AF837947

# Bioreactor Systems for Treating Perchlorate-Contaminated Water: Bench- and Pilot-Scale Investigations

*The mission of the Awwa Research Foundation (AwwaRF) is to advance the science of water to improve the quality of life. Funded primarily through annual subscription payments from over 1,000 utilities, consulting firms, and manufacturers in North America and abroad, AwwaRF sponsors research on all aspects of drinking water, including supply and resources, treatment, monitoring and analysis, distribution, management, and health effects.*

*From its headquarters in Denver, Colorado, the AwwaRF staff directs and supports the efforts of over 700 volunteers, who are the heart of the research program. These volunteers, serving on various boards and committees, use their expertise to select and monitor research studies to benefit the entire drinking water community.*

*Research findings are disseminated through a number of technology transfer activities, including research reports, conferences, videotape summaries, and periodicals.*

# Bioreactor Systems for Treating Perchlorate-Contaminated Water: Bench- and Pilot-Scale Investigations


Prepared by:
**Bruce E. Logan**, **Booki Min**, **Kijung Kim**, **Joel Miller**, and **Dana LaPoint**
Department of Civil and Environmental Engineering
The Pennsylvania State University, University Park, PA 16802

**Jacimaria Batista** and **Jian Liu**
Department of Civil and Environmental Engineering
University of Nevada, Las Vegas, NV

and

**Patrick J. Evans**, **Allyson Chu**, and **Steven Price**
Camp, Dresser and McKee, Inc., Denver, CO



Jointly sponsored by:
**Awwa Research Foundation**
6666 West Quincy Avenue, Denver, CO 80235-3098

**U.S. Environmental Protection Agency**
Washington D.C.

and

**East Valley Water District**
San Bernardino, CA

Published by:

Awwa Research Foundation
Advancing the Science of Water®

IWA Publishing

## DISCLAIMER

This study was funded by the Awwa Research Foundation (AwwaRF). AwwaRF assumes no responsibility for the content of the research study reported in this publication or for the opinions or statements of fact expressed in the report. The mention of trade names for commercial products does not represent or imply the approval or endorsement of AwwaRF. This report is presented solely for informational purposes.

**Published by IWA Publishing, Alliance House, 12 Caxton Street, London SW1H 0QS, UK**

Telephone: +44 (0) 20 7654 5500; Fax: +44 (0) 20 7654 5555; Email: publications@iwap.co.uk;
Web: **www.iwapublishing.com**

AwwaRF report number 90982F. Originally published by AwwaRF for its subscribers in 2003.
IWA Publishing version published 2004.

Copyright © 2003
by
Awwa Research Foundation

ISBN 1 84339 8893

# CONTENTS

LIST OF TABLES .......... ix

LIST OF FIGURES .......... xv

FOREWORD .......... xxiii

ACKNOWLEDGMENTS .......... xxv

EXECUTIVE SUMMARY .......... xxvii

CHAPTER 1: INTRODUCTION .......... 1
Project Approach .......... 2
Objectives .......... 2

CHAPTER 2: LITERATURE REVIEW .......... 5
Properties of Perchlorate .......... 5
Physical and Chemical Properties .......... 5
Health Effects of Perchlorate Exposure .......... 5
Characteristics of Perchlorate Reducing Microorganisms .......... 5
Early Studies on CRMs .......... 6
Abundances of Perchlorate Reducing Bacteria in Natural Systems .......... 6
Chlorate- and Perchlorate-Respiring Isolates .......... 7
Differences and Similarities of Denitrifying and Perchlorate Reducing Species .......... 8
Autotrophic Degradation of Nitrate Supported by Hydrogen .......... 10
Influence of Co-Contaminants on Treatment Efficiency .......... 13
Perchlorate and Chlorate Degradation in Bioreactors .......... 13
Methods of Hydrogen Production .......... 18
Hydrogen Production Using Fossil Fuels .......... 18
Electrolysis .......... 19
Other Methods of Hydrogen Production .......... 20

CHAPTER 3: ANALYTICAL METHODS .......... 23
Introduction .......... 23
Analytical Methods .......... 23
University of Nevada Las Vegas (UNLV) .......... 23
Penn State University .......... 25
Sample Exchange for Perchlorate Analysis .......... 26

CHAPTER 4: BIOLOGICAL PERCHLORATE REDUCING FIXED-FILM BIOREACTOR .......... 39
Introduction .......... 39
Materials and Methods .......... 39

Perchlorate Treatment in Pure Culture Sand Column Bioreactors at High Perchlorate Concentrations .......... 39
Perchlorate Removal at Low Perchlorate Concentrations in the Presence and Absence of Nitrate .......... 40
Perchlorate Treatment in Pure Culture Glass Bead Bioreactors .......... 41
Perchlorate Treatment in GAC-Packed Columns .......... 42
Batch Experiments With Suspended Pure and Mixed Cultures .......... 43
Results .......... 44
Pure Culture Sand Column Bioreactors at High Perchlorate Concentrations .......... 44
Sand Column Bioreactor Experiments at Low Perchlorate Concentrations in the Presence and Absence of Nitrate .......... 45
Perchlorate Removal in Glass Bead Bioreactors .......... 46
Perchlorate Treatment in GAC-Packed Columns .......... 47
Batch Experiments With Suspended Pure and Mixed Cultures .......... 49
Discussion .......... 49
Implications for Full Scale System Design and Operation .......... 50

CHAPTER 5: HYDROGEN-FED GAS PHASE UNSATURATED FIXED-FILM PERCHLORATE REACTORS .......... 65
Introduction .......... 65
Materials and Methods .......... 65
Selection and Isolation of Hydrogen-Oxidizing Bacteria Pure and Mixed Culture .......... 65
Hydrogen-Oxidizing Autotrophic Fixed Film Bioreactors: Design .......... 67
Column Operation .......... 68
Column Experiments .......... 70
Analytical Methods .......... 72
Results and Discussion .......... 72
Selection and Isolation of Hydrogen-Oxidizing Bacteria Pure and Mixed Culture .......... 72
Fixed Film Bioreactor Experiments for Column 1 .......... 73
Fixed Film Bioreactor Experiments for Column 2 .......... 75

CHAPTER 6: MEMBRANE-IMMOBILIZED BIOFILM .......... 103
Experimental Methods and Materials .......... 103
Set-Up of the Membrane-Immobilized Biofilm Reactor .......... 103
Microporous Membranes Selection .......... 103
Determination of Diffusion Coefficient for Perchlorate, Nitrate, and Sulfate .......... 104
Development of Perchlorate Degrading Enrichment Cultures .......... 105
SRT Studies for the Master Culture Reactor (MCR) .......... 106
Development of Membrane Immobilized Biofilms .......... 107
Perchlorate Biodegradation by Immobilized Biofilms .......... 108
Preliminary Kinetics Study of the "BALI" Culture .......... 108

Procedures for Characterization, Isolation and Identification of Perchlorate-Degrading "BALI" Cultures .... 109
Influence of Nitrate and Sulfate on Perchlorate Biodegradation by a Membrane-Immobilized Biofilm .... 111
Influence of Salinity on Perchlorate Biodegradation by the "BALI" Culture .... 112
Results and Discussion .... 112
Diffusivity Tests Through Membranes Without a Biofilm .... 112
Development of One-Liter Master Culture Reactor (MCR) .... 114
Development of Immobilized Biofilms on the Membranes .... 114
Biodegradation of Perchlorate by Membrane-Immobilized Biofilms .... 115
The "BALI" Enrichment Culture .... 118
SRT Studies .... 118
Preliminary Kinetics Study .... 119
Isolation, Characterization and Identification of Perchlorate-Degrading Bacteria From the "BALI" Mixed Culture .... 120
Influence of Nitrate on Perchlorate Biodegradation by the Membrane-Immobilized Biofilm .... 121
Influence of Sulfate on Perchlorate Biodegradation by the Membrane-Immobilized Biofilm .... 123
Influence of Salinity on Perchlorate Biodegradation by the "BALI" Culture .... 124

CHAPTER 7: PILOT SCALE TESTS .... 199
Introduction .... 199
Methods for the Pilot Study .... 199
Inoculum .... 199
Media .... 199
Bioreactors .... 200
Analytical Procedures .... 200
Results and Discussion of the Pilot Study .... 201
Sand Media Reactor .... 201
Plastic Media Reactor Operation .... 202

CHAPTER 8: ENGINEERING CONSIDERATIONS AND COST ESTIMATES .... 215
Introduction .... 215
Drinking Water Regulations .... 215
General Design Parameters .... 216
Conceptual Full-Scale Design .... 218
AFBF System Description .... 219
High-Rate Granular Filters Description .... 219
Ultraviolet Disinfection Description .... 220
Conceptual Cost Estimate .... 220

CHAPTER 9: CONCLUSIONS .......... 227
Packed-Bed Acetate-Fed Bioreactor (PSU-$O_4$ System) .......... 227
Unsaturated-Flow Hydrogen Gas Bioreactor .......... 228
Membrane Bioreactor .......... 228
Pilot Scale Tests With the Acetate-Fed Fixed Bed Bioreactor .......... 230
Recommendations .......... 231

APPENDIX A: QAPP .......... 233

APPENDIX B: ADJUSTMENTS TO QAPP .......... 249

APPENDIX C: PERCHLORATE ANALYTICAL PROCEDURE .......... 253

APPENDIX D: RAW DATA FOR SALINITY TESTS .......... 259

APPENDIX E: EXAMPLE CALCULATIONS FOR MEMBRANE-IMMOBILIZED BIOFILM REACTOR .......... 271

APPENDIX F: 16S rRNA ANALYSIS OF ISOLATE OF THE "BALI" MIXED CULTURE .......... 275

REFERENCES .......... 277

ABBREVIATIONS .......... 285

# TABLES


3.1 Calibration curve data for perchlorate at UNLV .......... 28

3.2 Calibration curve data for lactate at UNLV .......... 28

3.3 Calibration curve data for chloride at UNLV .......... 28

3.4 Calibration curve data for nitrate at UNLV .......... 29

3.5 Calibration curve data for sulfate at UNLV .......... 29

3.6 Amounts of DI water and NaOH withdrawn and added, respectively, to the eluent bottles for eluent preparation .......... 29

3.7 Calibration curve data for low level perchlorate (at Penn State University) .......... 29

3.8 Calibration curve data for high level perchlorate (at Penn State University) .......... 30

3.9 Calibration curve data for chloride (at Penn State University) .......... 30

3.10 Calibration curve data for acetate (at Penn State University) .......... 30

3.11 Calibration curve data for nitrate (at Penn State University) .......... 30

3.12 Exchange samples for perchlorate analysis between Penn State and UNLV for QA purposes (June 1999) .......... 30

3.13 Exchange samples for perchlorate analysis between Penn State and UNLV for QA purposes (June 2000) .......... 31

3.14 Integration corrections for perchlorate analysis in the presence of lactate or acetate .......... 31

3.15 Exchange samples for perchlorate analysis between Penn State and UNLV for QA purposes—Penn State samples (November 2000) .......... 32

3.16 Exchange samples for perchlorate analysis between Penn State and UNLV for QA purposes—UNLV samples (November 2000) .......... 32

4.1 The composition of Basal media .......... 52

4.2 Perchlorate profiles on selected days between days 38 to 77 .......... 52

4.3 The mass of perchlorate desorbed from GAC in the GAC packed column .......... 53

4.4 Acetate consumption for perchlorate reduction, when a perchlorate-reducing pure culture, KJ, was used ........ 54

4.5 Molar ratio of oxygen utilized per acetate consumed by KJ in batch experiment..... 54

4.6 TCE degradation by KJ and abiotic control........ 54

4.7 Statistical analysis of TCE degradation by KJ and abiotic control........ 54

5.1 Trace metal components in media used for perchlorate degradation experiments .... 78

5.2 Media components for column perchlorate experiments. Trace metals added at 5ml/L to achieve the concentration as noted in Table 5.1........ 78

5.3 Chemical composition of Redlands water as provided by CDM........ 79

5.4 Summary of conditions during chemostat experiment ........ 79

5.5 Summary of hydrogen-oxidizing mixed culture kinetic parameters for $ClO_4^-$........ 79

5.6 Hydraulic detention time at various flow rates before and after mixing........ 80

5.7 The statistical results from ANOVA (CI is the confidence interval)........ 80

6.1 Characteristics of the selected microporous membranes used in the membrane-immobilized biofilm reactor ........ 127

6.2 Concentration of buffer, nutrients and minerals used in the MCR media........ 127

6.3 Perchlorate and nitrate concentrations tested........ 128

6.4 Perchlorate and sulfate concentration tested........ 128

6.5 Contents of the culture tubes in the salinity test........ 128

6.6 Perchlorate diffusion testing by a BTS-55 membrane ........ 129

6.7 Perchlorate diffusion testing by a PVDF membrane ........ 130

6.8 Perchlorate diffusion testing by a FGLP membrane........ 131

6.9 Experimental determination of diffusion coefficient for perchlorate through three types of membranes ........ 131

6.10 Nitrate diffusivity testing by a BTS-55 membrane ........ 132

6.11 Sulfate diffusivity testing by a BTS-55 membrane .......... 133

6.12 The diffusion coefficients of perchlorate, nitrate, and sulfate in water and through Memcor BTS-55 membrane .......... 133

6.13 Perchlorate biodegradation in the first master culture reactor .......... 134

6.14 Biofilm development on a Millipore FGLP membrane .......... 134

6.15 The time sheet of perchlorate biodegradation testing .......... 134

6.16 Perchlorate biodegradation by a biofilm immobilized on a Millipore FGLP membrane (1st cycle) .......... 135

6.17 Perchlorate biodegradation by a biofilm immobilized on a FGLP membrane (2nd cycle) .......... 136

6.18 Perchlorate biodegradation by a biofilm-immobilized on a BTS-55 membrane (1st cycle) .......... 137

6.19 BTS-55 membrane: molar ratio of perchlorate to chloride and biodegradation rate (1st cycle) .......... 137

6.20 Perchlorate biodegradation by a biofilm immobilized on a BTS-55 membrane (2nd cycle) .......... 138

6.21 BTS-55 membrane: molar ratio of perchlorate to chloride and biodegradation rate (2nd cycle) .......... 138

6.22 Perchlorate biodegradation by a biofilm immobilized on a BTS-55 membrane (3rd cycle) .......... 139

6.23 BTS-55 membrane: molar ratio of perchlorate to chloride and biodegradation rate (3rd cycle) .......... 139

6.24 Estimation of thickness of biofilm immobilized on BTS-55 membrane .......... 140

6.25 Perchlorate biodegradation by a biofilm immobilized on a PVDF membrane .......... 140

6.26 PVDF membrane: molar ratio of perchlorate to chloride and biodegradation rate (1st cycle) .......... 141

6.27 Summary of relevant parameters calculated for the biofilms immobilized on the BTS-55 and PVDF membranes .......... 141

6.28 Mass balance of VSS in the reactor (1st cycle) .......... 141

6.29 Mass balance for perchlorate (1st cycle) ........ 142

6.30 Mass balance lactate (1st cycle) ........ 142

6.31 Mass balance for chloride (1st cycle) ........ 143

6.32 The daily molar ratio of perchlorate to chloride (1st cycle) ........ 143

6.33 Mass balance of VSS in the reactor (2nd cycle) ........ 143

6.34 Mass balance for perchlorate (2nd cycle) ........ 144

6.35 Mass balance for lactate (2nd cycle) ........ 144

6.36 Mass balance for chloride (2nd cycle) ........ 144

6.37 The daily molar ratio of perchlorate to chloride (2nd cycle) ........ 144

6.38 Perchlorate biodegradation in perchlorate limited bottles ........ 145

6.39 Perchlorate, lactate, and chloride concentration of preliminary kinetics study (KC1) ........ 145

6.40 Perchlorate, lactate, and chloride concentration in preliminary kinetics study (KC2) ........ 145

6.41 Perchlorate, lactate, and chloride concentration in preliminary kinetics study (KC3) ........ 146

6.42 Perchlorate, lactate, and chloride concentration in preliminary kinetics study (KC4) ........ 146

6.43 Perchlorate, lactate, and chloride concentration in preliminary kinetics study (KC5) ........ 146

6.44 Perchlorate, lactate, and chloride concentration in preliminary kinetics study (KL5) ........ 147

6.45 Perchlorate concentration of preliminary kinetics study in lactate-limited bottles (KL1-KL4) ........ 147

6.46 The characteristics of three isolates of "BALI" mixed culture ........ 147

6.47 The interference of nitrate on perchlorate biodegradation (batch testing) ........ 148

6.48 The interference of nitrate on perchlorate biodegradation (test 1) ........ 148

6.49 The interference of nitrate on perchlorate biodegradation (test 2) 149
6.50 The interference of nitrate on perchlorate biodegradation (test 3) 150
6.51 The interference of nitrate on perchlorate biodegradation (test 4) 151
6.52 Summary of nitrate interference on perchlorate biodegradation 152
6.53 Interference of sulfate on perchlorate biodegradation (batch testing) 152
6.54 The interference of sulfate on perchlorate biodegradation (test 1) 153
6.55 The interference of sulfate on perchlorate biodegradation (test 2) 153
6.56 The interference of sulfate on perchlorate biodegradation (test 3) 154
6.57 The interference of sulfate on perchlorate biodegradation (test 4) 155
6.58 Biodegradation rate for perchlorate in the presence of sulfate 155
6.59 Specific growth rate under different salinity conditions 156
7.1 List of measurements 204
8.1 General design parameters based on pilot testing 221
8.2 Design criteria for full-scale AFBF units 222
8.3 Design criteria for high-rate granular filter units 223
8.4 Comparison of conceptual costs for AFBF, FBBR, and ion exchange processes 224
A.1 Summary of critical and noncritical measurements 245
A.2 Summary of QA objectives for the critical parameter ($ClO_4^-$) 245
A.3 Summary of sample parameters 246
A.4 Method description of analytical techniques 247
D.1 The interference of different salinity levels on perchlorate biodegradation 260
D.2 The interference of salinity on perchlorate biodegradation (0%) 261
D.3 The interference of salinity on perchlorate biodegradation (0.5%) 262

D.4 The interference of salinity on perchlorate biodegradation (1%) ............................. 263

D.5 The interference of salinity on perchlorate biodegradation (1.5%) ........................... 264

D.6 The interference of salinity on perchlorate biodegradation (2%) ............................. 265

D.7 The interference of salinity on perchlorate biodegradation (2.5%) ........................... 266

D.8 The interference of salinity on perchlorate biodegradation (3%) ............................. 267

D.9 The interference of salinity on perchlorate biodegradation (3.5%) ........................... 268

D.10 The interference of salinity on perchlorate biodegradation (4%) ............................. 269

D.11 The interference of salinity on perchlorate biodegradation (6%) ............................. 270

# FIGURES


3.1 Calibration curve for perchlorate at UNLV ........ 33

3.2 Calibration curve for lactate at UNLV ........ 33

3.3 Calibration curve for chloride at UNLV ........ 34

3.4 Calibration curve for nitrate at UNLV ........ 34

3.5 Calibration curve for sulfate at UNLV ........ 35

3.6 Calibration curve for low level perchlorate (at Penn State University) ........ 35

3.7 Calibration curve for high level perchlorate (at Penn State University) ........ 36

3.8 Calibration curve for chloride (at Penn State University) ........ 36

3.9 Calibration curve for acetate (at Penn State University) ........ 37

3.10 Calibration curve for nitrate (at Penn State University) ........ 37

4.1 Setup used in sand column experiments for treating perchlorate at low concentrations ........ 55

4.2 Schematic of the GAC column system ........ 55

4.3 Influent and effluent perchlorate concentrations in pure culture sand column bioreactor ........ 56

4.4 Percent of effluent samples showing below each perchlorate concentration ........ 56

4.5 Critical hydraulic loading rate for the pure culture column experiment ........ 57

4.6 Molar ratios of acetate-consumed to perchlorate-removed in column experiments ........ 57

4.7 Cell concentrations measured along the pure culture column bioreactor ........ 58

4.8 Perchlorate concentrations in the effluent of the test and control columns during the 28 days of the column start-up period ........ 58

4.9 Perchlorate concentrations in the influent and effluents of the test and the control columns ........ 59

4.10 Perchlorate and nitrate concentrations in the pure culture column .......... 59

4.11 Tracer study using NaCl and NaClO4 in the control column (cleaned glass beads) prior to the biological degradation experiment .......... 60

4.12 Perchlorate concentrations in the bioreactor effluent for 23 days after the perchlorate was discontinued in the feed .......... 60

4.13 Perchlorate concentrations in the influent and effluent samples in the bioreactor, and the effluent from the abiotic column control (cleaned glass beds): (A) first trial; (B) second trial .......... 61

4.14 Intracolumn cell mass distribution measured at the conclusion of the pure culture (glass bead) column experiment .......... 62

4.15 Performance of the GAC packed column. The column was mixed on day 6 and 76 by different methods, and the acetate dosage was doubled on day 26. .......... 62

4.16 Perchlorate concentration profiles on days 15 and 16 in the GAC column .......... 63

4.17 Acetate concentration profiles on days 24, 28 and 30 in the GAC column .......... 63

4.18 Perchlorate concentration profiles on days 24, 28 and 30 in the GAC column experiment .......... 63

4.19 The reduction of perchlorate and nitrate in acetate-amended batch cultures with either perchlorate or nitrate, or in the presence of both compounds .......... 64

4.20 Time course of TCE degradation study by bacterial suspension KJ versus an abiotic control .......... 64

5.1 Schematic of Applikon Bioreactor used in chemostat experiment .......... 81

5.2 Schematic of the first hydrogen-oxidizing reactor (column 1) .......... 81

5.3 Set up for column 1 hydrogen bioreactor .......... 82

5.4 Schematic of the column 2 reactor .......... 83

5.5 Set up for the column 2 reactor .......... 84

5.6 Perchlorate reduction in hydrogen-oxidizing unsaturated column for high-perchlorate concentration feed .......... 85

5.7 Perchlorate removal in the hydrogen-oxidizing column at low influent perchlorate concentrations .......... 85

5.8 Tracer curves taken before and after mixing the bed to redistribute the biofilm-coated glass beads ........ 86

5.9 (A) Effluent concentration of perchlorate over time for various flow rates for column 1 at an influent perchlorate concentration of 700 μg/L. (B) Decrease in perchlorate concentration from the influent media as a function of flow rate using column 1 ........ 87

5.10 (A) Comparison of the tracer breakthrough curves in column 1 at flow rates of 1.3 mL/min and 11.3 mL/min. (B) Comparison of the effect of mixing on tracer curves in column 1 at a flowrate of 4.3 mL/min. ........ 88

5.11 Detention time as a function of flow rate before column mixing, and after column mixing in column 1 ........ 89

5.12 Tracer curve for column 2 at 2.3 mL/min ........ 89

5.13 (A) The influent, and effluent perchlorate concentration of column 2 during column startup and (B) The percent change in perchlorate at an average influent perchlorate concentration of 18 mg/L ........ 90

5.14 (A) The influent and effluent perchlorate concentrations of column 2 during column startup and (B) The percent change in perchlorate at an average influent perchlorate concentration of 1200 μg/L ........ 91

5.15 (A) Influent perchlorate concentrations effect on the perchlorate degradation rate for column 1 and column 2. (B) Percent perchlorate removal as a function of log mean perchlorate concentration in column 1 and 2 ........ 92

5.16 The percent perchlorate removal in column 2 as a function of pH ........ 93

5.17 (A) Perchlorate decrease from influent to effluent and (B) Percent perchlorate removal in the 64 mg/L sulfate and in the presence of 32.5 mg/L sulfate ........ 94

5.18 (A) Perchlorate decrease from influent to effluent in the absence and presence (after 9/1) of nitrate for the high perchlorate experiment. (B) Percent perchlorate removal in the absence and presence (after 9/1) of nitrate ........ 95

5.19 (A) Nitrate degradation in the bioreactor at an average influent perchlorate concentration of 826 μg/L. (B) Percent nitrate removal in the bioreactor at an average influent perchlorate concentration of 826 μg/L ........ 96

5.20 (A) Perchlorate decrease from influent to effluent in the absence (9/11-9/17) and presence (after 9/18) of nitrate for the low perchlorate experiment. (B) Percent perchlorate removal absence and presence (after 9/18) of nitrate ........ 97

5.21 (A) Nitrate degradation in the bioreactor at an average influent perchlorate concentration of 79 µg/L. (B) Percent nitrate removal in the bioreactor at an average influent perchlorate concentration of 79 µg/L........................................ 98

5.22 (A) Perchlorate decrease from influent to effluent for the Redlands water experiment. (B) Percent perchlorate removal for Redlands water............................ 99

5.23 (A) Nitrate degradation in the bioreactor at an average influent perchlorate concentration of 73 µg/L in Redlands water. (B) Percent nitrate removal in the bioreactor at an average influent perchlorate concentration of 73 µg/L in Redlands water........................................................................................................ 100

5.24 (A) Perchlorate degradation in the column 2 for sulfate and no sulfate situations. (B) Percent perchlorate removal in the column 2 at an average influent perchlorate concentration of 73 µg/L in Redlands water.......................................... 101

6.1 Schematic representation of the membrane-immobilized biofilm reactor ............... 157

6.2 Dimension of the membrane-immobilized biofilm reactor showing: (A) Top cover with three openings for sampling and monitoring; (B) Side view showing connecting plate and membrane opening................................................................ 158

6.3 Biofilm established on the microporous membrane at UNLV ................................. 158

6.4 Membrane biofilm growth chamber showing: (A) Top view with DO and nitrogen ports; (B) Front view showing holding grid and membrane plate, and (C) Side view showing membrane plate and holding grid....................................... 159

6.5 Construction of modified biofilm growth reactor .................................................... 159

6.6 The membrane-immobilized biofilm reactor during the biodegradation mode ......... 160

6.7 Preliminary kinetics study with the "BALI" culture performed under perchlorate and lactate limited conditions ............................................................. 160

6.8 Perchlorate diffusion coefficient by a Memcor BTS-55 membrane ......................... 161

6.9 Perchlorate diffusion through a Memcor PVDF membrane.................................... 161

6.10 Perchlorate diffusion through a Millipore FGLP membrane.................................. 162

6.11 Nitrate diffusivity testing through Memcor BTS-55 membrane.............................. 162

6.12 Sulfate diffusivity testing by a Memcor BTS-55 membrane ................................... 163

6.13 Anaerobic consumption of perchlorate and lactate in the BR ................................ 163

6.14 Anaerobic consumption of perchlorate and lactate in the DR .................................. 164

6.15 Perchlorate, lactate, and chloride concentrations in the DR and BR reactors during biodegradation of perchlorate by a biofilm immobilized on a FGLP membrane (1st cycle) .................................................................................... 164

6.16 Perchlorate, lactate, and chloride concentrations in the DR and BR reactors during biodegradation of perchlorate by a biofilm immobilized on a FGLP membrane (2nd cycle) .................................................................................... 165

6.17 Perchlorate, lactate, and chloride concentrations in the DR and BR reactors during biodegradation of perchlorate by a biofilm immobilized on a BTS-55 membrane (1st cycle) .................................................................................... 165

6.18 Perchlorate, lactate, and chloride concentrations in the DR and BR reactors during biodegradation of perchlorate by a biofilm immobilized on a BTS-55 membrane (2nd cycle) .................................................................................... 166

6.19 Perchlorate, lactate, and chloride concentrations in the DR and BR reactors during biodegradation of perchlorate by a biofilm immobilized on a BTS-55 membrane (3rd cycle) .................................................................................... 166

6.20 Perchlorate, lactate, and chloride concentrations in the DR and BR reactors during biodegradation of perchlorate by a biofilm immobilized on a PVDF membrane .................................................................................... 167

6.21 Master culture reactors containing the BALI red culture and another perchlorate degrading culture which does not have the red coloration .................................... 167

6.22 Kinetic studies with the "BALI" culture showing microbial growth in perchlorate limited bottles and no growth in lactate limited bottles ......................... 168

6.23 Perchlorate biodegradation in perchlorate limited bottles ...................................... 168

6.24 Microbial reduction of perchlorate to chloride when lactate/$ClO_4^-$ = 100 ............... 169

6.25 Microbial reduction of perchlorate to chloride when lactate/$ClO_4^-$ = 50 ................. 169

6.26 Microbial reduction of perchlorate to chloride when lactate/$ClO_4^-$ = 16.7 .............. 170

6.27 Microbial reduction of perchlorate to chloride when lactate/$ClO_4^-$ = 10 ................. 170

6.28 Microbial reduction of perchlorate to chloride when lactate/$ClO_4^-$ = 5 ................... 171

6.29 Microbial reduction of perchlorate to chloride when lactate/$ClO_4^-$ = 3 ................... 171

6.30 Gram-stain of "BALI" taken from growth reactor (TK1) showing the predominance of rod and helical gram-negative microorganisms ............................ 172

6.31 Gram-stain of "BALI" taken from kinetics study (KC5d) showing helical morphology of growing microorganisms ............................ 173

6.32 Micrograph of isolate KA2-1 ............................ 174

6.33 Micrograph of isolate KL1-A ............................ 175

6.34 Micrograph of isolate RAS-CCSD ............................ 176

6.35 Batch testing of nitrate interference on perchlorate biodegradation using "BALI" culture ............................ 177

6.36 The interference of nitrate on perchlorate biodegradation by BTS-55 membrane-immobilized biofilm (test 1) ............................ 177

6.37 Nitrate and perchlorate biodegradation rate (test 1) ............................ 178

6.38 The interference of nitrate on perchlorate biodegradation by a BTS-55 membrane-immobilized biofilm (test 2) ............................ 178

6.39 Nitrate and perchlorate biodegradation rate (test 2) ............................ 179

6.40 Interference of various concentration of nitrate on perchlorate biodegradation in BTS-55 membrane-immobilized biofilm reactors (test 1 & test 2) ............................ 179

6.41 The interference of nitrate on perchlorate biodegradation (test 3) ............................ 180

6.42 Biodegradation rates of perchlorate and nitrate (test 3) ............................ 180

6.43 The interference of nitrate on perchlorate biodegradation (test 4) ............................ 181

6.44 Biodegradation rates of perchlorate and nitrate (test 4) ............................ 182

6.45 Nitrate and perchlorate biodegradation in BTS-55 membrane-immobilized biofilm reactors (test 3 & test 4) ............................ 182

6.46 Interference of sulfate on perchlorate biodegradation (batch testing) ............................ 183

6.47 The interference of sulfate on perchlorate biodegradation (test 1) ............................ 183

6.48 The interference of sulfate on perchlorate biodegradation (test 2) ............................ 184

6.49 The interference of sulfate on perchlorate biodegradation (test 1 & test 2) ............................ 184

6.50 The interference of sulfate on perchlorate biodegradation (test 3) .......... 185
6.51 The interference of sulfate on perchlorate biodegradation (test 4) .......... 185
6.52 The interference of sulfate on perchlorate biodegradation (test 3 & test 4) .......... 186
6.53 Transmittance of different salinity concentrations .......... 187
6.54 Interference of salinity (0%) on perchlorate biodegradation .......... 188
6.55 Interference of salinity (0.5%) on perchlorate biodegradation .......... 189
6.56 Interference of salinity (1.0%) on perchlorate biodegradation .......... 190
6.57 Interference of salinity (1.5%) on perchlorate biodegradation .......... 191
6.58 Interference of salinity (2.0%) on perchlorate biodegradation .......... 192
6.59 Interference of salinity (2.5%) on perchlorate biodegradation .......... 193
6.60 Interference of salinity (3.0%) on perchlorate biodegradation .......... 194
6.61 Interference of salinity (3.5%) on perchlorate biodegradation .......... 195
6.62 Interference of salinity (4.0%) on perchlorate biodegradation .......... 196
6.63 Interference of salinity (6.0%) on perchlorate biodegradation .......... 197
7.1 Pilot-scale reactor with sand media .......... 205
7.2 Raw perchlorate data from influent and effluent of the sand media reactor .......... 206
7.3 Comparison of perchlorate removal before and after using a new backwash pump .......... 206
7.4 Data for days of acceptable operation (no mechanical failures) of perchlorate from influent and effluent at the sand media reactor .......... 207
7.5 Nitrate concentrations from influent and effluent at the sand media reactor .......... 207
7.6 Profile data from taps of the reactor at 1 gpm and 2 gpm .......... 208
7.7 Retention time test at 1 gpm and 2 gpm .......... 209
7.8 Pilot-scale reactor with plastic media .......... 210

7.9 Perchlorate raw data from influent and effluent in the plastic media reactor ........... 211

7.10 Comparison of perchlorate removal before and after using a new backwash pump depending on flow rates ........... 211

7.11 Data for days of acceptable operation (no mechanical failures) of perchlorate from influent and effluent in plastic media reactor ........... 212

7.12 Data for days of acceptable operation (no mechanical failures) of nitrate from influent and effluent in plastic media reactor ........... 212

7.13 Profile data from taps of the reactor at 1 gpm and 2 gpm ........... 213

7.14 Retention time test at 1 gpm and 2 gpm ........... 214

8.1 Existing Texas Street Well Facility process schematic ........... 225

8.2 Proposed Texas Street Well Facility conceptual process schematic ........... 225

8.3 Site plan of Texas Street Well Facility with proposed process facilities........... 226

8.4 Anaerobic fixed bed filter process schematic........... 226

# FOREWORD

The Awwa Research Foundation is a nonprofit corporation that is dedicated to the implementation of a research effort to help utilities respond to regulatory requirements and traditional high-priority concerns of the industry. The research agenda is developed through a process of consultation with subscribers and drinking water professionals. Under the umbrella of a Strategic Research Plan, the Research Advisory Council prioritizes the suggested project based upon current and future needs, applicability, and past work; the recommendations are forwarded to the Board of Trustees for final selection. The foundation also sponsors research projects through an unsolicited proposal process; the Collaborative Research, Research Applications, and Tailored Collaboration programs; and various joint research efforts with organizations such as the U.S. Environmental Protection Agency, the U.S. Bureau of Reclamation, and the Association of California Water Agencies.

This publication is a result of one of these sponsored studies, and it is hoped that its findings will be applied in communities throughout the world. The following report serves not only as a means of communicating the results of the water industry's centralized research program but also as a tool to enlist the further support of the nonmember utilities and individuals.

Projects are managed closely from their inception to the final report by the foundation's staff and large cadre of volunteers who willingly contribute their time and expertise. The foundation serves a planning and management function and awards contracts to other institutions such as water utilities, universities, and engineering firms. The funding for this research effort comes primarily from the Subscription Program, through which water utilities subscribe to the research program and make an annual payment proportionate to the volume of water they deliver and consultants and manufacturers subscribe based on their annual billings. The program offers a cost-effective and fair method for funding research in the public interest.

A broad spectrum of water supply issues is addressed by the foundation's research agenda: resources, treatment and operations, distribution and storage, water quality and analysis, toxicology, economics, and management. The ultimate purpose of the coordinated effort is to assist water suppliers to provide the highest possible quality of water economically and reliably. The true benefits are realized when the results are implemented at the utility level. The foundation's trustees are pleased to offer this publication as a contribution toward that end.

A broad spectrum of water supply issues is addressed by the foundation's research agenda: resources, treatment and operations, distribution and storage, water quality and analysis, toxicology, economics, and management. The ultimate purpose of the coordinated effort is to assist water suppliers to provide the highest possible quality of water economically and reliably. The true benefits are realized when the results are implemented at the utility level. The foundation's trustees are pleased to offer this publication as a contribution toward that end.

Edmund G. Archuleta, P.E.
Chair, Board of Trustees
Awwa Research Foundation

James F. Manwaring, P.E.
Executive Director
Awwa Research Foundation

# ACKNOWLEDGMENTS

The authors are grateful for the funding provided by the Awwa Research Foundation and the U.S. Environmental Protection Agency, which was obtained by the East Valley Water District through the efforts of Robert Martin and Congressman Jerry Lewis (CA).

The authors of this report are also indebted to the following individuals, water utilities and organizations for their cooperation and participation in this project: Douglas Headrick, Michael L. Huffstutler, and The City of Redlands, California for providing water samples and groundwater information; Camp Dresser and McKee Inc., for the engineering analysis and cost estimates on the perchlorate bioreactor system scale up. In addition, we thank Frank Blaha and the members of the PAC for their comments and guidance during the course of this project. In addition, we thank Dr. Kapal Madiereddi for his ideas on the design of the membrane-immobilized reactor; and UNLV students: Adriano Vieira for his help in setting up the IC and Peaknet software; Jon Becker for his assistance with the drawings; and Alex Markham and Hokanani for their valuable assistance with glassware preparation.

# EXECUTIVE SUMMARY

Widespread contamination of drinking water sources with perchlorate, a chemical used as the oxidizer in solid rocket propellant and in some munitions, is estimated to affect the drinking water of over 15 million people in the United States. Perchlorate interferes with iodide uptake by the thyroid gland and can adversely affect hormone production in humans. Although no federal drinking water standard exists yet for perchlorate, many states have enacted their own standards for this chemical ranging from 1 to 18 μg/L. In California, the action level for perchlorate ranged, during the time of this report, from 4 to 18 μg/L. The Environmental Protection Agency had not decided on what dose data to use, and projected drinking water standards were estimated to range from 1 to 32 μg/L.

Perchlorate is highly soluble and ordinarily chemically stable in water. However, perchlorate can be microbiologically degraded to chloride by some bacteria that use perchlorate as an electron acceptor during the oxidation of organic matter. High dissolved oxygen concentrations interfere with perchlorate reduction, and therefore anoxic conditions must be obtained for microbial perchlorate reduction. Nitrate can also interfere with perchlorate reduction. Thus, although it was known that perchlorate could be microbiologically degraded, little was known about how to design bioreactors to remove perchlorate from waters down to concentrations necessary to meet anticipated drinking water standards.

## RESEARCH OBJECTIVES

The purpose of this perchlorate project in Phase I was to conduct bench scale experiments on three different fixed-film biological treatment processes capable of being scaled up to treat large quantities of perchlorate-contaminated water to anticipated drinking water levels of less than 18 μg/L perchlorate. These three treatment systems were: a packed bed (slow sand or GAC filter) amended with soluble substrates (acetate or lactate); a hydrogen gas fed four-phase (hydrogen gas, water, biofilm, and support media), unsaturated trickle-type packed column; and a membrane-immobilized biofilm reactor. Pure and mixed enrichment microbial cultures were used in these reactors. It was also an objective of this research to investigate the effects of co-ions on perchlorate biodegradation. With information gained from the laboratory investigations, the costs of treating waters in a full scale system was estimated for the most successful treatment system, determined to be the packed bed acetate-fed bioreactor.

Based on these Phase 1 tests, an acetate-fed packed bed reactor (referred to as the PSU-$O_4$ system) was selected for further testing in pilot-plant tests at the Texas Street well field in Redlands, CA, for Phase 2 of this project. The purposes of Phase 2 were to: confirm laboratory-derived perchlorate removal efficiencies at pilot scale; to conduct sufficient tests to fully optimize the system over long operation periods in the field; and to continue to research the laboratory factors that can improve operation of full scale systems.

## APPROACH

In Phase 1, bench-scale reactors were constructed to test the ability of these different bioreactors to degrade perchlorate to low concentrations. The packed-bed reactors (saturated and unsaturated flow conditions) were tested in the Environmental Engineering laboratories at Penn State University, while the membrane-immobilized reactor experiments were conducted at the University of Nevada, Las Vegas. Most experiments were conducted using an artificial groundwater spiked with perchlorate at various concentrations. In the last series of experiments, nitrate was added to the water to study its effect on perchlorate degradation. These results were then compared to perchlorate removal using actual water samples taken from the Redlands-Crafton site containing perchlorate at concentrations of 60-80 ppb. In Phase 2, pilot scale reactors were tested at the Redlands Site in California to examine the ability of the system to reduce perchlorate to concentrations of less than 4 ppb.

## CONCLUSIONS

### Phase 1

All three systems successfully removed perchlorate at rates sufficient to achieve an acceptable level of perchlorate removal for subsequent treatment for potable use. A summary of the results for each system, and an economic analysis of the most feasible system, are summarized below.

#### *Packed Bed (Saturated-Flow) Reactors*

Bench scale testing using pure and mixed cultures, and sand and GAC as support media indicated that acetate-fed sand-bed reactors inoculated with a pure culture could provide the highest perchlorate removal rates. Using mixed cultures, perchlorate was reduced in sand column bioreactors from 20 mg/L to below the detection limit (<4 μg/L) at detention times as low as 18 minutes. GAC packed columns performed similarly for the first 10 days, but after the column was backwashed to re-distribute biofilm-coated particles in the reactor, reactor efficiency was substantially reduced due to perchlorate desorption. This indicated that the use of perchlorate-adsorbing media in packed beds was undesirable.

Perchlorate removal was found to be extremely rapid in a sand column bioreactor inoculated with a pure culture (bacterial isolate KJ). Perchlorate (25 mg/L) could be completely removed at empty bed contact times of less than 2.1 minutes by the pure culture reactor, achieving a maximum removal rate of 18,100 μg $L^{-1}$ $min^{-1}$. This detention time was approximately an order-of-magnitude smaller than those in mixed-culture tests. Sustained perchlorate removals were studied in this reactor system for 83 days after start up. Based on a total of 147 samples, perchlorate concentrations were not detectable (<4 μg/L) in 84.4% of effluent samples. Chloride measurements confirmed that all of the perchlorate was converted to chloride. The average molar ratio of chloride produced to perchlorate reduced was 1.0 for 91 samples.

A separate series of reactor experiments were conducted using water from the Redlands site that contained approximately 65 μg/L of perchlorate and 35 mg/L of nitrate. Perchlorate was reduced to below the detection limit for the Redlands water despite the presence of the nitrate. Nitrate was removed from a concentration of approximately 20 mg/L to effluent concentrations in the range of 9.3 to 11.3 mg/L during the first 6 days of testing a synthetic water similar to Redlands water. Nitrate was further reduced to 830 μg/L on day 12, and then remained below the detection limit (180 μg/L) after day 20 for all samples except one. It was concluded that this reactor was capable of continuously removing both perchlorate and nitrate from the Redlands water and that this reactor design was suitably advanced for further testing in Phase 2 of this project.

### *Hydrogen-Gas Phase (Unsaturated) Flow Reactors*

The hydrogen ($H_2$) reactor was tested at bench scale and determined to have perchlorate reduction rates suitable for scale up and field testing, but was not selected to allow field testing of the packed bed reactor. The biofilm reactor, operated in unsaturated-flow mode, was continuously fed water containing 20 mg/L $ClO_4^-$ and a gas mixture of $H_2$ (5%) and $CO_2$. After a 10-day startup period, the reactor feed was then switched to a lower influent perchlorate concentration of 740 μg/L. Over a 140 day period at a constant hydraulic loading rate of 0.45 cm/min, 39±9% of the perchlorate was removed in the reactor at detention times of 1.1 to 1.3 min producing an average removal rate of perchlorate of 230 μg $L^{-1}$ $min^{-1}$. These rates are comparable to other reactor types.

### *Membrane-Immobilized Biofilm Reactor*

Bench scale tests with this reactor, in which microbes are not in direct contact with the water to be treated, demonstrated successful removal rates of perchlorate from high concentrations to non-detectable levels. In laboratory tests using lactate as a growth substrate, perchlorate removal rates by a ~25 μm thick biofilm on a 78.5 $cm^2$ membrane surface area were found (triplicate tests) to be about 130 to 260 μg $ClO_4^-$ $L^{-1}$ $min^{-1}$ $mg^{-1}$ biomass. In addition to a rapid perchlorate removal rate, this reactor type produces treated water with consistently low perchlorate concentrations. A disadvantage of this reactor type was elevated chloride concentrations in the final product water. This was attributed to chloride diffusion back through the membrane and water driven by osmotic pressure from the untreated side of the membrane to the treated water side. The overall removal rates that could be achieved in a continuous flow reactor (i.e. information for scale up) were not determined, and therefore this reactor design was also not considered for field tests in Phase 2.

## Phase 2: Pilot-Plant Tests and Economic Analysis

Our preliminary economic and engineering analysis indicated that the least-expensive, most reliable system was an acetate-fed packed bed reactor. The packed-bed sand reactor achieved the highest perchlorate removal rates of the three systems in laboratory tests. In addition, there is a precedent for using an acetate-fed biological reactor for drinking water treatment in the U.S., making it likely that a reactor of this type would gain public acceptance. Nitrate has been treated using an acetate-fed packed bed reactor for drinking water pretreatment at a site in Coyle, Oklahoma. In addition, biological denitrifying systems have been successfully used for drinking water treatment in Europe for several years. The wider acceptance of biologically activated filters in the U.S. also points to new trends in the acceptance by water utilities to incorporate biological treatment into drinking water treatment trains. These factors, coupled with a national trend towards "green engineering" and sustainable technologies, suggested that an acetate fed bioreactor is a feasible perchlorate treatment technology.

The acetate-fed bioreactor was tested at the Redlands site in California. Two reactors were tested in parallel, a sand medium reactor and a plastic medium reactor. It was found that the sand media column could remove perchlorate from ~75 μg/L to the detection level (<4 μg/L) at hydraulic loading rates of 1 to 2 gpm/ft$^2$ when using a regular backwashing (weekly) cycle to clean and redistribute sand in the column. Salt tracer experiments indicated that the detention time for the cleaned medium was between 30 minutes (1 gpm/ft$^2$) and 18 minutes (2 gpm/ft$^2$). Prior to backwashing, the detention time was 30-45% lower than after backwashing. The plastic media bioreactor produced the most consistent perchlorate removals at 1 gpm/ft$^2$. Perchlorate removal below the detection level was possible at hydraulic loading rates of 1 or 1.5 gpm/ft$^2$ and was less sensitive to the backwashing frequency than the sand medium likely as a result of the higher porosity of the plastic medium. The detention times at a hydraulic loading rate of 1 gpm/ft$^2$ ranged from 60 minutes before backwashing to 70 minutes after backwashing.

The total estimated cost for perchlorate removal at the Redlands Site using the PSU-$O_4$ process (based on a fixed bed) was $22.6 million. The capital cost for perchlorate removal, calculated on a flow basis, was $0.91 per gpd of capacity. Operation and maintenance costs were $97 per acre-ft. It should be noted this cost is in addition to the existing capital expenditures at the site for the wells, reservoir and GAC units. The total cost of a fluidized bed reactor was approximately the same ($22.1 million) at this site because the fluidized bed requires fewer reactor tanks but has higher operating costs due to recycle. The capital cost of the fluidized bed reactor normalized to flow is slightly less ($0.76/gpd) than the fixed bed system, because the fluidized bed requires fewer reactor tanks, while the O&M costs are slightly larger ($104/acre-ft). The cost for an Ion Exchange system (ISEP±™) was estimated to be much larger at $30.9 million ($1.47/gpd). The ion exchange process has the advantage that downstream polishing processes are not required as for the biological processes. The operating cost for the perchlorate removal process consists primarily of the chemical costs for the acetate and ammonium phosphate, the power costs for the pumps and cost of additional staff to run the facility. Additional costs of repairs, equipment replacement and testing have not been included.

## RECOMMENDATIONS

Based on the laboratory, engineering and economic analysis performed during this study, it is recommended that full-scale systems to remove perchlorate from drinking water should be based on the acetate-fed, packed-bed reactor (PSU-$O_4$ system). In order to demonstrate this technology at full scale, we recommend that this system be constructed and tested at the Crafton-Redlands site in Redlands, CA. The full scale PSU-$O_4$ system would consist of a packed bed reactor fed an electron donor (acetate) and essential nutrients dissolved in the water, a post-filtration system to remove excess acetate and bacterial biofilm produced by the reactor, and a UV disinfection system. The bioreactor would use plastic media that would be backwashed regularly to maintain reactor performance.

# CHAPTER 1
# INTRODUCTION

Recent detection of perchlorate in several surface waters and ground water wells used to supply drinking water has created an unforeseen water contamination crisis in the Western states, although problems are likely to emerge at other sites where perchlorate is used. It is estimated that as many as 15 million people could be affected by perchlorate contamination of drinking water. Perchlorate is a health concern due to its interference with iodide in the production of hormones in the thyroid. In March of 1997, the California Department of Health Services (CDHS) developed a method that reduced the detection limit of perchlorate from 400 ppb to 4 ppb. Based on EPA work, they established a provisional action level of 18 ppb for drinking water. Subsequent monitoring of 232 groundwater wells by the CDHS indicated perchlorate was in 69 wells (30%) and at concentrations above the action level in 20 wells (9%) (AwwaRF 1997). Perchlorate concentrations in surface and groundwater range from detectable (4 ppb) to 0.37%, and endanger use of the Colorado River water in the western states. Samples taken from the Las Vegas Wash, which feeds Lake Mead and then flows into the Colorado River, contained 1,500 to 1,680 ppb (Urbansky 1998). The Los Angeles Metropolitan Water District measured 8 ppb in water at its intake in Lake Mead, and the Southern Nevada Water Authority found 11 ppb in its tap water. In January of 2002, California reduced the action level to 4 ppb, and it is now estimated that 292 out of 5,200 drinking water sources in California are contaminated with perchlorate. Eight states have advisory levels that vary between 1 ppb and 18 ppb.

Perchlorate contamination is thought to arise primarily from the manufacture and disposal of ammonium perchlorate, a highly oxidized compound produced for use as the oxidizer in solid rocket propellant. It is highly soluble and not easily removed from water. It was the consensus of a team of experts that met at a special workshop on perchlorate that "at this time there is no proven removal process available at the low concentrations being found in drinking water" (AwwaRF 1997). Conventional water treatment technologies such as air stripping and advanced oxidation processes have no effect on perchlorate removal or destruction. The costs associated with carbon adsorption and ion exchange can be high, and less expensive treatment technologies for drinking water were needed at the time this study was undertaken. Changes in solution redox potential alone (–200 mV) do not affect perchlorate because aqueous solutions are extremely stable (Bliven 1996); unlike nitrate, perchlorate does not react with zero-valent iron.

While perchlorate may be difficult to chemically treat or physically remove, it is very biodegradable. It is known that some microorganisms are able to respire using chlorate ($ClO_3^-$) and/or perchlorate ($ClO_4^-$): that is, they can use either of these compounds as an electron acceptor in the oxidation of many common substrates such as acetate, simple sugars and amino acids (Logan 1998a; Herman and Frankenberger 1998). It has been found that perchlorate does not naturally occur in the environment (in substantial concentrations), except in some fertilizers derived from Chilean saltpeter (Schilt 1979; Attaway and Smith 1993; Nzengung, Wang, and Harvey 1999). Both perchlorate and oxygen can be used as electron acceptors by some strains of microorganisms (Malmqvist, Welander, and Gunnarsson 1991; Hijnen et al. 1995; Korenkov et al. 1976; Wallace et al. 1996). Several strains of chlorate respiring microbes (CRMs) and perchlorate respiring microorganisms (PRMs) are thought to be related to denitrifying organisms, although some chlorate respiring cultures may lose the ability to reduce nitrate when cultivated on chlorate for long periods (Malmqvist et al. 1994). Recently, it has also been found that there are chlorate-respiring strains unable to degrade perchlorate (Logan et al. 2000).

One of the earliest proposed uses of chlorate reducing microorganisms (CRMs) was a test to evaluate the concentration of biodegradable organic matter in domestic wastewater by measuring the amount of $Cl^-$ produced from the reduction of $ClO_3^-$ (Bryan and Rohlich 1954, Bryan 1966). CRMs have since been proposed for use in a variety of engineered reactors. For example, chlorate produced during bleaching operations can be removed by CRMs in wastewater treatment lagoons (Malmqvist, Welander, and Gunnarsson 1991). Bacteria that have been identified to respire on chlorate have also been shown to reduce bromate (a carcinogen) with the addition of ethanol in batch cultures, and it was proposed to use microbial reduction of bromate (to bromide) for water treatment in water filters (Hijnen et al. 1995). Microbes identified to respire using chlorate and perchlorate include *Vibrio dechloraticans* Cuznesove B-1168 (Korenkov et al. 1976), *Ideonella dechloratans* (Malmqvist et al. 1994), and *Wolinella succinogenes* (Wallace et al. 1996). Since then, several other strains have been isolated that are capable of perchlorate reduction (Logan et al. 2000, Coates et al. 1999).

## PROJECT APPROACH

This project is directed at developing biological treatment systems to remove perchlorate from water so that it can subsequently be used as a source of drinking water. It is argued here that the most promising systems for microbial respiration of perchlorate are those based on fixed biofilms of microorganisms. A successful treatment system must have the ability to remove the perchlorate, leave very little oxidizable substrate in the water, and produce water relatively free of bacteria. At many sites where perchlorate has been measured, there is also contamination of the water by chlorinated aliphatics such as TCE. Therefore, the system should either be capable of degrading these compounds, or, other processes will be needed (such as air stripping) to remove residual chlorinated compounds.

In order to provide a rationale for the systems proposed in this study, the characteristics of perchlorate- and chlorate-reducing microorganisms, and existing processes using these microbes, are reviewed below. Since many of the attributes of nitrate and chlorate reducing microorganisms are similar, the literature on denitrifiers is also briefly summarized in order to hypothesize what native attributes might be characteristic of the microbial physiology of (per)chlorate reducing microorganisms. In addition, water containing perchlorate usually contains nitrate derived in part from ammonia oxidation to nitrate. There are substantial differences, however, between nitrate reduction and perchlorate reduction that will be noted. The main differences are that: nitrate can be more easily reduced abiotically than perchlorate, for example by zero valent iron; and nitrate is reduced to a gas (nitrogen) whereas chlorate is reduced to soluble ion (chloride). Other characteristics of (per)chlorate reducing microbes and denitrifying microbes are reviewed below.

## OBJECTIVES

Although perchlorate is biodegradable, there were no published reports at the time this research was undertaken of treatment technologies sufficient to remove perchlorate while maintaining water quality sufficient to meet other drinking water standards. The purpose of this project was therefore to explore three reactor designs in terms of their suitability for treating perchlorate contaminated water to produce drinking water. The objectives were to:

- Modify a sand filter to treat perchlorate contaminated water to drinking water standards;
- Design and test a gas (hydrogen) phase fixed film bioreactor for efficiency of perchlorate removal;
- Test a membrane-immobilized biofilm reactor for perchlorate removal to drinking water standards;
- To microbiologically characterize the perchlorate-reducing mixed enrichment culture and evaluate the effects of sulfate, nitrate and salinity on perchlorate biodegradation in the membrane-immobilized biofilm reactor;
- To conduct pilot scale tests (Phase 2) on one of the laboratory systems;
- Make an economic analysis of a reactor scaled up to capacities sufficient for treatment of large volume of water (millions of gallons per day).

Most of the laboratory studies were conducted using an artificial groundwater amended with perchlorate. To demonstrate perchlorate degradation for actual water sources, we also examined in the latter part of the study, perchlorate degradation using perchlorate-contaminated water samples from the Redlands-Crafton plume in California. The Redlands site was specifically indicated in the Awwa Research Foundation (AwwaRF) proposal guidelines. This site can treat 8.6 MGD of water, but has been closed due to perchlorate contamination of the water.

Preliminary sand column experiments completed before the start of this AwwaRF project demonstrated that perchlorate could be removed down to non-detectable levels. However, it was found that backwashing the sand in small laboratory columns was difficult. Therefore, in the first set of experiments conducted under this project, it was decided to try using granular activated carbon (GAC) instead of sand in order to make backwashing the column packing easier. Three different substrates—acetate, hydrogen gas, and lactate—were used in various reactors. Acetate had been shown to serve as an electron donor in studies demonstrating the removal of perchlorate to <18 ppb in batch and column experiments (Logan and Kim 1998). Using acetate, we conducted more detailed studies to better examine substrate (feed) concentrations and reactor detention times, biofilm substrate (sand, GAC, glass beads), inoculation strategy (pure versus mixed cultures), backwashing cycle frequency, and effect of a competing electron acceptor (nitrate). While the focus of this work was on bioreactor development, it was expected that the complete system would need to be a one to three compartment reactor system as described below. There was a patent issued on the first two reactor designs and on the concept of fueling these reactors using hydrogen gas. This patent was applied for prior to the start of this project, and during the project execution was granted to Bruce Logan and The Pennsylvania State University (Logan 1998b).

# CHAPTER 2
# LITERATURE REVIEW

## PROPERTIES OF PERCHLORATE

### Physical and Chemical Properties

Perchlorate salts, such as sodium perchlorate, have a relatively large water solubility (8 M). This is a result of the large, delocalized monovalent negative charge of the perchlorate ion, which leads to a decreased affinity for cations. The high solubility values mean that transport of perchlorate in contaminated groundwater is relatively easy (as compared to ions that may be more likely to sorb to organic material or precipitate from solution). Details of perchlorate physical properties, and its chemistry, is reviewed by Schilt (1979) and Urbansky (1998).

### Health Effects of Perchlorate Exposure

Perchlorate has been used in the past as a treatment option for individuals experiencing hyperthyroidism, a health condition more commonly referred to as Grave's disease. The thyroid gland of those people afflicted with Grave's disease over produces hormones (i.e. organic iodide), which negatively affects the body's metabolic activity. Ions are taken up by the thyroid based on their ionic volume. Therefore, because perchlorate has a greater molecular volume than iodide, perchlorate is preferentially taken from the blood stream (Chiovato, Santini, and Pinchera 1997; Foye 1989; Orgiassi and Mornex 1990). Individuals with Grave's disease benefit from the ingestion of perchlorate because fewer hormones are produced.

In 1992, EPA reviewed the effects of perchlorate given to patients with hyperthyroidism to evaluate to potential toxicity of perchlorate (Dollarhide 1992; Stanbury and Wyngaarden 1952). The review found an observable effect at a concentration of 0.14 mg/kg/day. Doses of 6 mg/kg/day over a period of at least two months resulted in fatal bone marrow alterations. Based on these findings, the CDHS established the 18 µg/L action level (CDHS, 1997b) which signifies the initiation of remediation activities and stoppage of water use. It is unknown what effect long term, chronic exposure to perchlorate will have on the human body (Urbansky 1998). There are ongoing studies by the EPA to set a new dose level of perchlorate, and therefore it is anticipated that the regulatory level of perchlorate will change (Urbansky and Schock 1999, Urbansky 2000).

## CHARACTERISTICS OF PERCHLORATE REDUCING MICROORGANISMS

Most chlorate-respiring microorganisms (CRMs) are also perchlorate reducing microorganisms (PRMs), and all PRMs are CRMs (Coates et al. 1999, Logan et al. 2000). Many strains of CRMs have many attributes in common with denitrifiers, but have some characteristics that are atypical of anaerobic microorganisms (Logan 1998a). This may be due (in part) to the fact that the theoretical energy yield of chlorate reduction is not only larger than for nitrate, sulfate or iron, but it is also larger than oxygen (Malmqvist, Welander, and Gunnarsson 1991). While this situation does not necessarily translate to more ATP production than for these other EAs, measured yields of 0.6 g-cell/g-acetate are larger than those typical of anaerobic processes and at the upper end (0.4 to 0.6) for aerobes (Grady and Lim 1980). Cell doubling times of 6 hr for CRMs are among the highest recorded for anaerobes (Logan et al. 2000, Logan 2000).

**Early Studies on CRMs**

Sodium chlorate was first used by Bryan and co-workers (Bryan and Rohlich 1954, Bryan 1966) as an alternative to dissolved oxygen in a BOD test. They found that chlorate concentrations of <1000 mg/L did not adversely affect calculated biochemical chlorate demands and that overall growth kinetics of chlorate reducers were only slightly less than those observed for aerobic microorganisms. Although chlorate has the potential to form chlorite, a toxic chemical, batch and continuous culture experiments have shown that the only end product of chlorate reduction is chloride ion, a non-toxic end product (Malmqvist, Welander, and Gunnarsson 1991). After the earlier studies of Bryan and coworkers, the use of chlorate as an alternate electron acceptor for the degradation of organic matter was largely ignored except for a patent by Korenkov et al. (1976). More recent studies of microbial growth on chlorate were conducted by Malmqvist and coworkers (Malmqvist, Welander, and Gunnarsson 1991, Malmqvist et al. 1994); these studies suggest the researchers thought that they were the first to show that chlorate could sustain microbial growth (they did not mention the earlier work by Bryan and co-workers).

**Abundances of Perchlorate Reducing Bacteria in Natural Systems**

It appears that chlorate respiring microorganisms (CRMs) are widely distributed in the natural environment, although it is not known if chlorate respiration is actually occurring in any of the environments sampled. Bliven (1996) tested different sources for chloride production from chlorate (500 mg/L) in BOD bottles amended with a glucose and glutamic acid solution (final concentration, 300 mg/L). The concentrations of chloride (mg/L) obtained after eight days of incubation by source was: anaerobic digester, 156; pulp and papermill wastewater, 63; primary clarifier effluent, 57; trickling filter effluent, 55; soil sample, 51.

van Ginkel, Plugge, and Stroo (1995) found chlorate reduction by river (Ijssel) samples, anoxic sediments from a ditch, surface soils (from a public garden), and a waste water treatment plant treating primarily domestic sewage. Microbial reduction of chlorate was supported by several many different organic chemicals including: carboxylic acids, alcohols (ethanol and propanols), some amino acids, and inorganic compounds including $H_2S$ and $H_2$. Oxygen inhibited chlorate reduction, but chlorate was completely converted to chloride in the presence of sulfate, Fe(III) and Mn(IV). Under denitrifying conditions, gas formation but not chloride production was observed implying that nitrate inhibited chlorate respiration (van Ginkel, Plugge, and Stroo 1995).

More recent work by Coates et al. (1999) and Wu (2000) demonstrate that both chlorate- and perchlorate-reducing microorganisms (CRMs and PRMs) are present in natural systems, although abundance varies as a function of the system. At sites not known to be contaminated with perchlorate, PRM abundances are highly variable. In order to account for differences in total bacterial concentrations in different samples, Wu (2000) normalized her results on the basis of $10^6$ or $10^9$ bacteria counted using acridine orange direct counts. They found more PRM in wastewater samples on a per-cell basis (15-350 PRM/$10^6$) than in water samples (0.02-0.4 PRM/$10^6$). There were also more PRM in soils from sites with direct evidence of perchlorate contamination (100-200 PRM/$10^9$) than from other sites (non-dectectable to 0.77 PRM/$10^9$). These results demonstrate that perchlorate-reducing bacteria are present at perchlorate-contaminated sites, and that perchlorate can be degraded by these microorganisms through the addition of different electron donors such as acetate and lactate. For comparison, a liter of creek water contains ~$10^9$ bacteria.

Measurements by Coates and coworkers (Coates et al. 1999) found that CRMs were quite abundant in most systems examined. They measured the abundance of CRMs, and assumed that all CRMs were PRMs based on findings that all of their CRM isolates were PRMs. Examples of the abundances of these microbes, based on cells per gram, were: aquatic sediments, $4.6\times10^3$; Mississippi river sediments, $4.3\times10^3$; pristine soil, $2.3\times10^3$.

## Chlorate- and Perchlorate-Respiring Isolates

Differing reports on strain size and morphology, spore formation, and chemicals that serve as reductants make it apparent that the ability to reduce perchlorate is not limited to a single bacterial species. Microbes known to respire both chlorate and perchlorate include: *Vibrio dechloraticans* Cuznesove B-1168 (Korenkov et al. 1976); *Ideonella dechloratans* (Malmqvist et al. 1994); GR-1, a strain identified to belong to the , subgroup of *Proteobacteria* (Rikken, Kroon, and van Ginkel 1996); and *Wolinella succinogenes* HAP-1, an obligate anaerobe (Wallace et al. 1996). Strains of *Pseudomonas fluorescens* have been found to reduce bromate (Hijnen et al. 1995). One chlorate-respiring isolate, AB-1, identified as most similar to *Comomonas testasteroni* using a Biolog test (as was *I. dechloratans*) was not tested for perchlorate reduction (Bliven 1996) but in all other cases tested, perchlorate reducers also reduced chlorate. All of these chlorate respiring microbes (CRMs) except HAP-1 are facultative anaerobes and are thought to be related to denitrifying organisms; it has been reported, however, that chlorate respiring cultures may lose the ability to reduce nitrate when cultivated on chlorate for long periods (Malmqvist et al. 1994).

Malmqvist and Welander (1992) obtained four chlorate reducing bacterial strains using streak plates and acetate/chlorate agar plates. All four isolates were Gram-negative, catalase- and oxidase-positive, motile rods. None of the isolates could use glucose, but Korenkov et al. (1976) indicated growth of their isolate only occurred on glucose in the presence of acetate. They grew aerobically or with nitrate as an electron acceptor. *I. dechloratans* is Gram-negative, motile, rod-shaped (straight or slightly curved, sometimes growing in filaments), and is capable of growth using oxygen or nitrate. It grew on acetate, alanine, asparagine, butyrate, fructose, glucose, lactate, propionate, pyruvate, and succinate as sole carbon sources, but did not grow on aminobenzoate, phenol and phenylalanine. A chlorate respiring isolate obtained by Bliven (1996), designated AB-1, was a slightly curved rod having a single polar flagellum. It grew aerobically on acetate or anaerobically on acetate and chlorate, but not anaerobically on phenol, benzene, toluene or xylene. Identification using Biolog microplates indicated a closest similarity to *Comamonas testosteroni* (as was *I. dechloratans*).

The perchlorate-respiring strain, GR-1, isolated by Rikken, Kroon, and van Ginkel (1996) was identified as a Gram-negative, oxidase positive, motile rod. It was isolated from activated sludge on plates containing acetate and sodium perchlorate with incubation under anaerobic conditions. GR-1 grew on acetate, propionate, caprionate, malate, succinate, and lactate, but was unable to grow on glucose, arabinose, mannose, mannitol, N-acetylglucosamine, maltose, gluconate, adipate, and phenyl acetate. GR-1 grew aerobically or on nitrate. It grew on perchlorate in the presence of nitrate, but nitrate increased measured doubling times from 3 hours to 9 hours. It could also respire on chlorate and Mn(IV), but not using sulphate, iodate, bromate, chlorite, selenate, or Fe(III).

An obligate anaerobic microorganism capable of perchlorate respiration at concentrations of 7 g/L of perchlorate, first designated as HAP-1, and then later classified as *Wolinella succinogenes* HAP-1, was isolated by Wallace et al. (1996) from an anaerobic sewage enrichment culture using agar plates. It was catalase-negative, with an optimum growth temperature of 40°C (range 20 to 45°C), and grew on $H_2$ and aspartate; fumarate; malate. It also grew on a mixture of $H_2$ and perchlorate on: pyruvate, succinate, acetate, whey powder, peptone, yeast extract, Brewers yeast, casamino acids or cottonseed protein. It did not grow on glucose, fructose, galactose, lactose, sucrose, butyrate, citrate, formate, propionate, benzoate, ethanol, methanol, 1-propanol, and starch. Earlier work reported by Attaway and Smith (1993) was conducted using suspended growth reactors that presumably were highly enriched with HAP-1. Aeration was found to inhibit perchlorate reduction, and completely inactivated mixed cultures after a 12 hour exposure (Attaway and Smith 1993). The mixed cultures reduced chlorate, nitrate, nitrite, and sulfate. Nitrate or sulfate did not affect reduction; chlorate (10 mM) reduced the rate of reduction, while nitrite and chlorite (10 mM) completely inhibited perchlorate reduction.

The bacterial strain perc1ace was isolated from a biosolids enrichment culture containing sodium acetate (1000 mg/L), perchlorate (500 mg/L), and other mineral salts (Herman and Frankenberger 1999). The facultative anaerobe was a Gram-negative, curved rod, capable of reducing between 0.1 to 1000 mg/L perchlorate to <0.005 mg/L. Optimum conditions for perchlorate reduction were between 25-30°C and pH 7.0-7.2. Perclace was able to grow on acetate (as electron donor) and perchlorate, nitrate, and oxygen. Nitrate did decrease the rate of perchlorate reduction, but within 48 hr both nitrate and perchlorate were reduced.

Coates and co-workers (1999) have obtained several perchlorate reducing isolates and have found that the majority of these isolates are closely related to one another and to the phototropic *Rhodocyclus* species. They have branched evolutionarily from two related subgroups: *Dechlorimonas* and *Dechlorisoma*.

Logan et al. (2000) isolated two microorganisms capable of both chlorate and perchlorate degradation. Comparative 16S rRNA sequence analysis indicated that these two isolates, KJ and PDX, were most closely related to "*Dechlorosoma suilla*" strain PS and "*Dechlorosoma*" spp. Iso1, Iso2 and SDGM. Three different methods of phylogenetic inference placed KJ and PDX with the genus *Dechlorosoma*. Two other isolates, PDA and PDB, were capable of chlorate respiration but not growth on perchlorate. These latter two isolates are the first known to respire chlorate, but not perchlorate.

**Differences and Similarities of Denitrifying and Perchlorate Reducing Species**

Based on the above characteristics of these isolates, there appear to be many common characteristics of these perchlorate respiring strains. For most strains, both chlorate and perchlorate could be reduced. Except for *W. succinogenes* (HAP-1), (per)chlorate reduction was partially or completely inhibited by high concentrations of either nitrogen and oxygen, and sulfate could not be used as a terminal electron acceptor. Chlorate-reductase has been isolated from microorganisms that also possess nitrate reductase, implying that chlorate-respiring strains may share many of the attributes of denitrifiers. While it may be that most (per)chlorate strains are facultative anaerobes that are denitrifiers, not all denitrifiers are chlorate reducers. There is also a variable effect of nitrate on (per)chlorate reduction that is interesting. The only exception to the inhibitory effect of nitrate on perchlorate reduction was reported by Attaway and Smith (1993); in their mixed cultures (presumably containing *W. succinogenes* HAP-1), perchlorate was reduced in the presence of nitrate. Although the facultative anaerobe (GR-1) isolated by Rikken, Kroon, and van Ginkel (1996) was also able to reduce perchlorate in the presence of nitrate, cell doubling times increased in the presence of nitrate implying an inhibitory effect of nitrate on perchlorate respiration.

Little is known about the biochemical pathways involved in bacterial utilization of chlorate or perchlorate as an electron acceptor. Although chlorate reductases have been isolated, these enzymes have been obtained from denitrifying strains known to reduce, but not necessarily shown to respire, chlorate (Oltmann, Reijnders, and Stouthamer 1976; DeGroot and Stouthamer 1969). Studies on the induction of chlorate-reducing enzymes suggest that in order for cells to respire using chlorate, both nitrate and oxygen would have to be absent, an observation which is not true for all chlorate respiring strains. In batch cultures, the presence of oxygen has been found to be detrimental to cell growth for all species examined except HAP-1. Because these chlorate reducing isolates obtained thus far have been found to be facultative anaerobes, as long as $NO_3^-$ is absent, dissolved oxygen in a water sample is not a problem if sufficient oxidizable substrate is present. In such a case, oxygen would be fully removed by cell growth prior to growth supported by chlorate or perchlorate respiration.

There are important differences in the physiology and biochemistry of denitrifying and perchlorate reducing species. Denitrification does not progress from $NO_3^-$ to $N_2$ in one step, but rather follows the sequence of:

$$NO_3 \rightarrow NO_2 \rightarrow NO \rightarrow N_2O \rightarrow N_2 \tag{2.1}$$

Thus, overall it is a five-electron transfer process. In contrast, the conversion of chlorine in perchlorate ($Cl^{+7}$) to chloride ($Cl^-$) requires the overall transfer of eight electrons—and it can be accomplished completely by one bacterium without the detection of toxic intermediates in solution (such as chlorite). The reason for this in at least one strain was determined by van Ginkel et al. (1996). They were able to isolate a novel enzyme (chlorite dismutase, from strain GR-1) capable of converting chlorite ($ClO_2^-$) to oxygen. This implies a sequence of perchlorate reduction of :

$$ClO_4^- \rightarrow ClO_3^- \rightarrow ClO_2^- \rightarrow O_2 + Cl^- \tag{2.2}$$

GR-1 is a facultative anaerobe and can therefore use oxygen as well as other compounds as electron acceptors including chlorate, nitrate, and Mn(IV). Oxygen produced from chlorite was not found to accumulate in solution, and therefore oxygen was probably used directly by GR-1 as an electron acceptor. It is not known if chlorite dismutase is present in all other strains of chlorate and perchlorate reducing microorganisms, but tests by Coates et al. (1999) have found evidence of this enzyme in all strains isolated so far in their laboratory.

Perchlorate reduction is significantly different from nitrate in that perchlorate is extremely stable in water. Many inorganic chemists use perchlorate under highly reducing conditions in order to maintain ionic strengths while studying the reduction of more easily reduced compounds (Espenson 1997). Perchlorate should, from a strictly overall thermodynamic perspective, react with many metal complexes but in fact it is stable with almost all complexes except methyl rhenium dioxide suspensions at low pH (where it has very high reaction kinetics) (Espenson 1997). In contrast, nitrate is quite reactive. For example, nitrate was found to be completely reduced by granular metallic iron and hydrogen with a palladium catalyst (with nitrite as an intermediate) within 14 minutes (Siantar et al. 1997). However, both chlorate and perchlorate are stable in water in the presence of zero-valent iron.

The reduction of perchlorate, shown in equation 2.2, is also coupled to an increase in biomass (Rikken, Kroon, and van Ginkel 1996; Logan et al. 2000). An increase in biomass demonstrates that perchlorate reduction is an energy yielding reaction. Perchlorate reduction occurs only under anaerobic conditions (Attaway and Smith 1993; Rikken, Kroon, and van Ginkel 1996; Mulvaney 1999). The following redox reactions describe perchlorate reduction with the oxidation of acetate to chlorite versus that of oxygen (Milazzo and Caroli 1978):

$$CH_3COO^- + 2O_2 \rightarrow 2HCO_3^- + H^+ \qquad (\Delta G° = -844\ kJ/mol\text{-}acetate) \qquad (2.3)$$

$$CH_3COO^- + ClO_4^- \rightarrow 2HCO_3^- + H^+ + Cl^- \qquad (\Delta G° = -966\ kJ/mol\text{-}acetate) \qquad (2.4)$$

$$½\ CH_3COO^- + ClO_4^- \rightarrow HCO_3^- + ½\ H_2 + ClO_2^- \qquad (\Delta G° = -801\ kJ/mol\text{-}acetate) \qquad (2.5)$$

The Gibbs free energy based on oxygen (eq. 2.3) is less favorable than that for perchlorate (eq. 2.4) only if perchlorate is completely converted to chloride. However, when the equation is written in terms of what actually occurs in the microorganisms, i.e. perchlorate conversion to chlorite (eq. 2.5), it can be seen that aerobic respiration would be more energetically favorable that perchlorate respiration. The reasoning that energy generation is only associated with the reduction of perchlorate to chlorite is supported by the purification and identification of chlorite dismutase. This enzyme is responsible for the nonenergy-yielding conversion of chlorite into chloride and oxygen (van Ginkel et al. 1996). The presence of chlorite dismutase enables cells to rapidly convert toxic chlorite to harmless chloride and oxygen. The functioning of the enzyme would also explain why chlorite does not accumulate in solution during perchlorate reduction. Bacteria capable of anaerobic respiration on perchlorate must possess chlorite dismutase as a detoxifying mechanism.

### Autotrophic Degradation of Nitrate Supported by Hydrogen

Bacteria that utilize hydrogen as an energy source are characterized as chemolithotrophs (organisms that obtain energy from the oxidation of an inorganic compound). Many of these bacteria are also mixotrophs, meaning that they concomitantly use inorganic and organic compounds as energy and/or carbon sources. The half-cell oxidation reaction of hydrogen is:

$$H_2 \rightarrow 2e^- + 2H^+ \qquad (E_o = -0.41\ V) \qquad (2.6)$$

Bacterial cells able to carry out the oxidation of hydrogen possess either one or two hydrogenase enzymes. All cells that oxidize hydrogen contain a membrane-bound hydrogenase. Some bacteria contain a second type of cytoplasmic hydrogenase, although many cells that undergo aerobic respiration lack this enzyme. The presence of hydrogen most likely does not induce the synthesis of hydrogenases; instead, a decrease in the availability of an energy source triggers the synthesis of the hydrogenase enzyme. Temperature is also a key parameter in the synthesis of hydrogenase;

synthesis of both hydrogenase enzymes ceases at temperatures above 33°C (Friedrich and Friedrich 1983).

The membrane bound hydrogenase is responsible for the oxidation of hydrogen, and hydrogen atoms are transported directly from the hydrogenase to a quinone in the electron transport chain. The electron transport chain (ETC) of cells able to oxidize hydrogen is quite similar to that of cells that use organic compounds as an energy source. Following the transport to quinone, electrons are shuttled to a series of cytochromes (either cytochrome c→ cytochrome $aa_3$, or directly to cytochrome b) and finally to the terminal electron acceptor (TEA). Most hydrogen oxidizing bacteria use oxygen as the TEA, but some also respire anaerobically (sulfate reducers, homoacetogenic bacteria, and most methanogens also use hydrogen as an electron donor). As a result of the transfer of electrons through the ETC, protons are generated and passed across the cytoplasmic membrane to create a proton gradient. This proton motive force is responsible (through the enzyme ATP synthase) for the production of ATP.

The cytoplasmic hydrogenase is responsible for the oxidation of hydrogen and reduction of $NAD^+$. The reduced form of $NAD^+$, NADH, is used as reducing power in the Calvin Cycle, which will be discussed in more detail below. The soluble hydrogenase has also been found to transfer electrons to the ETC directly.

Most hydrogen-utilizing bacteria are microaerophilic, meaning that they prefer environments with relatively low oxygen concentrations. This is a result of hydrogenases being oxygen-sensitive enzymes. These cells appear to grow best when oxygen concentrations are half of typical aerobic environments (i.e., 0.1 atm oxygen partial pressure). The growth rate of cells growing under autotrophic conditions have been decreased significantly, or growth has ceased altogether, in environments with oxygen concentrations greater than 40% (vol/vol) (Bowien and Schlegel 1981). Another characteristic of the hydrogenase enzyme is that it requires a nickel cofactor; therefore, nickel should be included in media used to grow the bacteria. Nickel is believed to be coordinated in the active site of the metalloenzyme, but the structure of the Ni center is unknown (Friedrich and Schwartz 1993).

Most bacteria that are hydrogen oxidizing are autotrophic, meaning that carbon dioxide is fixed to produce a carbon source necessary for anabolism. However, these bacteria are generally characterized as facultative chemolithotrophs. This description signifies that the bacteria are able to fix carbon dioxide through the Calvin Cycle (i.e. autotrophs) or use organic compounds for biosynthetic purposes (Madigan, Martinko, and Parker 1997). In the presence of organic compounds capable of being used as a carbon source, the synthesis of the inducible enzymes responsible for carbon dioxide fixation is repressed. When the organic substrate is depleted, the synthesis of carbon dioxide fixing enzymes begins.

For the purposes of the research presented here, the hydrogen-oxidizing bacteria of interest are autotrophs. These microorganisms utilize the Calvin Cycle to produce fructose 6-phosphate, which is used in biosynthesis. Relatively large energy (18 ATP) and reductant (12 NADPH or NADH) input is required for the production of 1 mole of fructose 6-phosphate. The overall stoichiometry of the Calvin Cycle is:

$$6CO_2 + 12\ NADPH + 18\ ATP \rightarrow C_6H_{12}O_6(PO_3H) + 12\ NADP^+ + 18\ ADP + 17\ P_I \qquad (2.7)$$

The reducing power (i.e., NADPH or NADH) originates from cytoplasmic hydrogenases or from reverse electron flow. Use of reverse electron flow is necessary if cells do not possess the hydrogenase enzyme responsible for the direct oxidation of hydrogen and subsequent reduction of $NAD^+$ or $NADP^+$. Reverse electron flow describes the transport of electrons from the "quinone pool" against the electron gradient, eventually resulting in the reduction of $NAD^+$ to NADH (or $NADP^+$ to NADPH). A membrane potential, resulting from the breakdown of ATP molecules is required because electrons are being forced against a gradient. Bacteria that undergo this process characteristically have lower growth rates and cell yields. Removal of the cytoplasmic hydrogenases resulted in an increase in the doubling time of *Alcaligenes eutrophus* from 3.6 to more than 12 hours. The loss of the membrane-bound hydrogenase activity affected the growth of *A. eutrophus* only slightly, which indicated the production of NADH by means of reverse electron transport is the growth-rate limiting step in autotrophic growth (Hogrefe, Romermann, and Friedrich 1984; Komen and Schmidt 1992).

When autotrophic hydrogen-oxidizing bacteria use oxygen as the TEA, the following reaction describes the general metabolic reaction:

$$6H_2 + 2O_2 + CO_2 \rightarrow (CH_2O) + 5H_2O \qquad (2.8)$$

where $(CH_2O)$ represents new cell material. The actual generation of ATP originates from the oxidation of hydrogen by oxygen

$$H_2 + 0.5O_2 \rightarrow H_2O \qquad (\Delta G^o_{rxn} = -237 \text{ kJ/mole}) \qquad (2.9)$$

Expression of only a few genes is necessary for the hydrogen oxidation process to occur. These genes have been located in plasmids of several hydrogen-oxidizing species. Genetic information located within plasmids can be passed "horizontally" between individuals of the same species or between different species (Perry and Staley 1997). According to thermodynamic calculations, the reduction of perchlorate to chloride (an eight electron transfer) appears to be very favorable. The following reduction reactions show the standard electrode potentials (E°; Sawyer, McCarty, and Parkin 1994) for perchlorate and oxygen reduction:

$$1/8ClO_4^- + H^+ + e^- \rightarrow 1/8Cl^- + 1/2H_2O \qquad (E^\circ = 1.37 \text{ V}) \qquad (2.10)$$

$$1/4O_2 + H^+ + e^- \rightarrow 1/2H_2O \qquad (E^\circ = 1.229 \text{ V}) \qquad (2.11)$$

The above equations indicate that perchlorate reduction could be coupled to the oxidation of water to oxygen since the E° of the perchlorate reaction is greater than that of the oxygen reaction. However, this event does not occur naturally (i.e. abiotically) at an appreciable rate, therefore it was concluded that the reaction is kinetically limited (Harris 1991). The inherent stability of the perchlorate ion is demonstrated by its use as a counter ion in studies of metal cation chemistry, and as an ionic strength adjuster (Urbansky 1998).

## Influence of Co-Contaminants on Treatment Efficiency

It has been found that phenol can serve as the sole substrate for growth of mixed cultures under chlorate reducing conditions (Logan et al. 1998). Both anaerobic and aerobic pathways are known for phenol degradation. Under denitrifying conditions, phosphate is added to phenol and then carboxylation of phenylphosphate occurs by phenol carboxylase to produce 4-hydroxybenzoate. Since CRMs may have had access to low concentrations of oxygen in chemostat work, it is possible that microaerophilic processes may be necessary for pollutant degradation. Mono and dioxygenases are important in a number of aromatic compound degradation pathways. Phenol degradation under aerobic conditions by *Pseudomonas putida* occurs by oxygen insertion into the ring leading to ring cleavage and mineralization of phenol. The work by van Ginkel et al. (1996) demonstrating that oxygen is produced by chlorite dismutase, suggests that oxygen could be available for oxygenases to use for ring cleavage.

The potential of using (per)chlorate respiring microbes for degradation of persistent chemicals such as benzene, toluene, and xylene have been investigated by several researchers (Bliven 1996, Olsen 1997, Wu 2000). In all cases, these researchers were unable to obtain batch cultures that could oxidize these chemicals while reducing chlorate. However, Wu (2000) has shown that the addition of chlorate can enhance the rate of toluene degradation. It was concluded that this increase in toluene degradation was due to leakage of oxygen by chlorate respiring bacteria into a biofilm that resulted in aerobic toluene degradation. Coates et al. (1999) have shown that a naphthalene-degrading microorganism can degrade naphthalene in the presence of a chlorate-respiring microbe only when chlorite is added into the culture. The addition of chlorite resulted in the production and release of oxygen, which allowed the aerobic naphthalene degradation. In the absence of chlorite addition, however, there was no naphthalene degradation.

In pilot scale tests on perchlorate-contaminated water containing volatile organic carbon (VOC) compounds such as trichloroethylene (TCE), it was found that VOC's were removed during treatment in a fluidized bed bioreactor fed ethanol. Breakdown products such as vinyl chloride, however, was found, likely as a result of adsorption to the granular activated carbon media in the fluidized bed. A subsequent shut down of the reactor did show some evolution of vinyl chloride over five days, but no such compounds thereafter.

There is great potential for the degradation of chlorinated aliphatics in mixed cultures under hydrogen oxidizing conditions for two reasons. First, a bacterium has been isolated that is capable of reductively dechlorinating tetrachloroethylene (PCE) to ethylene when grown with $H_2$ (Maymó-Gatell et al. 1997). It may be possible to bioaugment perchlorate-respiring cultures and biofilms with these microbes in order to facilitate reductive dechlorination of some chemicals. Second, many other chemicals can be reduced under methanogenic conditions (Ballapragada et al. 1997). Although we will not intentionally try to develop a methanogenic consortium, some growth of methanogens is likely in hydrogen fed reactors. The growth of these cells might be sufficient to degrade chlorinated chemicals in the water.

## Perchlorate and Chlorate Degradation in Bioreactors

Reactors for perchlorate degradation are needed for two different types of treatment: wastewater treatment, and drinking water treatment. Microbiological degradation of chlorate and perchlorate to low levels has been reported in the literature, but most of the literature has concerned engineered bioreactors suitable for wastewater treatment. The distinction between water and wastewater reactors is that for water treatment, there must be no toxic chemicals and little organic matter remaining in solution. For wastewater treatment, microbes and organic matter can remain in the water if it is to be discharged to a receiving water body or a POTW.

In order to specify design characteristics of a new biological treatment process, in terms of reactor detention time or loading rate, kinetic constants and growth yields for mixed or pure cultures need be known or pilot or full-scale operation data would need to be available. Important kinetic constants include: $\mu_m$, the maximum growth rate, $K_s$ (the half saturation constant), and $b$, an endogenous decay coefficient. In order to microbiologically reduce (per)chlorate, an oxidizable substrate will need to be added to a bioreactor. Any remaining substrate will need to be removed to limit subsequent biological growth in the water distribution system. Due to cell maintenance and energy requirements, there is a minimum achievable substrate concentration that is a function of the reactor configuration and microbial kinetic constants. For example, the minimum substrate concentration achievable in a completely mixed, constant flow suspended growth bioreactor is $S_{min}=bK_s/(u_m-b)$ (Logan 2000). In order to calculate this minimum substrate concentration, one needs to know values of these kinetic constants.

### *Treatment of Perchlorate and Chlorate in Reactors Using Organic Substrates*

To determine growth rate constants for chlorate-respiring mixed cultures, Logan and co-workers (Patnaik 1996, Bliven 1996, Olsen 1997, Logan et al. 1998) conducted laboratory-scale chemostat experiments using nitrogen-air purged reactors fed different substrates: acetate, glucose-glutamic acid, and phenol. Reactors were small capacity (0.5 to 1 L working volume) run at room temperature with no recycle. Chlorate was added in excess in order to measure kinetic parameters for growth on the oxidizable substrates. Maximum growth rates using acetate, glucose-glutamic acid, and phenol were 0.56, 0.12 and 0.040 $h^{-1}$, with cell yields of 0.12, 0.41, and 0.12 g-cell/g-substrate, respectively (Logan et al. 1998). These results indicate that growth rates of CRMs are quite high, and indicate that CRMs might be able to out compete other anaerobic microorganisms for oxidizable substrates in water treatment bioreactors. However, some minimum concentration of substrate will remain in solution. Assuming an endogenous decay coefficient of $b$=0.005 $d^{-1}$, acetate would only be reduced to $S_{min}$=0.26 mg/L based on the kinetic constants given above for acetate.

There have been a few other published reports on reactors suitable for treating chlorate and perchlorate, but much of the information is incomplete with respect to engineering design information such as microbial growth constants, cell yields, and reactor configuration details. For example, Malmqvist and Welander (1992) reported that fixed film reactors could be used to eliminate chlorate from paper mill wastewaters, but the type of media packing and other reactor configuration details were not provided (Malmqvist and Welander 1992). In laboratory tests, chlorate (50 mg/L) in a kraft mill wastewater (COD=1600 mg/L) was 100% removed (no detection limit noted) at detention times larger than 0.6 hr under conditions of 37°C and at a wastewater pH=7. In pilot tests, in a 20 $m^3$ reactor operating at a pH range of 6.2-7, a temperature range of 36 to 38°C, and influent chlorate concentrations of 40 to 75 mg/L, removals were complete at a detention time of 12.8 h, with removals of >90% at detention times of 1.6 h.

Malmqvist, Welander, and Gunnarsson (1991) conducted laboratory-scale chemostat experiments at 37°C at pH=7, using a 0.36 L working volume in a 0.4 L reactor, and no recycle. The reactor (but not the feed) was sparged with nitrogen to maintain anoxic conditions, and only one detention time of 25 hours was examined. Chlorate removal at concentrations of 83 to 1250 mg/L (1-15 mM) was complete with stoichiometric conversion to chloride. Cell yields were high, ranging from 1.9 to 3.8 g-VSS per g-equivalent electrons.

Researchers at the Tyndall Air Force Base have developed a suspended growth reactor to remove perchlorate from high concentrations (3000 mg/L) down to relatively lower concentrations (<0.5 mg/l). Some information on their system is contained in two publications (Attaway and Smith 1993, Wallace et al. 1996) while other information is available through Internet postings (*www.brooks.af.mil*, and *www.afcesa.af.mil*). Their skid-mounted system consisted of two reactors, a 1300 L anaerobic reactor and a 2700 L aerobic reactor, complete with automatic pH control, clarifier and feed and product holding tanks. Their anaerobic reactor used primarily cultures of *Wollinella succinogenes* HAP-1 with Brewer's yeast as a nutrient source. In a pilot test that began in May of 1995, and ran for >600 hours, they examined the effect of reactor detention times of 8-36 hours on perchlorate reduction; perchlorate was reduced to <0.5 mg/L even at the lowest detention times. Although this system was successful at treating perchlorate-contaminated waste water, it is unlikely that this system could be used to treat drinking water to acceptable levels (<18 ppb) due to high concentrations of organic matter used in the system.

An up-flow anaerobic fixed bed reactor able to treat up to 1,000 gal/day of effluent was also designed, fabricated, and successfully tested at Tyndall Air Force Base, Florida (Wallace et al. 1998). In this process, a mixed anaerobic bacterial culture containing mostly (29%-48%) *Wolinella succinogenes* HAP-1 was used. The mixed culture formed a biofilm on the packing material (diatomaceous earth pellets), and reduced perchlorate to chloride at the rate of 1 g of perchlorate reduced per hour per liter at the residence times of 1.17 and 0.46 hour.

After this AwwaRF project was begun, several other biological perchlorate treatment systems were developed and described in the literature (Coppola 1998, Logan and Kim 1998, Green and Pitre 1999, Catts 1999, Kim 1999, Logan 2000). It was reported by Coppola (1998) that Applied Research Associates had developed, for the Thiokol Corporation, a prototype reactor to treat wastewater containing perchlorate salts, corrosion inhibitors and other contaminants. In addition to perchlorate, it was also found that nitrate, nitrite, chlorate, and chromium VI were reduced by this process.

It was shown by Logan and Kim (1998) that perchlorate could be removed to below detectable levels in fixed bed sand filters inoculated with perchlorate-degrading enriched cultures. They developed a perchlorate degrading biofilm in 14.2-cm long sand columns by injecting enriched cultures into the column. After incubating them overnight (30°C), the column was switched to continuous flow and fed an artificial groundwater amended with acetate, trace minerals and nutrients (ammonia and phosphorus), and 20 mg/L of perchlorate. Oxygen was not removed from the feed water, but instead was removed within the column by the microbial consortia, resulting in anoxic conditions. Perchlorate concentrations in the column effluent were below detectable levels (<4 μg/L) at loading rates of <0.11 gpm/ft$^2$. Higher loading rates (>0.12 gpm/ft$^2$) resulted in perchlorate breakthrough producing high effluent concentrations (>150 μg/L) that rapidly increased with hydraulic loading.

A fluidized bed reactors (FBRs) was used to treat perchlorate-contaminated water to low levels at a site in Rancho Cordova, California. The system consisting of four FBRs each 4.27 m in diameter, treated groundwater contaminated with perchlorate using ethanol as an electron donor (Greene and Pitre 2000). The system treated ground water at a flow rate of 4,000 gal/min, and had a perchlorate loading rate of approximately 0.7 kg/m$^3$-day. The FBR used granular activated carbon (GAC) as a support medium for biofilm growth, and ethanol as a carbon source and electron donor for perchlorate reduction. When the carbon source changed to methanol or acetate, the biofilm growth was reduced and its characteristic changed. The full scale FBR system was shown to reduce perchlorate from 8 mg/L to less than 18 μg/L.

An FBR pilot test was conducted for the Baldwin Park Operable Unit (BPOU) Steering Committee in San Gabriel Basin, California (Catts 1999). The FBR, filled with granular activated carbon as a biofilm support medium, removed perchlorate from concentrations of 100 μg/L to less than 4 μg/L. The bioreactor also removed nitrate from 50 mg/L to 0.1 mg/L. Performance of the bioreactor was strongly dependent upon the dissolved oxygen in the influent, the hydraulic retention time in the bioreactor, and the ethanol concentration in the reactor influent.

A critical factor in the design of reactors for water treatment is the choice of the substrate (electron donor) added to the reactor. While acetate has been used extensively in laboratory work as the electron donor, it remains to be seen if acetate will be cost effective for larger systems. Alternative candidates include: acetic acid, ethanol, methanol, and hydrogen gas. Of these, hydrogen is particularly appealing and is the subject of a recent patent (Logan 1998b). Chlorate respiring microorganisms are known to oxidize hydrogen. This gas is sparingly soluble, and therefore would not be likely to persist in water in appreciable quantities. Methanol and ethanol have proven to work, but cost efficiency is only one aspect of the system operation. Acetic acid reduces the pH compared to acetate and therefore cannot be introduced at high concentrations without buffering the water.

Most of the bioreactors described above have been based on packed bed or fluidized bed bioreactors (Catts 1999, Kim and Logan 2000, Greene 1999, Giblin et al. 2000b, Wallace et al. 1998). In biological packed-or fluidized-bed reactors, the perchlorate-contaminated water passes directly over the perchlorate-reducing culture. Bacterial cells and soluble microbial products can contaminate the treated water. A carbon source (e.g., acetate, lactate, and ethanol) is added directly to the water to be treated. Excess organic carbon in treated water, if not removed, will stimulate microbial growth in the water distribution system, perhaps leading to the spread of pathogens.

A membrane-immobilized biofilm bioreactor can be designed so that the water to be treated is kept separate from a microbial biofilm by a microporous membrane. In this reactor the water contaminant is allowed to diffuse from the contaminated water through the membrane to the biofilm growing on the other side of the membrane. The biofilm-side of the membrane is kept in contact with the electron donor, nutrient and minerals needed for microbial growth. The membrane and biofilm act as a barrier to prevent the carbon source from mixing with the water being treated, minimizing the contamination of treated water by carbon source, bacterial cells, and soluble microbial products.

Membrane-immobilized biofilm reactors have been investigated for their effectiveness in converting nitrate to nitrogen gas (denitrification), but not for perchlorate reduction. In the early 1980s, Nilsson and coworkers began the work on denitrification by immobilized cells (Nilsson et al. 1980, Nilsson and Ohlson 1982). Later on, Lemoine, Jouenne, and Junter (1988, 1991a) used sandwiched immobilized cells between microporous membranes to separate the immobilized culture from the water being treated. They were able to obtain decreased concentrations of organics in the treated waters (Lemoine, Jouenne, and Junter 1991b).

For each of these reactors, it is necessary to determine optimum feed rates, reactor detention times, and reliability. Each of these systems may also require additional unit processes. For example, water polishing by conventional filtration or further treatment in additional biological reactors to remove any residual substrate in the water may be necessary for the non-membrane based systems. For systems treating water containing low concentrations of perchlorate (<100 μg/L), it may also be necessary to include a system for regenerating perchlorate degrading capacity of the biofilm by taking the system off line and feeding the reactor high concentrations of chlorate and oxidizable substrate (Logan 1998b). The fact that perchlorate respiration proceeds in the absence of substrate in the source water (i.e. the microbes can use stored substrates) may make it possible to develop reactors based solely on endogenous decay. Laboratory and pilot scale testing will be necessary to determine which of these approaches will ultimately prove to be the most stable and economical.

***Potential for Treatment of Perchlorate in Reactors Using Hydrogen Gas***

The potential for using hydrogen gas as an energy source in a biological reactor degrading perchlorate can be see from studies using hydrogen gas for denitrification. Microorganisms carrying out biological denitrification have been shown to be able to use hydrogen ($H_2$) as a source of electrons both in bench scale reactors, and reactors operating in full-scale water treatment facilities (Rittmann 1995). When $H_2$ is used, an inorganic source of carbon such as carbon dioxide gas or natural alkalinity must be supplied to allow for bacterial growth. A typical primary-substrate reaction coupling $H_2$ oxidation and nitrate reduction is

$$2.5\ H_2 + NO_3^- + H^+ \rightarrow 0.5\ N_2 + 3\ H_2O \qquad (2.12)$$

This equation demonstrates that a strongly buffered system is required since $H^+$ is consumed. Typical kinetic values when $H_2$ was the limiting substrate in biological denitrification processes were found to be the following: cell yield Y=0.85 g-cell/g-substrate, $K_s$= 0.125 g/m$^3$, and the minimum substrate concentration necessary to support a steady-state biofilm $S_{min}$=0.01 g/m$^3$ (Rittmann 1995).

A bench-scale, fluidized sand bed contained in a conical column was used by Kurt, Dunn, and Bourne (1987) to measure denitrification rates under $H_2$ oxidizing conditions. Hydrogen gas was delivered to the column at a rate of 7.5 L/hr, and a modified polarographic oxygen electrode was used for measurement of dissolved $H_2$. Kinetic parameters were calculated using a double Michaelis-Menten equation where nitrate ($NO_3^-$) and $H_2$ were assumed to be the limiting substrates in the first denitrification step ($NO_3^-$ to $NO_2^-$), and nitrite ($NO_2^-$) and $H_2$ were assumed limiting in the second step ($NO_2^-$ to $N_2$). The rate of biological denitrification seemed to be limited more by the concentration of electron acceptor (nitrate) than the electron donor (hydrogen). The Monod half-saturation constant ($K_s$) value for $H_2$ was found to be less than 0.1% of saturation (approximately 1 mg/L). Twenty-five mg/L $NO_3^-$-N was removed in the bench-scale reactor with a 4.5 h detention time, and the optimum pH for biological dentrification was found to be 7.5. The pH tended to rise as denitrification proceeded if insufficient buffering was provided.

A full-scale biological denitrification plant using $H_2$ gas as the electron donor was put into operation at Rasseln near Monchengladbach, Germany (Gros, Schnoor, and Rutten 1986). The treatment process was given the trade name Denitropur. Raw water is saturated with $H_2$ and amended with phosphate and carbon dioxide. Following the introduction of the electron, carbon, and buffering sources, nine fixed-bed denitrification reactors in series, packed with a commercial support structure (Mellapack), were used to biologically remove nitrate. After the denitrification step, water was aerated to remove the aqueous $H_2$ that was not utilized by microorganisms. Biologically treated water was filtered through a dual-media filter, and was finally disinfected by ultraviolet radiation. The plant had been designed to treat 50 m$^3$/hr, and to reduce an influent nitrate concentration from 75 to less than 5 mg/L $NO_3^-$ with a residence time in the bioreactors of 1 hour. At an operating temperature of 10°C, the plant was capable of removing 0.25 kg $NO_3^-$-N/m$^3$/hr. The microbial growth rate was determined to be in the range of 0.1-0.3 day$^{-1}$.

A gas-lift bioreactor (packed with pumice particles) which was inoculated with activated sludge was used by van Ginkel et al. (1998) to biologically reduce chlorate to chloride under $H_2$ oxidizing conditions. Complete chlorate reduction, along with the evolution of chloride in stoichiometric quantities, was observed after 2 weeks of incubation. A decrease in hydraulic detention time from 24 to 6 hours did not affect the bioreduction process. Following the establishment of a biofilm on the pumice particles, these particles were packed into a reactor and studied for the ability of this consortium to accomplish perchlorate removal. An influent concentration of 2 mg/L perchlorate was fed to the reactor, and a 99% removal of perchlorate occurred with a detention time of 2 hours. Reduction occurred under anaerobic, $H_2$ oxidizing conditions. A maximum conversion rate of perchlorate was calculated as 1 g-perchlorate/m$^3$-reactor/h. These results suggest that it could be possible to support perchlorate-degrading biofilms by using perchlorate to "regenerate" the reactor by developing deep biofilms throughout the reactor.

## METHODS OF HYDROGEN PRODUCTION

The proposal to use hydrogen in a drinking water treatment system is relatively novel for a treatment system in the United State. However, there are two considerations for using hydrogen: is this system sustainable; and what will it cost? Organic substrates are more commonly produced and used, but methods of hydrogen production are less familiar. Therefore, we review below different methods that can be used to produce hydrogen: several of these methods are environmentally desirable (more sustainable) but can be more costly.

The future use of hydrogen in drinking water treatment systems will depend in large part on the economics of hydrogen production. On a global scale, it is anticipated that hydrogen gas will become increasingly popular and more economical as a means of transferring energy. The potential use of hydrogen gas as an energy carrier is beneficial for several reasons. As an energy carrier, hydrogen serves as a medium through which the energy source (e.g. solar or nuclear energy) is stored and transmitted. Hydrogen gas can be produced safely, is a nonpolluting gas, and has various potential energy uses including powering nonpolluting vehicles and heating homes. Increased use of domestically produced hydrogen could also allow the United States to decrease its dependence on importation of oil. The following sections highlight methods of large-scale hydrogen production currently used, as well as several developing technologies that could prove economical and efficient.

### Hydrogen Production Using Fossil Fuels

Carbonaceous materials (e.g. natural gas, coal, or biomass) can be used in the production of hydrogen gas by five different methods (Cox and Williamson 1977). In the first method, methane is directly converted to carbon (coke) and a gas stream that is as much as 95% pure hydrogen gas. In the second method, called the steam-iron process, there are separate oxidation and reduction steps: in one reactor, steam is transformed into hydrogen, while in a separate reactor a gas is used to reduce the product $Fe_3O_4$ to $Fe_{0.947}O$. The remaining three processes are: catalytic steam reforming, catalytic partial oxidation, and non-catalytic partial oxidation. These processes produce hydrogen and carbon monoxide (CO) as synthesis gases in the first stage of production. A subsequent water-gas shift reaction (Eq 2.13) converts the CO to additional hydrogen and carbon dioxide.

$$CO + H_2O \rightarrow CO_2 + H_2 \tag{2.13}$$

Commercial hydrogen production processes separate product hydrogen and carbon dioxide by counter-current absorption using liquids (e.g. water, methanol, and potassium carbonate solutions). The liquid absorbs the unwanted hydrogen sulfide and/or carbon dioxide, leaving hydrogen gas of increased purity.

Steam reforming of natural gas (methane) is currently the most commonly used process for large-scale hydrogen production, and is also the lowest-cost method (Basye and Swaminathan 1997). This process involves the reaction between methane and steam at elevated temperatures in the presence of metal catalyst(s). The popularity of natural gas-derived hydrogen originated from the high-hydrogen content, relatively low cost, and vast quantities of natural gas (Cox and Williamson 1977). However, a shift to increased use of noncatalytic partial oxidation processes is occurring. The reasons for the shift are decreasing natural gas supplies, the resulting increase in costs, and the variety of fossil fuels able to be used in hydrogen production by noncatalytic partial oxidation (NCPO). NCPO, along with the steam-iron process, are the only commercial processes which can use both solid (e.g. coal, char) and liquid (e.g. heavy oils, petroleum residuals) fuel sources (Cox and Williamson 1977). The NCPO process involves the mixture of steam, oxygen, and fuel in necessary proportions to only partially oxidize the fuel of CO. As is the case in natural gas steam reforming, a water-gas shift is required to produce higher purity hydrogen gas. Until rationing of methane is necessary, the steam reforming process will remain popular due to its lower capital and operating costs (Cox and Williamson 1977).

The use of solid fuels in partial oxidation systems is termed gasification. The fuel is broken down into highly reactive vapors, and subsequently condensed into oils capable of being steam reformed in the ultimate production of hydrogen gas (NREL 1995). Coal is the most commonly used fuel in gasification, although the use of biomass (e.g. wood, grass, agricultural and municipal waste) offers a economically feasible, sustainable alternate fuel source. Either the use of municipal solid waste, which must be dried before use, or wood may be a potentially low-cost fuel for gasification.

**Electrolysis**

The production of hydrogen gas by electrolysis is achieved by passing a direct current between two electrodes immersed in an electrolytic solution so that water decomposes to hydrogen and oxygen according to:

$$2H_2O \rightarrow 2H_2 + O_2 \tag{2.14}$$

Both alkaline and acidic electrolytes can be used in the production of hydrogen and oxygen by electrolysis (Cox and Williamson 1977).

In order for the splitting of water to occur, sufficient energy (the quantity of the free energy of formation of water) must be supplied. The source of this energy (i.e. electricity) may be derived from fossil fuels, solar, wind, and hydropower sources. The result of the necessary energy source is high operational costs, and therefore, electrolysis is not used for large-scale hydrogen production in the United States and Europe (Basye and Swaminathan 1997).

In the late 1970's, large electrolysis plants were located near large rivers where relatively cheap hydroelectric power was available to supply electricity to the electrolysis process (Cox and Williamson 1977). Recent modifications to the electrolyzer design have allowed for increased efficiency and lowering of operational costs. In the past, commercial electrolyzers were operated at 70-90°C to minimize the corrosion of electrode materials and loss of water through evaporation (Cox and Williamson 1977). Steam electrolysis is a current advancement to the electrolysis process in which a portion of the energy required to split the water originates from heat, thereby making the process more efficient. In addition, proton exchange membranes (PEM) are used to separate the cathode and anode in the electrolyzer cell, resulting in the direct separation of the two produce gases. This separation is a cost-intensive processing step that is avoided by the use of PEM. Electrolysis systems using PEM also operate at higher current densities which result in decreased capital costs and increased hydrogen generation efficiency (Basye and Swaminathan 1997).

**Other Methods of Hydrogen Production**

The previous methods of hydrogen production required a relatively large energy input, and the source of the energy came from electricity, natural gas, or some other fossil fuel. Energy from the sun can also be utilized by both biological and chemical systems in the production of hydrogen.

Certain bacteria and algae are capable of transforming solar energy into chemical energy, in the form of hydrogen. Within these microorganisms, chlorophyll absorbs the solar photons and enzymes subsequently use the energy to dissociate water into hydrogen and oxygen. A substantial limitation of these photobiological systems is that many of the enzymes able to dissociate water are inhibited by the presence of oxygen. To overcome this limitation and the relatively low efficiency of photobiological production (only 5-6% of the sun's energy is converted into hydrogen energy), two system manipulations are currently being studied (NREL 1995).

Water dissociating enzymes found to be more oxygen tolerant have been genetically introduced to *Chlamydomonas*, a green algae that possesses oxygen-generating enzymes (NREL 1995). The result of the introduction is an alga that concurrently produces hydrogen and oxygen. Hydrogen-producing and oxygen-producing enzymes have also been extracted from microorganisms and placed on opposite sides of a conducting surface. This surface can conduct the flow of electrons from one side to the other. The oxygen-generating enzymes use solar energy to oxidize water, which results in a flow of electrons to the hydrogen-producing enzyme. The hydrogen that is produced could theoretically occur at efficiencies of nearly 25% (NREL 1995), which mimic the thermal efficiencies common to electrolysis systems (24-35%; Cox and Williamson 1997).

There exist several methods of hydrogen production by chemical systems that utilize solar energy. However, the use of solar energy in these systems falls into two general descriptions. The sun's energy can be used as a source of heat to thermally decompose water or to achieve necessary temperatures for other processes (i.e. producing steam that undergoes electrolysis). An example of the second method is photoelectrochemical (PEC) production, a system that directly uses the sun's energy in electrolysis systems. PEC production of hydrogen gas incorporates photovoltaic cells, which conduct electrical current when exposed to light, and an electrolyzer in a single device. The electrical current within the photovoltaic can create a voltage capable of splitting water by electrolysis. Current research into photovoltaics is focusing on materials that can more efficiently transform solar energy into an electrical voltage at a quantity high enough to dissociate water. PEC technology is potentially an excellent option for sustainable production of low-cost hydrogen (NREL 1995).

# CHAPTER 3
# ANALYTICAL METHODS

## INTRODUCTION

The analytical procedures used in this research follow the QA/QC plan submitted to the Awwa Research Foundation and included in Appendix A with modifications in Appendix B. Various aqueous sample measurements were made at Penn State University and the University of Nevada, Las Vegas (UNLV). In this section, we report analytical methods used at the two universities. More detailed information is given on the UNLV procedures; differences are noted for the PSU methods. Detailed methods related to the individual bioreactor experiments are contained in subsequent chapters.

## ANALYTICAL METHODS

### University of Nevada Las Vegas (UNLV)

#### *Perchlorate, Nitrate, Sulfate, Lactate, and Chloride Analysis*

Perchlorate, nitrate, sulfate, lactate, and chloride were measured using a DX-120 ion-chromatograph with a Dionex IonPac AS11 4mm (10-32) separation column and IonPac AG-11 4mm (10-32) guard column. The IC was connected to a computer equipped with the Dionex Peaknet Software (Version 5.1). Varying concentrations of sodium hydroxide (NaOH) were used as the eluent for all analyses; 49 mM for perchlorate; 5 mM for nitrate and sulfate; and either 5 mM or 1 mM for chloride and lactate. The procedure used for perchlorate analysis is similar to that developed by the California Department of Health Services (Appendix C). For perchlorate analyses, a 1000 μL injection loop was used while a 25 μL injection loop was used for the other anion analysis. The procedures for the analysis of lactate, chloride, nitrate and sulfate were developed in the UNLV Environmental Engineering laboratory and used the same columns as for perchlorate, but the eluent concentrations were different. Several trials were needed to determine the eluent concentration that allowed the detection of the anions at the desired levels and produced calibration curves with linear least square regression coefficient ($R^2$) greater than 0.995.

Calibration curves for all anions used either five or seven different concentration levels. After the calibration procedure was completed, $R^2$ values of 0.997 or greater for perchlorate, and 0.995 or greater for the other anions were achieved. A mid-standard and QC check consisting of a calibration standard and a sample prepared by another member of the research group, respectively, were added after every 20 samples. Samples with anion concentrations outside the calibration range were diluted with DI water so that the analyses were performed within the calibration range. Sample calibration curves are shown in Figures 3.1 to 3.10. The data used to build the figures are shown in Tables 3.1 to 3.11.

#### *Preparation of the Eluent for IC Runs*

Prior to measuring anion concentrations, it was necessary to prepare different concentrations of NaOH solutions to be used as the eluent. Proper degassing of the DI water to be used in the eluent and addition of the correct amount of NaOH were very important steps in obtaining a good calibration curve. The following procedure was used to prepare the eluent solutions for anions analysis:

- Four liters of DI water were added into an eluent bottle.
- A desired amount of DI water (depending on the eluent concentration—Table 3.6) was withdrawn from the bottle using a pipette.
- The DI water contained in the bottle was then purged with helium gas for about one hour.
- After the water was degassed, different volumes of 50% (w/w) NaOH was added to the bottle (Table 3.6).
- Next, the bottle was capped tightly and turned upside down gently at least three times to assure the solution was well mixed.

***Ion Chromatograph Operating Conditions***

For all the anion analyses, the IC operating flow rate was 1 mL/min and it operated with external water mode for auto suppression. The system operating pressure was maintained below 1500 psi.

***Total Suspended Solids Measurement***

Suspended solids analysis was performed gravimetrically to determine the concentration of microbes in the reactors. Samples were filtered into an Erlenmeyer flask, using a filter support, funnel and a vacuum pump. For some tests, a GF/C Whatman glass-fiber filter (APHA, AWWA, and WEF 1995) was used, and for others, 0.45 μm and 0.2 μm membrane filters were used. The filter type and size used in every experiment is described in each individual experimental procedure.

***Calibration of the Dissolved Oxygen (DO) Meter***

The DO meter was calibrated before being used to monitor anaerobic conditions in the reactors. Calibration was performed as follows:

- The YSI 54A Dissolved Oxygen Instrument (YSI Models 54 ARC and 54 ABP) was switched to the "OFF" position and the meter mechanical zero was adjusted.
- The instrument was turned on, and the knob was adjusted to RED Line.
- The probe (YSI 5720) was plugged into the instrument for 10-15 minutes for probe polarization.
- The instrument was switched to the "ZERO" position and adjusted to "0" on the mg/L scale.
- The instrument was switched to "TEMP" and the probe (YSI 5720) was placed in a BOD bottle, containing about 1 inch of water, for about ten minutes for stabilization.
- The temperature on °C scale was obtained from the instrument.
- Calibration values were determined by using the probe's temperature and the true local atmospheric pressure or the altitude above sea level. The altitude of the UNLV

Environmental Engineering Laboratory is 2000 ft above sea level (the correction factor is 0.93).

- The instrument was switched to the 0-10 mg/L D.O. range and was adjusted with the "CAL control" until the calibration value determined by the above step was reached.

The membrane of the DO meter probe (YSI 5720) was changed before every experiment. Model 5775 Oxygen Probe Service Kit containing membrane booklets and one bottle of $O_2$ Probe Solution was used to replace the membrane.

***Calibration of the pH Meter***

The pH meter used in the experiment was a Corning pH/ion meter 450 with an Orion combination pH 91-56 probe. Three points calibration (pH = 4.0, 7.0 and 10.0) was used.

**Penn State University**

The concentrations of the anions perchlorate, chlorate, acetate, and chloride were measured using two different ion chromatography (IC) systems: a Dionex DX 100 and a Dionex DX 500 system (Dionex Corp., Sunnyvale, CA). The DX 100 was equipped with an AS4A 4mm column, isocratic pump, and a conductivity detector, while the DX 500 system was equipped with an IonPac AS11 4mm column, IonPac AG11 4 mm guard column, ED40 electrochemical detector, GP50 gradient pump, and AS40 automated sampler. Data was recorded using PeakNet software (Version 5.1).

The AS-11 column (Dionex) was used for perchlorate measurements. Determination of perchlorate concentrations in the 1-20 mg/L range was accomplished using a 25 μL injection loop and in the 4-100 μL range using a 1000μL sample loop. Independent of the sample perchlorate concentration, 100 mM sodium hydroxide served as the eluent. Samples known to have perchlorate concentrations outside the above ranges were diluted with either a solution containing 15.5 mg/L $K_2HPO_4$, 8.5 mg/L $NaH_2PO_4$, 5.0 mg/L $NH_4H_2PO_4$, 0.1 mg/L $MgSO_4 \cdot 7H_2O$, and a trace metals solution (hydrogen reactor measurements) or deionized water (fixed-bed column experiment). The concentrations of the above components are the same as they appear in the growth medium used as the nutrient/buffer media in the biological growth experiments. This solution or deionized water was used to dilute samples because phosphate peak elutes before perchlorate from the AS-11 column. Thus, if phosphate is present at high concentrations, separation of the phosphate and perchlorate peaks cannot be accomplished.

Perchlorate standards were prepared with the same solution stated above. Three perchlorate concentrations of 1, 10, and 20 mg/L (as $ClO_4^-$) were used to generate a standard curve for the determination of perchlorate in the mg/L (ppm) range. Four different perchlorate standards within the 5-100 μg/L range (5, 20, 50, and 100 μg/L) were used for determination in the μg/L (ppb) range. The diluting solution was periodically analyzed for perchlorate to verify that it had not been contaminated. Either new IC vials and caps were used, or they were rinsed 5 times with DI water and reused. The detection limit of perchlorate by IC (DX-500) was determined by running 5 ppb samples (n=7) (APHA, AWWA, and WEF 1995; Standard Method 1030-E). The detection limits were determined to be 1.5 μg/L for the samples in deionized water, and 3 μg/L for the samples made in the column feeding water.

The AS-4 and AS-11 analytical columns (Dionex) were both used to measure chloride and acetate concentrations. The eluent for these anions was a 10mM NaOH solution. Procedures were the same as those described above, except the system was calibrated with 0.5, 4 and 10 mg/L chloride, and 1, 10 and 20 mg/L acetate. A 25 μL sample injection loop was used for these analyses.

For the measurement of nitrate either the IC or a nitrate analyzer (Technicon™ AutoAnalyzer™ II, Technicon Industrial Systems, Tarrytown NY) were used. This colorimetric method is described in Standard Methods (APHA, AWWA, and WEF 1995), according to Method 4500-$NO_3^-$E. To measure nitrate, the solution is passed through a copper-cadmium column to reduce nitrate to nitrite. Nitrite ion reacts with a color reagent (sulfanilamide) and is detected with a spectrophotometer. The nitrate is calculated as the difference between the original and converted solution. The detection limit by this instrument is 0.180 mg/L of nitrate. The system was calibrated using 1, 10 and 20 mg/L samples of nitrate.

## SAMPLE EXCHANGE FOR PERCHLORATE ANALYSIS

As part of the Quality Assurance/Quality Control (QA/QC) Plan, Penn State and UNLV exchanged samples to determine the accuracy of the data being collected for the project. Three sample exchanges were performed for QA/QC purposes. The exchanges took place in June 1999, June 2000 and November 2000.

Penn State and UNLV exchanged blind samples containing varying amounts of perchlorate. Standards prepared in DI water and samples taken from the biological reactors were exchanged. Perchlorate standards were prepared using sodium perchlorate (Aldrich chemicals, >99% purity). The samples were taken from perchlorate-containing bioreactors and filtered through a 0.2 or 0.45 μm membrane filter and placed in polyethylene vials. The samples were then placed on a cooler with ice-packs and shipped over night to Penn State and to UNLV. Upon arrival, samples were stored in the refrigerator and analyzed within 72 hrs. The IC was calibrated so that calibration curves for perchlorate analysis should have an $R^2$ (from linear least square analysis) of at least 0.997 and the percent difference between measured samples should not be more than 20%.

The results of the perchlorate analyses by both Penn State and UNLV, in June 1999, are shown in Table 3.12. Very similar results were found for the samples exchanged and the maximum difference between measurements was 7%. The data were very good for samples Perch-1 to Perch-3 with a DI matrix and for samples Perch 4-5 with a high TDS matrix. These are good results for the QA of both labs.

The results of the second QA/QC sample exchange (June 2000) are shown in Tables 3.13 and 3.14. Out of 20 samples exchanged, sixteen analyses met the quality assurance criterion with an $R^2$ greater than 0.9997 and analyzed concentrations differed by less than 20%. However, four out of 20 samples did not meet the quality assurance standard. It was observed that for these samples the chromatogram generated when running some of these samples had a very unusual baseline. Investigation of the irregular baseline revealed the irregular curves were the result of the presence of large amount of either lactate or acetate used in the bioreactors. The irregular baseline causes the Dionex PeakNet software to integrate a larger peak area than that resulting from the perchlorate peak. The Peaknet Software contains provisions to correct such integration problems by changing the threshold of the peak to be integrated. Such corrections were performed in reruns of samples Penn-4 and Penn-5 to evaluate the causes of the difference in results. The results are shown in Table 3.14.

Notice in Table 3.14 that after integration corrections of the baseline, the analyses are consistent and the re-run results all fall within the 20% difference stipulated by the Quality Assurance Plan. Because future perchlorate treatment plants using biological reduction may add lactate/acetate or other similar carbon sources to the water, perchlorate analysis for operational and regulatory purposes should take into account the effects of baseline deviation on peak area integration. The corrections can be easily done by inspection of the chromatograms' baselines.

Tables 3.15 and 3.16 show the results of the third sample exchange between Penn State University and UNLV for QA/QC purpose. Notice that UNLV analyses matched the target perchlorate concentrations very well except for sample Penn-4. Sample Penn-4 was re-run by UNLV and the value of 209.44 ppb was found with an error in relation to the target value of about 4.2%. The samples analyzed by Penn State during this sample exchange failed the internal QA/QC. It is suspected that a dilution error was made when diluting samples. Unfortunately, these samples could not be re-run by Penn State because the initial volumes of sample taken from the reactors was very small. The amounts of sample left were not sufficient for the second measurement.

Table 3.16 shows the UNLV analysis and Penn State re-run analysis for the UNLV prepared samples. The target concentration for samples UNLV-1 to UNLV-5 is given as a range because these samples were generated by taking samples from perchlorate-reducing bioreactors currently running in the UNLV laboratory. Thus, a percent error as compared to the target value cannot be calculated for these samples. For samples UNLV-1 to UNLV-5, the percent errors were calculated by comparing the UNLV and Penn State measurements. Notice that both UNLV and Penn State analyses have very small errors as compared to the target values and the errors are within the QA/QC range of 20% except for sample UNLV-4. UNLV re-ran this sample and found 98.16 ppm and 100.24 ppm, with errors of 4.5% and 0.24%, respectively, compared to the 1st run.

These QA/QC procedures confirmed the accuracy of the in-house procedures. When our IC systems were running and passing our in-house procedures, the inter-laboratory checks were sufficient to identify measurement problems. IC systems do require frequent checks and maintenance. For samples that did not meet our QA/QC procedures for the inter-laboratory comparison, in-house checks routinely alerted researchers of routine measurement problems and these problems were immediately corrected.

**Table 3.1**
**Calibration curve data for perchlorate at UNLV**

| Perchlorate concentration (ppb) | Peak area |
|---|---|
| 5 | 6833 |
| 10 | 14864 |
| 15 | 21675 |
| 30 | 42888 |
| 60 | 87623 |
| 80 | 118565 |
| 100 | 149887 |

**Table 3.2**
**Calibration curve data for lactate at UNLV**

| Lactate concentration (ppb) | Peak area |
|---|---|
| 30 | 1533 |
| 50 | 2404 |
| 60 | 3336 |
| 120 | 6547 |
| 200 | 11504 |
| 400 | 23807 |
| 600 | 35380 |

**Table 3.3**
**Calibration curve data for chloride at UNLV**

| Chloride concentration (ppb) | Peak area |
|---|---|
| 20 | 2937 |
| 25 | 3347 |
| 30 | 4177 |
| 50 | 7416 |
| 100 | 13698 |
| 150 | 20467 |
| 200 | 27431 |

**Table 3.4**
**Calibration curve data for nitrate at UNLV**

| Nitrate concentration (ppb) | Peak area |
|---|---|
| 25 | 1199 |
| 30 | 1625 |
| 50 | 2987 |
| 100 | 6423 |
| 300 | 20844 |
| 500 | 35822 |
| 1000 | 73613 |

**Table 3.5**
**Calibration curve data for sulfate at UNLV**

| Sulfate concentration (ppb) | Peak area |
|---|---|
| 30 | 2471 |
| 50 | 4646 |
| 80 | 6513 |
| 100 | 9009 |
| 500 | 49512 |
| 300 | 28880 |
| 1000 | 102032 |

**Table 3.6**
**Amounts of DI water and NaOH withdrawn and added, respectively, to the eluent bottles for eluent preparation**

| NaOH eluent (mM) | DI water withdrawn (mL) | 50% (w/w) NaOH added (mL) |
|---|---|---|
| 1 | 0.2092 | 0.2092 |
| 5 | 1.048 | 1.048 |
| 10 | 2.096 | 2.096 |
| 49 | 24 | 10.4 |

**Table 3.7**
**Calibration curve data for low level perchlorate (at Penn State University)**

| Perchlorate concentration (ppb) | Peak area |
|---|---|
| 5 | 7750 |
| 20 | 30998 |
| 50 | 75638 |
| 100 | 168843 |

**Table 3.8**
**Calibration curve data for high level perchlorate (at Penn State University)**

| Perchlorate concentration (ppm) | Peak area |
|---|---|
| 1 | 63430 |
| 10 | 770189 |
| 20 | 1627637 |

**Table 3.9**
**Calibration curve data for chloride (at Penn State University)**

| Chloride concentration (ppm) | Peak area |
|---|---|
| 0.5 | 63266 |
| 4.0 | 624094 |
| 10 | 1657282 |

**Table 3.10**
**Calibration curve data for acetate (at Penn State University)**

| Acetate concentration (ppm) | Peak area |
|---|---|
| 1 | 54163 |
| 10 | 565621 |
| 20 | 1131242 |

**Table 3.11**
**Calibration curve data for nitrate (at Penn State University)**

| Nitrate concentration (ppm) | Peak area |
|---|---|
| 1 | 127648 |
| 10 | 1034954 |
| 20 | 2175251 |

**Table 3.12**
**Exchange samples for perchlorate analysis between Penn State and UNLV for QA purposes (June 1999)**

| Sample ID | Measured by Penn State | Measured by UNLV | % difference |
|---|---|---|---|
| Percl-1, ppb | 10.6 | 10.0 | 6.0 |
| Percl-2, ppb | 31.6 | 30.0 | 5.3 |
| Percl-3, ppb | 81.8 | 80.0 | 2.3 |
| Percl-4, ppm | 205.0 | 208.3 | 1.7 |
| Percl-5, ppm | 147.0 | 136.9 | 7.6 |
| $R^2$ of calibration | 0.997 | 0.999 | |

**Table 3.13**
**Exchange samples for perchlorate analysis between Penn State and UNLV for QA purposes (June 2000)**

| Sample ID | Penn State analysis | UNLV analysis | % difference |
|---|---|---|---|
| UNLV-1, ppb | 33.61 | 31.02 | 8.35 |
| UNLV-2, ppb | 297.57 | 255.05 | 16.67 |
| UNLV-3, ppb | 114.92 | 100.52 | 14.33 |
| UNLV-4, ppb | 150.88 | 150.37 | 0.34 |
| UNLV-5, ppb | 68.9 | 59.98 | 14.87 |
| UNLV-6, ppm | 100.15 | 101.5 | 1.33 |
| UNLV-7, ppm | 87.75 | 112.58 | 22.06 |
| UNLV-8, ppm | 192.08 | 209.06 | 8.12 |
| UNLV-9, ppm | 244.07 | 246.8 | 1.11 |
| UNLV-10, ppm | 209.02 | 233.54 | 10.50 |
| Penn-1, ppm | 22.66 | 20.31 | 11.57 |
| Penn-2, ppm | 3.12 | 2.96 | 5.41 |
| Penn-3, ppm | 1.5 | 1.57 | 4.46 |
| Penn-4, ppb | 734.52 | 576.45 | 27.42 |
| Penn-5, ppb | 984.07 | 673.65 | 46.08 |
| Penn-6, ppb | 247.16 | 208.81 | 18.37 |
| Penn-7, ppb | 125.53 | 121.74 | 3.11 |
| Penn-8, ppm | 12.06 | 11.83 | 1.94 |
| Penn-9, ppb | 690.35 | 375.5 | 83.85 |
| Penn-10, ppm | 24.09 | 21.66 | 11.22 |
| $R^2$ of calibration | 0.9997 | 0.9997 | |

**Table 3.14**
**Integration corrections for perchlorate analysis in the presence of lactate or acetate**

| Sample ID | No integration correction | First re-run w/correction | Second re-run w/correction | Third re-run w/correction |
|---|---|---|---|---|
| Penn-4 | 245.75 | 552.05 | 552.95 | 576.45 |
| Penn-5 | 32.85 | 690.05 | 673.65 | 671.40 |

**Table 3.15**
**Exchange samples for perchlorate analysis between Penn State and UNLV for QA purposes—Penn State samples (November 2000)**

| Sample I.D. | Target | Penn State analysis | Error % (Penn State) | UNLV analysis | Error % (UNLV) |
|---|---|---|---|---|---|
| Penn-1, ppm | 200 | 501.7 | 150.9 | 206.8 | 3.4 |
| Penn-2, ppm | 20 | 22.6 | 13 | 20.58 | 2.9 |
| Penn-3, ppm | 2 | 2.2 | 10 | 2.07 | 3.5 |
| Penn-4, ppb | 200 | 120 | -40 | 4.12 ppm * | 97.94 |
| Penn-5, ppb | 20 | 12.7 | -36.5 | 22.22 | 11.10 |
| Penn-6, ppb | 50 | 40.7 | -18.6 | Sample spilled | N/A |
| Penn-7, ppb | 20 | 15.9 | -20.5 | 19.32 | -3.4 |
| Penn-8, ppb | 100 | 92 | -8 | 93.45 | -6.55 |
| Penn-9, ppm | 10 | 15.2 | 52 | 13.00 | 30.00 |
| Penn-10, ppb | 92 | 200 | 117.4 | 96.40 | 4.78 |
| QA/QC (UNLV) | 100 | | | 98.23 | 1.77 |

* UNLV re-run this sample three times and found a concentration of 209.44 ppb. The error in relation to the target value is then 4.72%.

**Table 3.16**
**Exchange samples for perchlorate analysis between Penn State and UNLV for QA purposes—UNLV samples (November 2000)**

| Sample I.D. | Target | UNLV analysis | Error % (UNLV) | Penn State Re-run | Error % (Penn re-run) |
|---|---|---|---|---|---|
| UNLV-1, ppm | 30-50 | 43.3 | - | 40.08 | -7.4* |
| UNLV-2, ppm | 20-40 | 24.46 | - | 23.44 | -4.17* |
| UNLV-3, ppm | 80-100 | 79.52 | - | 76.49 | -3.81* |
| UNLV-4, ppm | 80-100 | 102.79† | - | 2026 | 1871.0* |
| UNLV-5, ppm | 3000-5000 | 3459.1‡ | - | N/A | — |
| UNLV-6, ppb | 20 | 21.49 | 7.45 | 22.42 | 12.1 |
| UNLV-7, ppb | 150 | 148.41 | -1.06 | 203.87 | 1.94 |
| UNLV-8, ppb | 200 | 202.87 | 1.44 | 222.61 | 11.31 |
| UNLV-9, ppb | 80 | 77.48 | -3.15 | 76.35 | -4.56 |
| UNLV-10, ppb | 50 | 48.34 | -3.32 | 48.97 | -2.06 |
| QA/QC (UNLV) | 100 | 97.93 | 2.07 | | |

* the error (%) was obtained comparing UNLV and Penn State data.
† UNLV re-run this sample in duplicate and found 98.16 ppm and 100.24 ppm, with errors f 4.5% and 0.24% respectively compared to the 1st run.
‡ UNLV re-run this sample in duplicate and found 2771.2 ppm and 2822.2 ppm, with errors of 19.9% and 18.4% compared to the 1st run.

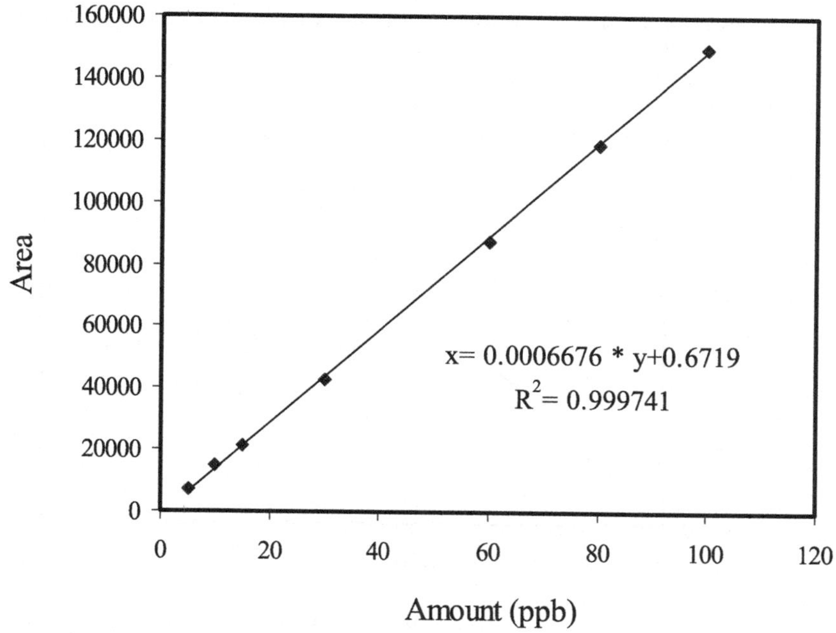


**Figure 3.1 Calibration curve for perchlorate at UNLV**

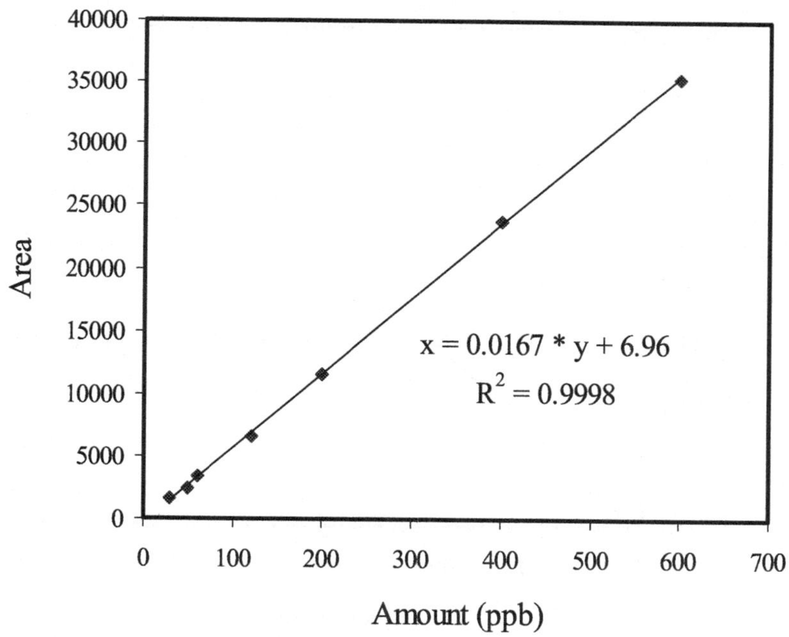


**Figure 3.2 Calibration curve for lactate at UNLV**

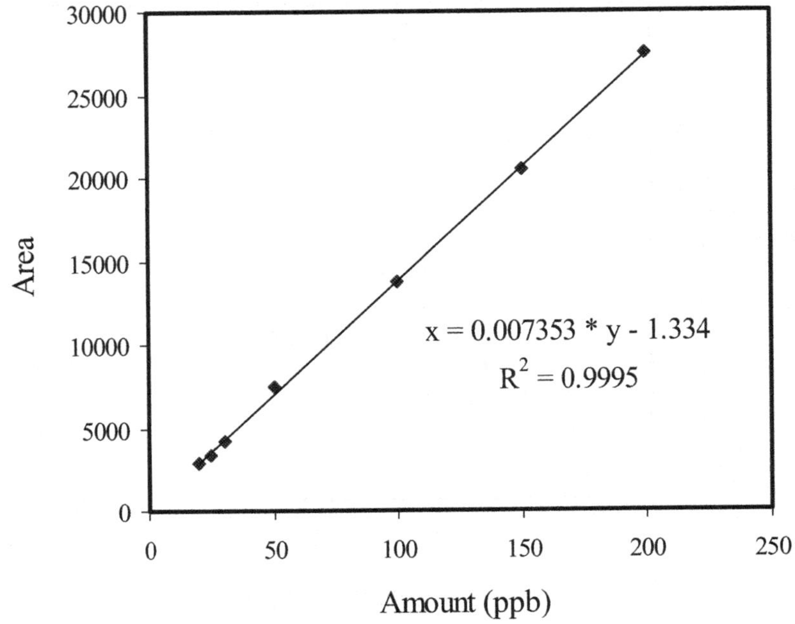


**Figure 3.3 Calibration curve for chloride at UNLV**

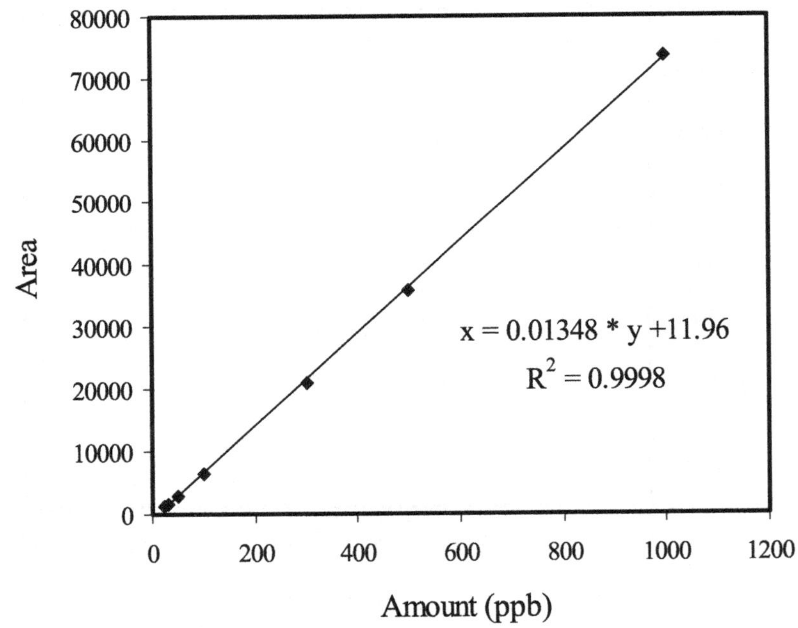


**Figure 3.4 Calibration curve for nitrate at UNLV**

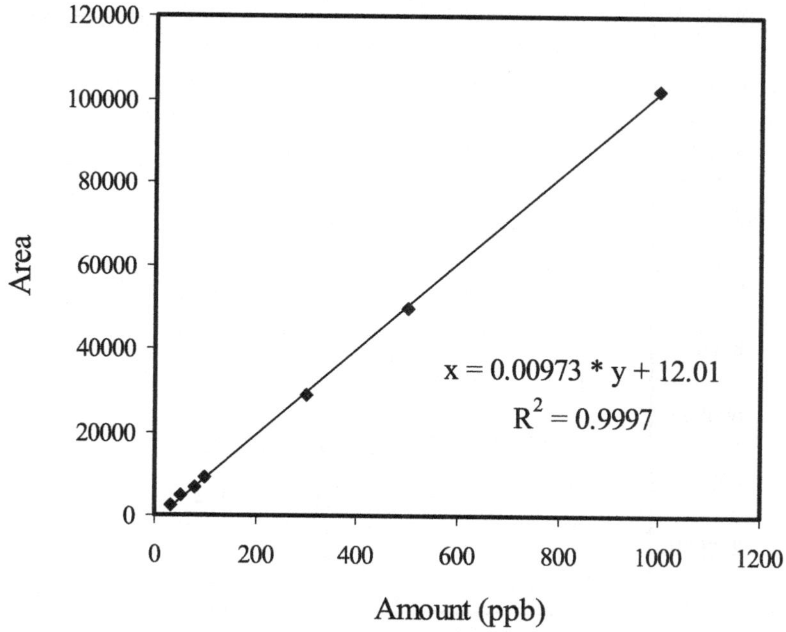


**Figure 3.5 Calibration curve for sulfate at UNLV**

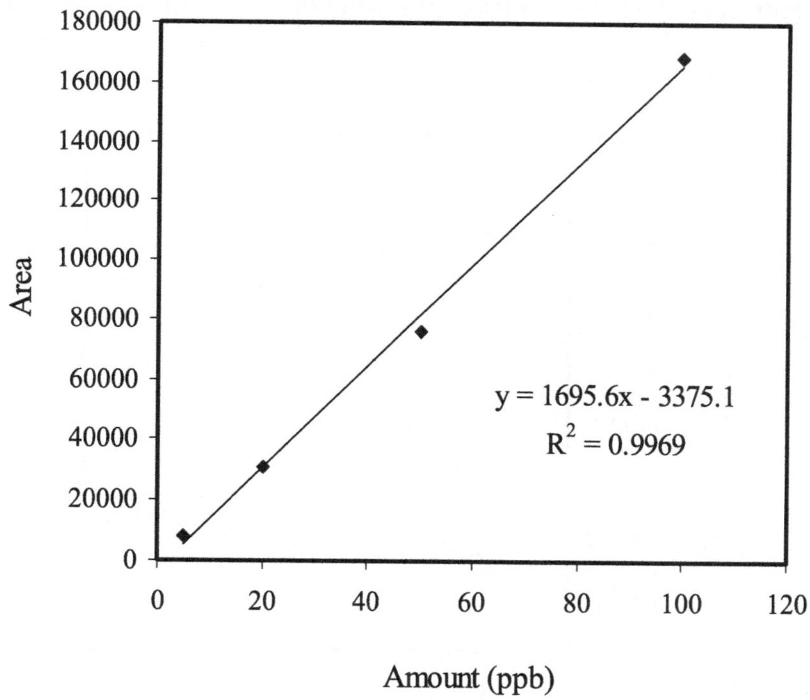


**Figure 3.6 Calibration curve for low level perchlorate (at Penn State University)**

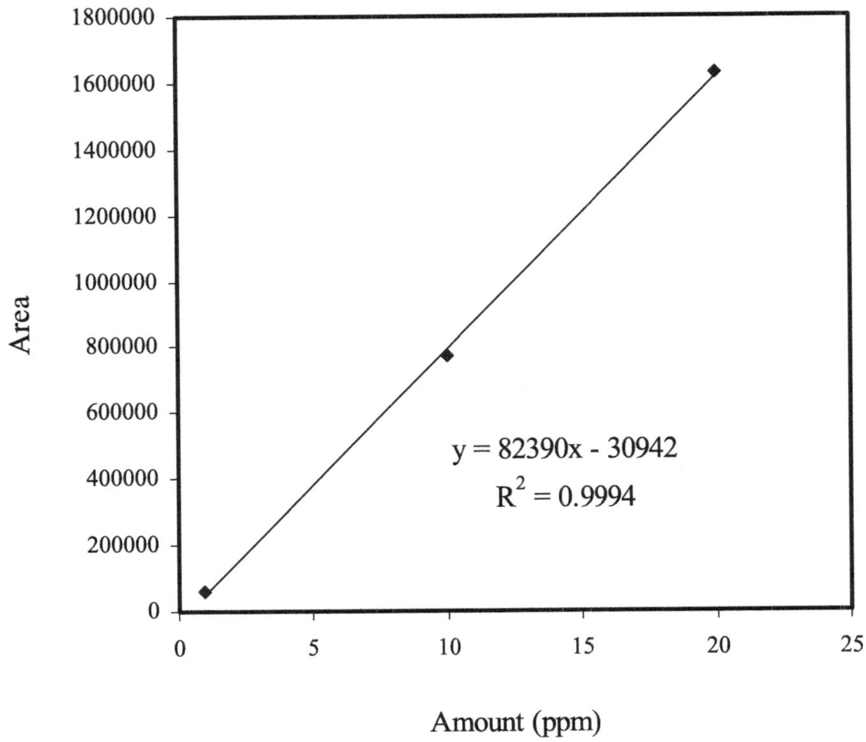


**Figure 3.7 Calibration curve for high level perchlorate (at Penn State University)**

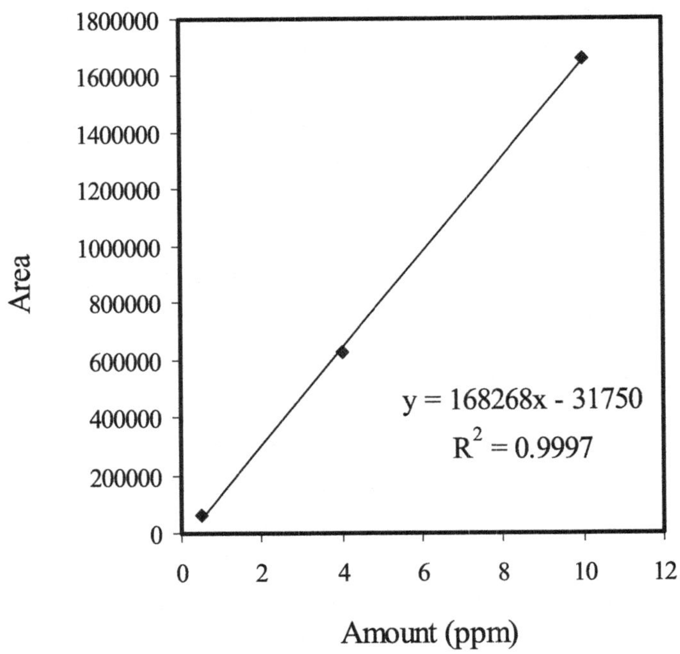


**Figure 3.8 Calibration curve for chloride (at Penn State University)**

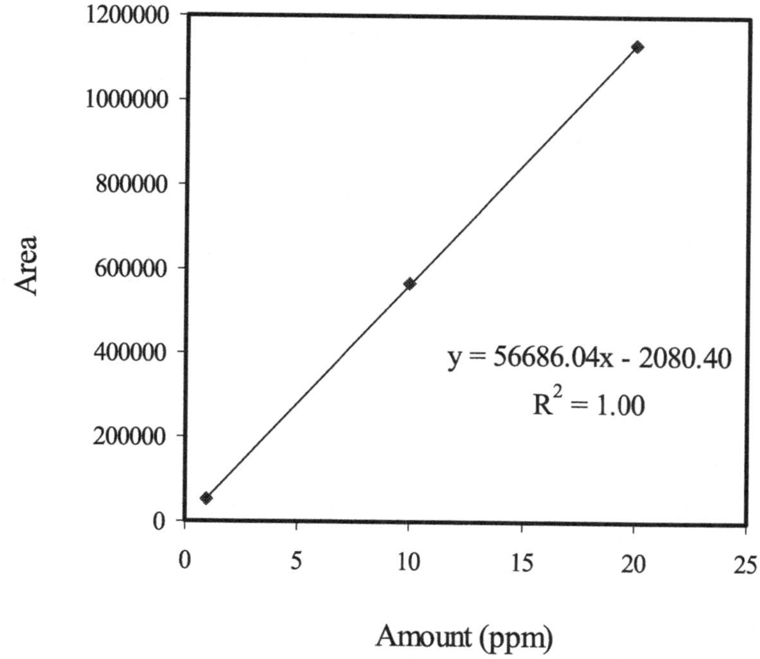


**Figure 3.9 Calibration curve for acetate (at Penn State University)**

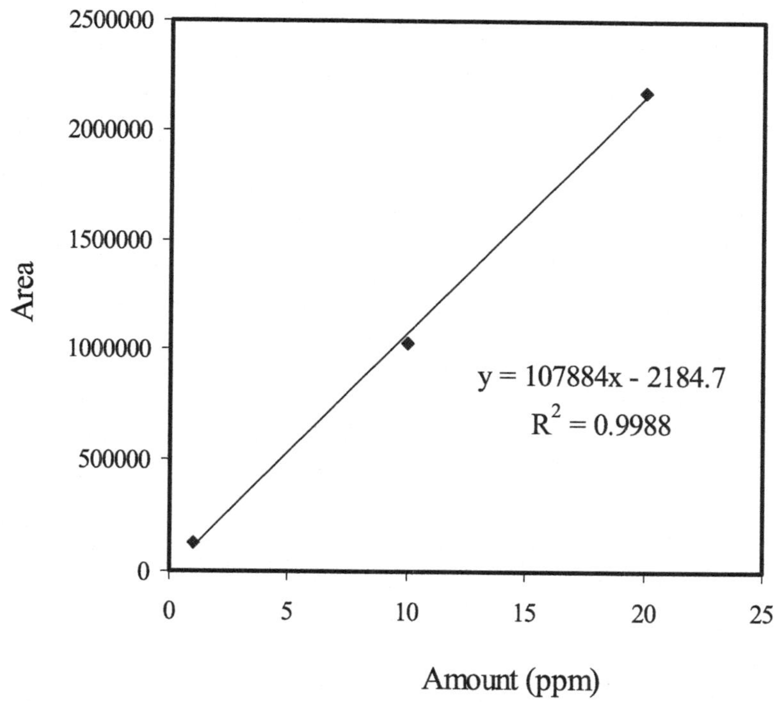


**Figure 3.10 Calibration curve for nitrate (at Penn State University)**

# CHAPTER 4
# BIOLOGICAL PERCHLORATE REDUCING FIXED-FILM BIOREACTOR

## INTRODUCTION

Biological reduction of perchlorate in fixed-film bioreactors was examined using reactors inoculated either with acclimated mixed cultures or with a pure culture, *Dechlorisomas* spp. KJ (Logan et al. 2000). Perchlorate removal from 20 mg/L to <18 μg/L in acetate-fed sand columns had already been proven in our laboratory prior to the start of this project (Logan and Kim 1998) for an empty bed contact time (EBCT) of 31 minutes. Shorter EBCTs (≤28 minutes) resulted in perchlorate breakthrough producing high effluent concentrations (>150 μg/L) that further increased at greater hydraulic loadings. Acetate concentrations remained relatively high in the column effluent during treatment and were not optimized to limit effluent concentrations.

In order to further examine and optimize perchlorate removal in packed bed reactors, a series of bioreactor tests were conducted using different packing materials including silica sand, granular activated carbon (GAC), and glass beads. In these tests, perchlorate removal was examined as a function of the concentration of acetate and the minimum EBCT was again tested for pure and mixed cultures. Experiments were then conducted to more fully examine treatment of perchlorate contaminated groundwaters such as the water in the Redlands Crafton site. Many perchlorate-contaminated waters such as these are also contaminated with nitrate, and nitrate can compete with perchlorate as an electron acceptor. Therefore, the effect of nitrate on perchlorate reduction was also examined. The perchlorate concentrations in the water were also decreased to <100 μg/L, and perchlorate removal tested with reactors inoculated with a pure culture (KJ). Finally, actual groundwater from the Redlands site was tested side-by-side in a bioreactor with an identical bioreactor treating a simulated groundwater containing a similar amount of perchlorate but no nitrate. The Redlands water contained approximately 18 mg/L nitrate and 65 μg/L of perchlorate. The control column was operated in parallel in order to assess the long-term performance of the bioreactor with water contaminated with low-levels of perchlorate in the absence of competing electron acceptors such as nitrate. Running these columns over a long period allowed examination of the potential for column clogging. However, larger pilot plant tests will be necessary to more fully examine potential long term operational success.

## MATERIALS AND METHODS

### Perchlorate Treatment in Pure Culture Sand Column Bioreactors at High Perchlorate Concentrations

The effect of using a pure culture (KJ) on perchlorate removal in a fixed-film column was studied. KJ was originally obtained from the consortium of mixed perchlorate column effluent and isolated using a streak plate technique (Mulvaney 1999; Logan, Wu, and Unz 2001). The column reactor was 28 cm long and 2.5 cm in diameter. Silica sand (Unimin Corp.) was packed in the column, producing a final porosity of 0.40. A bacterial suspension of KJ was inoculated into the column and was allowed to remain in the column for one day to ensure microbial attachment to the column support media. Thereafter, water containing approximately 20 mg/L perchlorate and 60 or 100 mg/L of acetate in a basal media (trace metals with phosphate buffer) was pumped through the column. This media was pumped into the column at the EBCT (empty bed contact time) of 117 min (0.059 gpm/ft$^2$) for the first 19 days and then progressively lowered to an EBCT of 2.1 min (3.345 gpm/ft$^2$) in order to find a critical hydraulic loading rate (breakthrough of perchlorate).

### *Acetate Consumption*

Acetate was used as a carbon source and electron donor for bacteria. Acetate concentrations in the influent and effluent of the column were measured using an ion chromatograph (Dionex-500). Acetate consumption was expressed on the basis of a molar ratio.

### *Cell Mass Distribution Within the Column*

Following an experiment, the distribution of biomass in the column was measured by extruding the column and obtaining sand samples (in triplicate) from various depths of the column triplicate. These samples were dried, weighed, and then combusted again at 550°C. The weights before and after combustion were defined as the cell mass. Potential interferences caused by organic matter on the sand were eliminated because the sand was pre-combusted to remove organic matter prior to using it in the column.

## Perchlorate Removal at Low Perchlorate Concentrations in the Presence and Absence of Nitrate

Another set of experiments were performed to: (a) demonstrate that perchlorate could be completely removed from water when it was present at very low concentrations (10-100's ppb) using a pure culture inoculum (isolate KJ) as was done at higher influent perchlorate concentrations (20 ppm); (b) to investigate the effect of nitrate on perchlorate reduction.

Two columns (one for the treatment and one as a control) were used, each one was the same as described above (28 cm in length and 2.5 cm in diameter, constructed of plexiglass) as shown in Figure 4.1. The columns were filled up to a height of 20 cm with silica sand that was combusted in a furnace at a temperature (550°C) sufficient to oxidize any organic matter on the sand. The columns were not sealed at the top so that plugging in the column (if it occurred) could be studied. Therefore, each column was run in an up-flow mode, and they were designed to be operated with 8 cm of water column above the sand bed.

The columns were inoculated with perchlorate-reducing pure culture, KJ, and one day was allowed for the microbial attachment to the column support media before pumping the feed solution. The cell concentration used for inoculation had an optical density at 600 nm (OD600) of 1.07, and was measured to contain 54 µg-DW/mL of bacteria. The feed solution initially contained ~20 ppm perchlorate and 100 ppm acetate. After 28 days of operation to build up biomass in the column (start-up period), the feed solution was switched to water containing 100 g/L of perchlorate. For the treatment (test) column, the media was switched to the Redlands water on day 19, when the perchlorate removal had stabilized below the detection level of perchlorate (<4µg/L) for 11 days.

Because Redlands water contains low-concentrations of perchlorate, it was thought that these low concentrations might not be sufficient to sustain a sufficiently thick biofilm of perchlorate-respiring bacteria over time. The amount of time that the reactor could successfully operate cannot be predicted, and must therefore be determined experimentally. Failure of the column to completely remove perchlorate, as evidenced by perchlorate breakthrough, would require that the biofilm in the column be regenerated. This could be accomplished using water containing several hundred milligrams per liter of chlorate (which is less toxic than perchlorate) since KJ can grow on chlorate. This water would be recycled until the chlorate was removed. In the event that the perchlorate-reducing bacteria in the column were lost, the columns would be bioaugmented (reinoculated). Following biofilm regeneration or bioaugmentation, the columns would be put back into service again to treat low-levels of perchlorate contaminated water.

The effect of nitrate on perchlorate degradation rates was tested by pumping water containing both perchlorate and nitrate through the column. As a control for the effect of nitrate, perchlorate-contaminated water (no nitrate) was fed into a control column. The concentrations of perchlorate and nitrate to be tested were set as follows: feed solution 1 contained 100 μg/L perchlorate and 20 mg/L nitrate in Basal medium. These concentrations were chosen to simulate the Redlands ground water

The flow rates of the media pumped into both columns were 1mL/min producing a residence time of 40.5 minutes. On day 28, the media used for both columns was changed to contain a lower concentration of perchlorate. On day 19 after the pumping media was switched to low-level perchlorate, it was again switched to the Redland water in the test column, and the water was amended with 3 mM sodium acetate, 0.26 mM $(NH_4)H_2PO_4$, autoclaved before use.

### Perchlorate Treatment in Pure Culture Glass Bead Bioreactors

Glass beads are often used in bioreactors so that the packing is uniform and has a known specific surface area. In order to compare sand and glasss bead reactors, the performance of a reactor filled with glass beads was examined with the reactor inoculated with the a pure culture (KJ). The same columns as those described above were used (28 cm high with a 2.5 cm inside diameter) except the column was packed with glass beads 0.5 mm in diameter (BIOSPEC Products, Bartlesville, OK). The porosity of the glass bead filled column was 0.40.

To begin the experiment, one pore volume of a suspension of KJ bacteria was inoculated into the reactor at a cell concentration of 85 μg/mL ($OD_{600}$=0.195) and this suspension was left in the reactor undisturbed for one day prior to pumping a feed solution into the system An abiotic column (control) was operated in parallel to examine if there was perchlorate adsorption to the glass beads. The feed solution to these columns contained 20 mg/L perchlorate and 30 to 60 mg/L acetate in basal media (trace metals and phosphate buffer). This solution was initially pumped at a rate of 1 mL/min, corresponding to an empty bed contact time (EBCT) of 142 minutes for the hydraulic loading rate of 0.20 cm/min (or 0.05 gpm/ft$^2$).

#### *Tracer Study*

Adsorption/desorption of perchlorate to the glass beads was studied in the biotic and abiotic columns as part of a tracer study. In both columns, an inert tracer consisting of 500 ppm NaCl was first used to measure column detention time. Effluent samples were collected with time and the conductivity was measured using a conductivity meter (CDM230 conductivity meter, MeterLab™, Copenhagen).

In the abiotic column, a 500 mg/L perchlorate solution was added to the column and its concentration in the effluent measured using the conductivity meter. The tracer curves from the

NaCl and perchlorate experiments were compared in order to determine if there was retardation of perchlorate produced by its adsorption to the glass beads. The NaCl tracer study was also conducted on the biofilm reactor after 23 days of column operation in order to see the adsorption/desorption behavior of perchlorate in the presence of the biofilm.

### *Cell Mass Distribution*

The distribution of bacteria according to total cell dry weight was measured in the column at the conclusion of the pure culture experiment (after 23 days). The glass bead samples taken at various height of the column (in triplicate) were dried in an oven at 110°C, and then the dry weight of the beads and cell mass were measured. The dry samples were then combusted at 550°C in a furnace to remove the organic matter, which was assumed to be equal to the weight of the cell mass. As a control for changes in the bead mass during combustion, the weight loss of clean glass beads were also measured. The biofilm mass was assumed to be the difference between that measured for the biofilm reactor beads versus those of the clean beads.

## Perchlorate Treatment in GAC-Packed Columns

Biological treatment of perchlorate contaminated water was evaluated in a fixed-bed column packed with granular activated carbon (GAC; 1240, Norit, Oklahoma) having a bulk density of 0.44 g/cm$^3$ and porosity of 0.41. The objective of using this GAC reactor was to assess the biological reduction of perchlorate with GAC used as a column support media, under conditions where we could fluidize the bed. The column was operated for a total of 77 days. The column was mixed intermittently for two purposes: to prevent the clogging caused by excessive microbial growth near the influent, and to redistribute biomass more evenly within the column.

### *Column Configuration*

The length of the GAC bed was 91.4 cm (3ft), in a column 244 cm (8 ft) long. The top 61 cm (2 ft) of the column had a larger diameter (15 cm) to accommodate the bed expansion during fluidization. The column was operated up-flow mode in order to minimize the introduction of air and maintain anaerobic condition in the column. Under normal operation (i.e. not during fluidization), the water level was at the exit valve (Figure 4.2). This column was equipped with 8 sampling ports with 6 ports located within the GAC bed. Samples were collected from the ports and used to obtain the chemical or biomass concentration profile. The sampling ports were 15 cm (0.5 ft) distant from the adjacent ports and the sample from port 7 was taken as an effluent sample.

### *GAC Adsorption Isotherm*

The adsorption of perchlorate onto GAC was quantified using a Freundlich isotherm. Five samples of different GAC doses were equilibrated for 40 h on a shaker table with 100 mg/L of perchlorate in deionized water.

### *Inoculation of the GAC Column With a Perchlorate-Reducing Mixed Culture*

The inoculum was obtained from a previous perchlorate-reducing column experiment in our laboratory, where sand was used as a column support media in a column that could not be fluidized (Kim 1999). Sand was taken from the column when the operation ended and stored for future use. A culture for GAC column inoculation was obtained by the sand samples (about 10% by volume) and adding them to Basal media (Table 4.1) containing 500 mg/L of perchlorate and 750 mg/L acetate. After a week, the solution became turbid and the supernatant was used to inoculate the GAC column.

After the inoculation, one day was allowed before the column operation for cell attachment to GAC. The HLR was 0.16 cm/min (0.04 $gpm/ft^2$) for the first day and was increased to 0.18 cm/min (0.055 $gpm/ft^2$) on day 2 and then to 0.35 cm/min (0.11$gpm/ft^2$) on day 58.

### *Column Mixing*

The column mixing was attempted twice, on day 6 and on day 73. The first mixing was performed by fluidizing the column using tap water in the laboratory. Since tap water was used to fluidize the column, unfortunately this water contained a chlorine residual. The concentration of chlorine was measured using a spectrometer (HACH DR/2000). Further attempts to fluidize the column using laboratory-scale pumps were not successful. Therefore, the second column mixing, which was on day 73, was achieved by pouring out the GAC and gently mixing (by hand) the GAC in order to minimize biomass loss.

## Batch Experiments With Suspended Pure and Mixed Cultures

In order to understand the effect of nitrate on perchlorate reduction, batch perchlorate degradation experiments were conducted using mixed cultures in the presence and absence of nitrate. Perchlorate acclimated mixed cultures were developed using wastewater (primary digester; 1% by volume) obtained from a local wastewater treatment plant added to the basal media (trace metals and phosphate buffer) containing 300 mg/L perchlorate and 450 mg/L acetate. Four days later, the acclimated suspension was transferred (8% by volume) to fresh media. The sample became turbid in two days and was used as the inoculum for three batch reactor tests containing, in basal media: nitrate only (3mM $NO_3^-$), perchlorate only (3mM $ClO_4^-$), and both nitrate and chlorate (3mM each). The batch reactors were inoculated, and kept stirred, in an anaerobic chamber during the experiment. The samples were collected over time and filtered through 0.2 μm pore-diameter syringe filters (VWR Scientific). The concentrations of nitrate and perchlorate were measured by ion chromatography (Dionex DX-500).

### *TCE Degradation*

Trichloroethylene (TCE) is present in the City of Redlands perchlorate-contaminated water. Batch experiments were conducted using isolate KJ to determine if TCE could be degraded under perchlorate-reducing, acetate-oxidizing conditions. KJ cultures were grown in basal medium autoclaved at 121°C for 1 hour and degassed in an anaerobic glove box (Coy Laboratories, Ann Arbor, MI) overnight before inoculation. Tubes (55 mL anaerobic culture tubes) were inoculated (5% v/v) with a 48-hour old culture maintained by weekly transferring on fresh medium (final optical density of 0.33). Each tube was spiked with 100 μL pure TCE (99.9%, Fisher Chemical, Fair Lawn, NJ). Controls for adsorption and abiotic losses (no cells) were spiked with the same amount of TCE. Both the treatment and control tubes were sacrificed periodically for TCE measurements. Basal medium contained, per liter of deionized water: 1.55 g $K_2HPO_4$, 0.85 g $NaH_2PO_4$, 0.5 g $NH_4H_2PO_4$, 50 mg $MgSO_4{\cdot}H_2O$, 3 mg EDTA, 1 mg $CaCl_2{\cdot}2H_2O$, 0.4 mg $CoCl_2{\cdot}6H_2O$, 4 mg $FeSO_4{\cdot}H_2O$, 0.4 $NaMoO_4{\cdot}H_2O$, 0.1 mg $NiCl_2{\cdot}6H_2O$, $NaSeO_3{\cdot}5H_2O$, 0.6 mg $H_3BO_3$, 2mg $ZnSO_4{\cdot}7H_2O$, 0.2 mg $CuCl_2{\cdot}2H_2O$, 1 mg $MnCl_2{\cdot}4H_2O$, 500 mg sodium perchlorate, and 500 mg sodium acetate.

Using a Henry's constant for TCE of 9.655 atm•mL/mmol (La Grega, Buckinham, and Evans 1994), the partial pressure of TCE in the gas phase is: p = HC = (9.655 atm•mL/mmol) (C mmol/mL) = 9.655C atm, where C is the aqueous TCE concentration. From the ideal gas law, this produces 9.88 mmol of TCE in the gas phase, and from a mass balance 30 mmol in the liquid phase. Thus, the aqueous concentration was 3657 mg/L. Due to an error in the original calculation, this was 1000× higher than originally planned.

TCE was analyzed using an SRI 9610 gas chromatograph system (SRI, Menlo Park, CA) equipped with a dry electrolytic conductivity detector (DELCD) and appropriate column (5% AT-1200 + 1.75% Benlone 34, 6′ x 0.125′ x 0.085′ stainless steel). Samples were withdrawn from tubes using a gas-tight syringe. During analysis, the oven was held at 40°C for 1 minute, ramped at 10°C/min to a final temperature of 120°C, and then held constant for 10 minutes. The DELCD detector was operated at 100°C. Helium was used as the carrier gas at a flow rate of 10 mL/min.

## RESULTS

### Pure Culture Sand Column Bioreactors at High Perchlorate Concentrations

After one day of operation, perchlorate was reduced from 25.6 ppm to non-detectable (<4 μg/L) at an EBCT of 117 min (Figure 4.3).

Effluent perchlorate concentrations increased on day 7 to 14.9 mg/L, but the acetate was found to be limiting perchlorate removal. Acetate in the feed solution was increased from an average of 60 mg/L to 100 mg/L on day 9, and this succeeded in reducing the effluent perchlorate concentration to 15 μg/L by the next day. Perchlorate concentrations in the effluent fluctuated during the next 83 days of the operation but for many of the samples perchlorate was below the detection level. The percent of samples below successive effluent concentrations were calculated using all column data between days 11 to 66 as shown in Figure 4.4. Based on a total of 43 samples (averages of measured values), perchlorate concentrations were not detectable in 67.4% of effluent samples, and all the samples contained perchlorate concentrations less than 53 μg/L for the high-perchlorate concentration feed.

Chloride measurements confirmed that all of the perchlorate was converted to chloride. The average molar ratio of chloride produced to perchlorate reduced was 1.0 for 91 samples.

#### *Minimum Detention Time for Complete Perchlorate Removal*

The critical hydraulic loading rate for complete perchlorate removal was determined by increasing the flow rates into the column until the perchlorate concentrations in the effluent were consistently high. A critical hydraulic loading rate was found to be 3.35 gpm/ft$^2$, corresponding to an EBCT of 2.1 min (detention time = 50 sec) (Figure 4.5) This EBCT is much lower than that previously measured for mixed culture column experiments of 31 minutes. This low detention time for the column inoculated with KJ suggests that the pure culture could be used to provide more efficient removal of perchlorate than reactors inoculated with mixed cultures. Thus, KJ should be used as the inoculum for perchlorate reducing bioreactors in pilot studies.

### *Acetate Consumption*

The amount of acetate necessary for perchlorate removal was measured based on influent and effluent perchlorate and acetate concentrations from the column. The molar ratio of acetate to perchlorate ($Ac^-/ClO_4^-$) averaged 6.6 ± 2.0 but varied over a wide range of 3.5 to 10.8 (Figure 4.6). The fluctuation in this ratio could be due to uptake of acetate without cell growth, acetate uptake by non-perchlorate respiring bacteria in the column, or variations due to fluctuations in the concentrations of dissolved oxygen entering the column (which were not monitored).

### *Biomass Distribution Concentration Within the Column*

The distribution of biomass in the column measured at the end of the experiments showed that the cell concentrations were highest near the inlet of the column (13.7 mg-DW/g-sand) and decreased toward the outlet of the column (Figure 4.7). This distribution in biomass in the column is not surprising since the highest biomass concentrations were found near the inlet of the column where the substrate concentrations were also the highest.

## Sand Column Bioreactor Experiments at Low Perchlorate Concentrations in the Presence and Absence of Nitrate

Separate experiments were conducted to determine the effect of low influent perchlorate concentrations on perchlorate removal rates. Two columns were first operated in parallel with a perchlorate concentration of 21.6 mg/L for a 28-day period defined as a “startup” period. Two days after the columns were inoculated, perchlorate was reduced from 21.6 mg/L in the influent to an effluent concentration of 158 μg/L in the treatment column, and to <4 μg/L in the control column. Effluent perchlorate concentrations from the treatment column were not consistent during the first 10 days of operation, with effluent concentrations ranging from <4 μg/L to 11.1 mg/L (Figure 4.8).

After this 28-day period, the medium was changed to contain a lower influent perchlorate concentration (Figure 4.9) and the low-perchlorate treatment phase initiated. Perchlorate was reduced below the detection limit in the control column within two days, while it took eight days to remove perchlorate below the detection limit in the test column (Figure 4.10).

Nineteen days into the treatment period, the pumping medium to the treatment column was switched to the Redlands water while the control column continued to receive water without nitrate. The Redlands water contained approximately 65 μg/L of perchlorate and 35 mg/L of nitrate on day 19. Perchlorate was reduced to below the detection limit for the Redlands water despite the presence of the nitrate. Nitrate was removed from a concentration of approximately 20 mg/L to effluent concentrations in the range of 9.3 to 11.3 mg/L during the first 6 days (Figure 4.10). Nitrate was further reduced to 830 μg/L on day 12, and then remained below the detection limit (180 μg/L) after day 20 for all samples except one on 39.

## Perchlorate Removal in Glass Bead Bioreactors

Tracer studies were conducted with the column packed with clean glass beads to determine if perchlorate would adsorb to the column packing. Conductivities of the NaCl and perchlorate-spiked water samples were recorded and normalized against the influent conductivities (Figure 4.11). The results show that the breakthrough curves generated by two salts match very closely indicating little dispersion in the reactor. Initially, the measured average pore volumes based on the NaCl and perchlorate spikes were 55 and 48 mL, where the difference in values is probably due to the frequency and accuracy of measurements. The calculated pore volume was 57 mL. The lack of perchlorate retardation relative to the NaCl tracer indicates a negligible amount of perchlorate adsorption to clean glass beads.

The tracer study conducted after 23 days of continuous bioreactor operation (see below) was also obtained and produced similar results (Figure 4.12). The desorption curve (perchlorate concentrations after the perchlorate additions in the feed were discontinued) shows that perchlorate did not adsorb to the biofilm coated media and therefore that perchlorate adsorption to the biofilm is not a factor in perchlorate reduction. Measured pore volumes for the abiotic and biofilm columns were 58 and 62 mL (Figure 4.11).

The glass bead columns were inoculated with the pure culture KJ, and after one day, fed perchlorate-contaminated water. Perchlorate removal to non-detectable concentrations occurred within 67 hours, but after 167 hours perchlorate removal ceased (Figure 4.13A). The experiment was repeated, and again perchlorate removal could not be sustained with the pure culture reactor using glass beads (Figure 4.13B). The detention time in the column was 57 to 101 min, which should have been sufficient to completely and consistently remove perchlorate, as found for the sand column reactors even after accounting for differences in surface area.

The intra-column cell concentration shows that the highest cell mass was found at 4 cm from the influent forming a peak (Figure 4.14). This biomass peak coincides with the region of the column where dissolved oxygen (present in the feed) was expected to be consumed; this assumption is based on the fact that oxygen is preferentially utilized as an electron acceptor over perchlorate. A peak in biomass was also exhibited at ~19 cm along the column. It is not known why this second peak appeared, but it is possible that it was due to bacteria re-distributed along the length of the column in the direction of flow.

It was concluded that the glass beads did not provide an adequate support medium for the pure culture. It may be that the biofilm development was insufficient to control dissolved oxygen concentrations in the column. If dissolved oxygen is not completely removed, perchlorate reduction will be inhibited. Control of dissolved oxygen was apparently not a problem in mixed culture experiments or those experiments using the pure culture and sand. When using a mixed culture, other microbes in the reactor can help to lower dissolved oxygen concentrations in the reactor and maintain perchlorate-reducing conditions. When sand was used, we hypothesize that more biofilm was kept in the reactor, providing enough biofilm to lower the dissolved oxygen and consistently remove perchlorate. Because it was unlikely that glass beads would be used in larger reactors, we did not further explore the reasons for the failure of this reactor beyond the factors examined above related to adsorption and biomass in the reactor.

**Perchlorate Treatment in GAC-Packed Columns**

The use of GAC in bioreactors was investigated as GAC is a good medium for bacterial attachment and therefore biofilm development. Following inoculation of the GAC column with a perchlorate-degrading mixed culture, effluent perchlorate concentrations decreased to non-detectable levels in four days. This low effluent concentration was expected as complete removal was previously accomplished at the same hydraulic loading in a much shorter sand column. In order to see the effect of the column mixing and to redistribute the biomass within the column, the column was fluidized on day 6 using tap water. Unfortunately, tap water was not aerated to remove chlorine and it was later determined that the tap water likely contained as much as 0.7 mg/L of a chlorine residual. Therefore, in addition to the biomass lost by fluidization of the column, additional biomass in the column was likely killed by the chlorine in the tap water. As a result of these two factors, after fluidization the column operation did not return to its original efficiency within four days (Figure 4.15).

Perchlorate concentrations in the reactor effluent remained high, especially from day 9 to day 16. On days 12 and 13, the effluent concentrations exceeded the influent concentrations, indicating that perchlorate was being desorbed from the GAC, originally present near the column influent, faster than it could be degraded by the biofilm in the reactor. It was suspected that the lack of sufficient perchlorate degradation by the biofilm was due in part to insufficient oxidizable substrate (acetate) available for the biofilm. The acetate in the column feed was therefore doubled from 30 to 60 mg/L on day 26. Thereafter, the perchlorate concentration in the effluent rapidly decreased from 5040 μg/L (day24) to 28.5 μg/L (day 28).

***Gentle Column Mixing***

Perchlorate removal in the column had sufficiently improved, and therefore it was decided to again study the effect of column fluidization on performance. The GAC in the column was mixed on day 73 by pouring out the GAC from the column, and mixing it by hand. This was hoped that this procedure would minimize the loss of biomass, but allow us to study the effect of media redistribution on reactor performance. This time, the effluent perchlorate concentrations after the mixing were not affected by mixing, indicating that gentle mixing will be essential to performance of the bioreactor. However, perchlorate concentration remained in detectable range suggesting perchlorate desorption continued to reduce bioreactor performance. This potential role of perchlorate desorption was further investigated as described below.

***Perchlorate and Acetate Concentration Profile in the Column***

Perchlorate and acetate concentration profiles across the length of the column were obtained on 12 different days. On days 15 and 16, following the first mixing of the column, perchlorate was reduced from the influent concentration of 45 mg/L to about 5 mg/L within a short distance of the column (15.2 cm), but the concentration increased to about 20 mg/L at 30.5cm (Figure 4.16). After that, perchlorate levels remained in the range of 25 to 37 mg/L along the column.

Acetate concentration profiles were also obtained on days 24, 28 and 30 (Figure 4.17). As in the perchlorate profile, most of the acetate was utilized in the first 15cm of the column. The rapid reduction of acetate confirmed that the most of the microbial perchlorate removal occurred within the short distance of the column (15cm).

Once the applied acetate concentration was doubled (day 26), the perchlorate effluent concentrations decreased over time. Perchlorate concentrations after changing influent acetate concentration decreased to very low levels within the first 15 cm (0.8, 0.2, and 0.009 mg/L on day 24, 28 and 30, respectively) (Figure 4.18). The perchlorate concentrations down the column on the two days after increasing the acetate concentration (day 28 and 30) were stable throughout the column.

The perchlorate concentration profiles were obtained seven times between days 38 and 77 (Table 4.2). On day 58, when the hydraulic loading rate was doubled, the perchlorate removal efficiency did not apparently decline. The perchlorate concentration in the effluent was less than 3 µg/L on day 58.

On day 77, which was the last day of the operation, the variations of perchlorate concentrations with depth were still observed indicating continued perchlorate desorption. For example, the perchlorate concentration in effluent (48µg/L) was higher than the concentration at the first port of the column (3 µg/L).

The variable perchlorate concentration profiles indicated that GAC was not a good column support medium due to desorption of perchlorate back into the water. Also, it was very difficult to estimate microbial perchlorate reduction occurring in the column because of the interference by adsorption/desorption behavior of perchlorate. The result of the slow desorption of perchlorate is that it was not possible to accurately predict perchlorate concentration in the effluent.

#### *Perchlorate Adsorption Isotherm*

A mass balance indicated that the mass of perchlorate that desorbed over the 71 days of operation (days 6 to 77) was less than that which could have been adsorbed to the first 15.2 cm of GAC. Using the measured Freundlich isotherm ($q=0.022\ C^{2.93}$ [mg/L]; n=5, $R^2$=0.83), and assuming GAC in equilibrium with a linear reduction of perchlorate from 20 to 0 mg/L over 15.2 cm, the mass of adsorbed perchlorate was 44 g. Based on 33 days of intra-column perchlorate profiles, the amount of perchlorate desorbed during 77 days of operation was calculated (Table 4.3). From the data in the table, the total mass of perchlorate desorbed was calculated to be 4966 mg, which accounts for 5.8% of perchlorate that could have been adsorbed to GAC (86266 mg) based on the isotherm and the influent concentrations. Thus, the erratic column performance can be tied to perchlorate desorption through intra-column profiles and a perchlorate mass balance.

#### *The Effect of Nitrate on Perchlorate Reduction*

Nitrate concentrations in water were found to affect perchlorate degradation rates in the acetate-fed packed bed bioreactors. In order to further examine the effect of nitrate on mixed cultures, batch experiments were performed to determine whether nitrate would be preferentially used by mixed cultures. It was found that for acetate-amended suspended cultures, nitrate reduction occurred without a lag phase in the presence or absence of perchlorate even though these cultures were acclimated to perchlorate (Figure 4.19).

## Batch Experiments With Suspended Pure and Mixed Cultures

In the absence of nitrate, perchlorate was degraded very slowly for approximately 40 h for unknown reasons. However, perchlorate reduction was rapid after this 40 hr period. When both nitrate and perchlorate were added to the perchlorate-acclimated culture, perchlorate reduction occurred much sooner than when nitrate was absent. However, perchlorate was not removed until the nitrate had been completely removed.

These results suggest that nitrate is a preferred electron acceptor to perchlorate in this mixed culture because perchlorate removal always took longer than that of nitrate. However, these results are inconclusive. It is possible that different microbes mediated perchlorate and nitrate degradation since a mixed culture was used. Further batch experiments with pure cultures will be necessary to better understand the effect of nitrate on perchlorate reduction.

### *Acetate Requirements During Microbial Growth on Perchlorate and Oxygen*

Batch experiments with isolate KJ were conducted to determine the overall acetate requirements during aerobic and anaerobic growth of isolate KJ. An overall molar ratio of acetate consumed to perchlorate reduced was determined to be 2.6±0.9 (n=4) under perchlorate reducing conditions (Table 4.4). However, more acetate was consumed on the first day than on the second, suggesting that acetate uptake was uncoupled from cell respiration and that the cells were able to store acetate. Work by Mulvaney (1999), however, has failed to find any discernable polyhydroxybutryrate (PHB) granules in cells of KJ. The overall molar ratio of acetate to oxygen was determined to be 1.0±0.1 (n=5) for KJ suspensions (Table 4.5).

### *TCE Degradation Study*

TCE did not appear to be degraded to a measurable extent over an 11-day period in batch culture tests. Figure 4.20 shows the time course of aqueous TCE concentrations in KJ cell suspension (open circles) and abiotic control (solid circles). The aqueous TCE concentration varied from 3039 to 4198 mg/L in the control tubes (no bacteria), and from 2759 to 3522 mg/L in treatment tubes (KJ cell suspension). The average aqueous TCE degradations by KJ and control were normalized to 100% based on the average starting concentration, and removals calculated on this basis as summarized in Table 4.6. While there appears to be a difference between the control and cell suspension, this difference was not significant. TCE was degraded 13.6% by the KJ suspension, while there was only an average of 2.6% degradation by the abiotic control. These differences were not significant based on a t-test (n=4, P=0.12) as summarized in Table 4.7. due to the high concentrations of TCE used in this study relative to those in typical field conditions, further field investigations of TCE degradation should be conducted as a part of the Phase 2 investigation.

## DISCUSSION

It has been shown in this chapter that an acetate-fed, packed bed reactor can be used to successfully remove perchlorate to below detectable concentrations. Reactor detention times of 10.2 min (EBCT of 31 min) are necessary for mixed cultures, but when a reactor was first inoculated with perchlorate-respiring KJ it was possible to get perchlorate removal using a sand column reactor with a detention time of only 0.8 min (EBCT of 2.1 min). Sand column reactors successfully treated artificial groundwater initially containing ~20 mg/L and 80 µg/L of perchlorate. Perchlorate removal was also complete when an actual groundwater from the Redlands Crafton site in California was treated.

Acetate-fed bioreactor experiments were also conducted with columns packed with glass beads and granular activated carbon (GAC). Glass beads were not found to be an acceptable support medium for saturated flow experiments, likely as a result of insufficient biofilm development. GAC was also not an acceptable biofilm support medium because perchlorate adsorbed to the GAC. When the column was mixed, the perchlorate desorbed resulting in perchlorate concentrations in the effluent well above the detection limit.

**Implications for Full Scale System Design and Operation**

The laboratory-scale reactor filled with sand is typical of the type of reactor used to filter water in a water treatment plant following conventional coagulation/ sedimentation treatment. The main differences are that acetate is added as an electron donor for bacteria growing on the sand, and nutrients are included to ensure bacterial growth on the added substrate. In a pilot or full-scale system, the sand bioreactor would be constructed in the same manner as a dual media filter, and operated in either up-flow or down-flow mode, with most of the filter composed of sand with the bottom media consisting of gravel. The reactor would be inoculated, and operated in a recycle mode until a biofilm had been developed as indicated by a loss of perchlorate in the system. The system could be fluidized (i.e. backwashed) to remove accumulated bioflim (as needed), but backwashing would be done very gently for the sole purpose of homogenizing the column media. The reactor could then be operated again in plug flow mode again with fresh feed water containing one of the supplemental carbon sources and perchlorate-amended artificial groundwater.

There are two important components to this reactor that require additional discussion: the ability to regenerate the biofilm; and the ability to fluidize and mix the filter media. Because of the low perchlorate concentrations present in many drinking water sources, there might not be sufficient perchlorate in the water to continuously support a perchlorate respiring biofilm in competition with other anaerobes (such as methanogens) even though the cell yields and growth rates of these other anaerobic communities are so low compared to perchlorate respiring strains. Thus, it is anticipated that it will be necessary to periodically *regenerate a thick perchlorate respiring biofilm* by infusing the column with high concentrations of chlorate (or if necessary, perchlorate) and substrate (such as acetate). The biofilm in this reactor would therefore be regenerated by: temporarily halting the flow of contaminated water through the reactor; recycling water containing relatively high concentrations of electron donor and electron acceptor (chlorate, at 10-1000 mg/L levels) to those bacteria so that they may develop into a thick biofilm; the reactor would then be rinsed with clean water, and placed back into service. Thus, during regular water treatment the bacteria in the biofilm would scavenge perchlorate while in an overall state of endogenous decay. The biofilm would be regenerated as often as necessary, but it should certainly be on a less frequent time scale than needed for a water treatment filter for backwashing.

Second, the reactor would have to be periodically fluidized for two reasons. One, it may be necessary to periodically redistribute bacteria that preferentially grow near the column effluent to the whole column. Two, it may be necessary to periodically dislodge old biofilms (or other material that may accumulate on the media packing) to prevent the sand bed from clogging. The need to only periodically fluidize the bed, versus continuously fluidize the bed (as proposed by others) means that operation costs are substantially reduced for normal operation compared to a fluidized bed.

Contaminated water from any biofilm regeneration cycle (possibly containing high concentrations of chlorate or electron donor) would be held for subsequent treatment. Excess chlorate would rapidly be removed in the presence of excess electron donor. Excess electron acceptor could be removed through ordinary anaerobic treatment processes such as methanogens. In our laboratory, we find that water in batch culture can be disposed of after a day or so because it no longer contains perchlorate. In a water treatment plant, we expect such water could be added back into the flow entering the perchlorate reactor or sent to a conventional wastewater treatment plant.

**Table 4.1**
**The composition of Basal media**

| Buffer | Concentration (g/L) | Metals | Concentration (mg/L) |
|---|---|---|---|
| $K_2HPO_4$ | 1.55 | $MgSO_4 \cdot 7H_2O$ | 55 |
| $NaH_2PO_4 \cdot H_2O$ | 0.98 | $H_3BO_3$ | 0.6 |
| $NH_4H_2PO_4$ | 0.5 | $CaCl_2 \cdot 2H_2O$ | 1 |
| | | $CoCl_2 \cdot 6H_2O$ | 0.4 |
| | | $CuSO_4 \cdot 5H_2O$ | 0.2 |
| | | EDTA | 3 |
| | | $FeSO_4 \cdot 7H_2O$ | 4 |
| | | $MnCl_2 \cdot 4H_2O$ | 1 |
| | | $NiCl_2 \cdot 6H_2O$ | 0.1 |
| | | $NaMoO_4 \cdot 2H_2O$ | 0.4 |
| | | $NaSeO_3 \cdot 5H_2O$ | 0.15 |
| | | $ZnSO_4 \cdot 7H_2O$ | 2 |

**Table 4.2**
**Perchlorate profiles on selected days between days 38 to 77**

| Sample | Height (cm) | Perchlorate concentration (mg/L) | | | | | | |
|---|---|---|---|---|---|---|---|---|
| | | Day38 | Day44 | Day47 | Day56 | Day58 | Day75 | Day77 |
| Influent | 0 | 64.8 | 39.6 | 44.7 | 40.2 | 27.2 | 24.9 | 24.9 |
| Port 1 | 15.2 | 0.004 | 0.004 | 0.004 | 0.015 | 1.019 | 0.005 | 0.004 |
| Port 2 | 30.5 | 0.004 | 0.004 | 0.004 | 0.007 | 0.041 | 0.004 | 0.022 |
| Port 3 | 45.7 | 0.336 | 0.004 | 0.004 | 0.011 | 1.045 | 0.004 | 0.056 |
| Port 4 | 61.0 | 0.666 | 0.004 | 0.012 | 0.004 | 1.634 | 0.004 | 0.111 |
| Port 5 | 76.2 | 1.248 | 0.009 | 0.004 | 0.018 | 0.004 | 0.004 | 0.073 |
| Port 6 | 91.4 | 1.116 | 0.004 | 0.004 | 0.004 | 0.004 | 0.004 | 0.069 |
| Effluent | 106.7 | 1.136 | 0.004 | 0.004 | 0.010 | 0.004 | 0.004 | 0.048 |

**Table 4.3**
**The mass of perchlorate desorbed from GAC in the GAC packed column**

| Day | Perchlorate desorbed (mg/day) | Day | Perchlorate desorbed (mg/day) |
|---|---|---|---|
| 1 to 6 | 0 | 38 | 16.6 |
| 7 | 6 | 39-40 | 18.0 |
| 8 | 19.2 | 41 | 1.3 |
| 9 | 65.5 | 42-43 | 0.7 |
| 10 | 78.2 | 44 | 0.0 |
| 11 | 241.9 | 45-46 | 0.0 |
| 12 | 1272.1 | 47 | 0.0 |
| 13 | 896.1 | 48 | 0.0 |
| 14 | 702.7 | 49 | 0.0 |
| 15 | 459.5 | 50-55 | 0.0 |
| 16 | 320.1 | 56 | 0.0 |
| 17 | 201.3 | 57 | 0.0 |
| 18 | 82.5 | 58 | 0.0 |
| 19 | 91.3 | 59 | 0.6 |
| 20 | 100.2 | 60 | 0.3 |
| 21 | 82.0 | 61 | 0.0 |
| 22 | 63.9 | 62 | 0.0 |
| 23 | 62.8 | 63 | 0.0 |
| 24 | 61.8 | 64-66 | 0.0 |
| 25-27 | 92.7 | 67 | 0.0 |
| 28 | 0.0 | 68-69 | 1.4 |
| 29 | 0.0 | 70 | 1.4 |
| 30 | 0.1 | 71-72 | 1.4 |
| 31-32 | 0.1 | 73 | 0.0 |
| 33 | 0.0 | 74 | 0.0 |
| 34 | 1.1 | 75 | 0.0 |
| 35 | 2.3 | 76 | 0.6 |
| 36-37 | 18.9 | 77 | 1.3 |

**Table 4.4**
**Acetate consumption for perchlorate reduction, when a perchlorate-reducing pure culture, KJ, was used**

| Sample | Perchlorate (mg/L) | | | Acetate (mg/L) | | | Molar ratio ($Ac^-/ClO_4^-$) | |
|---|---|---|---|---|---|---|---|---|
| | Day 0 | Day 1 | Day 2 | Day 0 | Day 1 | Day 2 | Day 1 | Day 2 |
| 1 | 305.1 | 204.0 | 0.025 | 513.4 | 369.9 | 247.0 | 2.4 | 1.5 |
| 2 | 302.7 | 178.9 | 0.155 | 517.0 | 244.6 | 45.7 | 3.7 | 2.6 |
| | | | | | Average | | 2.6 (±0.9) | |

**Table 4.5**
**Molar ratio of oxygen utilized per acetate consumed by KJ in batch experiment**

| Sample | Molar ratio ($O_2/Ac^-$) |
|---|---|
| 1 | 0.8 |
| 2 | 1.0 |
| 3 | 1.0 |
| 4 | 1.1 |
| 5 | 1.1 |
| Average | 1.0±0.1 |

**Table 4.6**
**TCE degradation by KJ and abiotic control**

| | Control | KJ |
|---|---|---|
| Starting TCE concentration (mg/L) | 3731 | 3522 |
| The average of four following measurements of TCE concentrations (mg/L) | 3636 | 3056 |
| Degradation of TCE (mg/L) | 95 | 466 |
| Percent degradation | 2.6% | 13.2% |

**Table 4.7**
**Statistical analysis of TCE degradation by KJ and abiotic control**

| Days | 1 | 2 | 3 | 4 |
|---|---|---|---|---|
| Group I: TCE degradation by KJ (%) | 12.5 | -18.5 | 5 | -9.2 |
| Group II: TCE degradation by Control (%) | -6.6 | -6.9 | -21.7 | -17.9 |
| P value of t-test between Group I and II | | 0.12 | | |

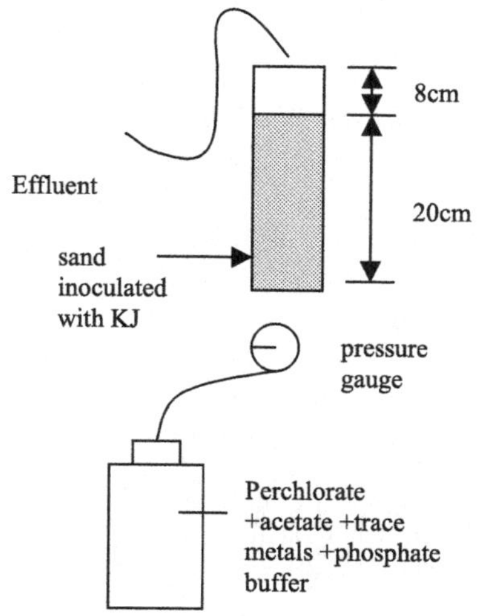


**Figure 4.1 Setup used in sand column experiments for treating perchlorate at low concentrations**

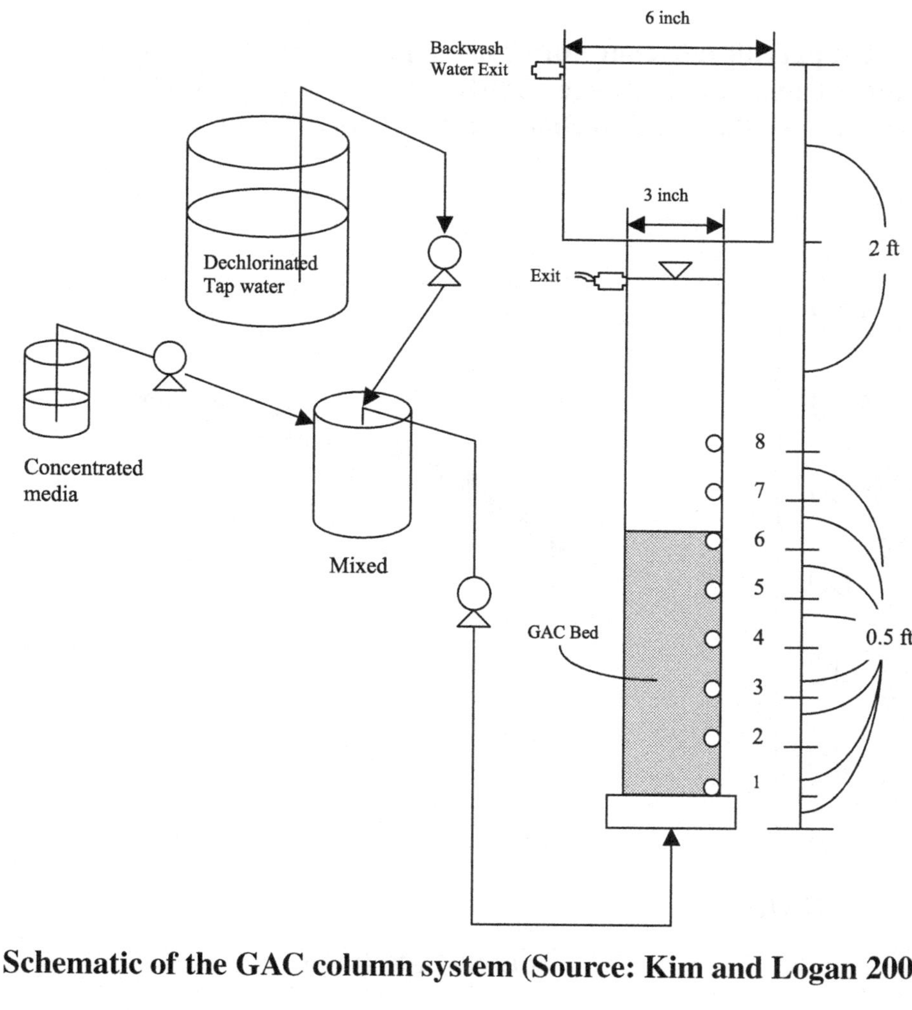


**Figure 4.2 Schematic of the GAC column system (Source: Kim and Logan 2000)**

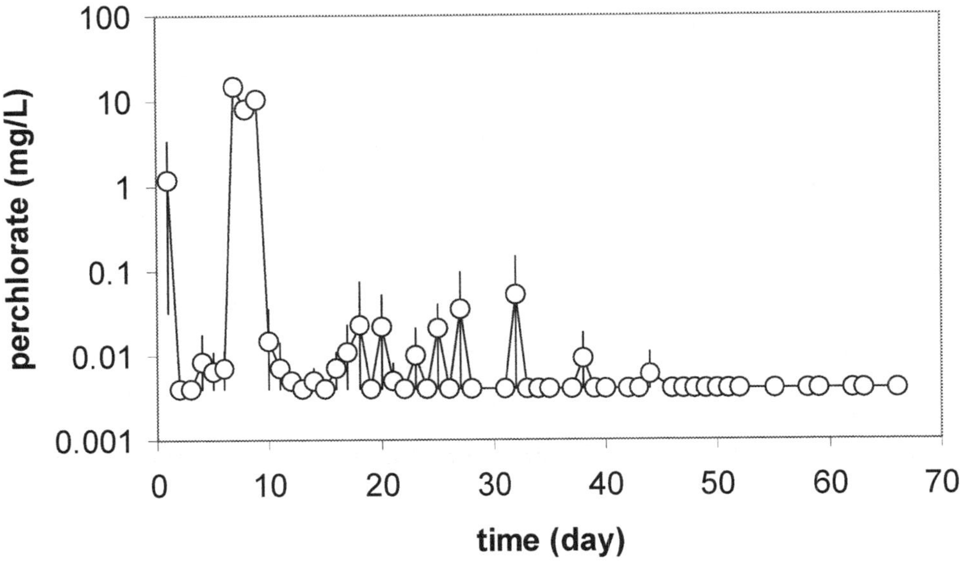


**Figure 4.3 Effluent perchlorate concentrations in pure culture sand column bioreactor (Source: Reprinted from Kim and Logan 2001. "Microbial Reduction of Perchlorate in Pure and Mixed Culture Packed-Bed Bioreactors." *Water Research.* With permission from Elsevier)**

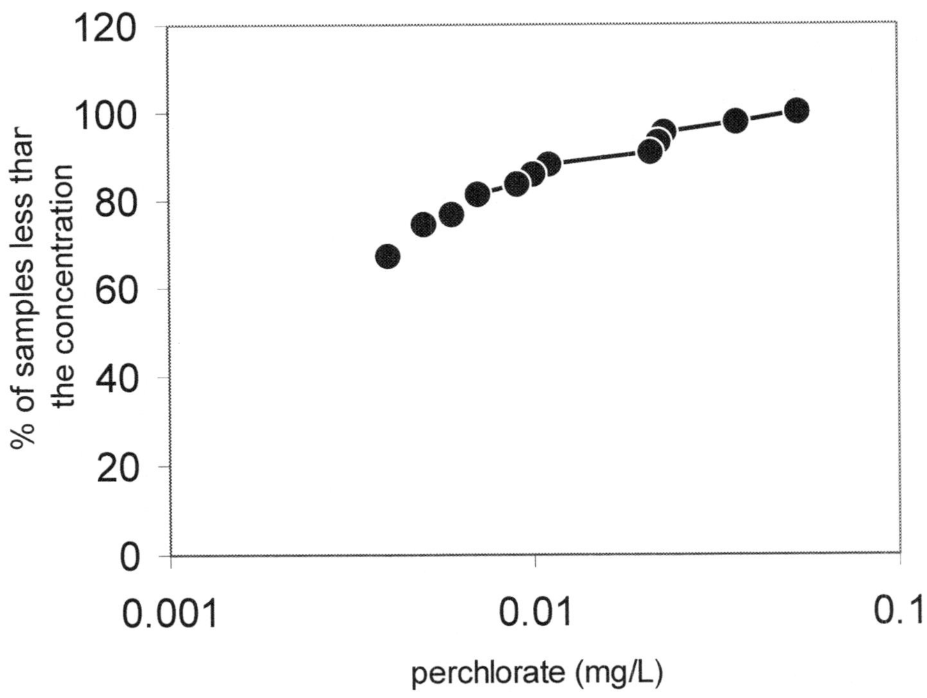


**Figure 4.4 Percent of effluent samples showing below each perchlorate concentration**

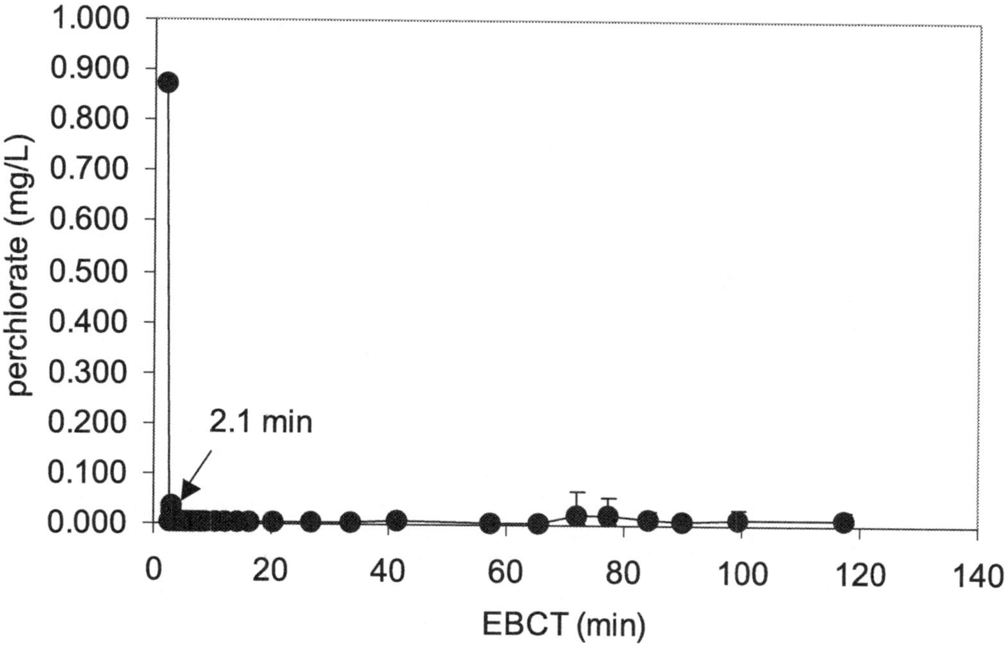


**Figure 4.5 Critical hydraulic loading rate for the pure culture column experiment**

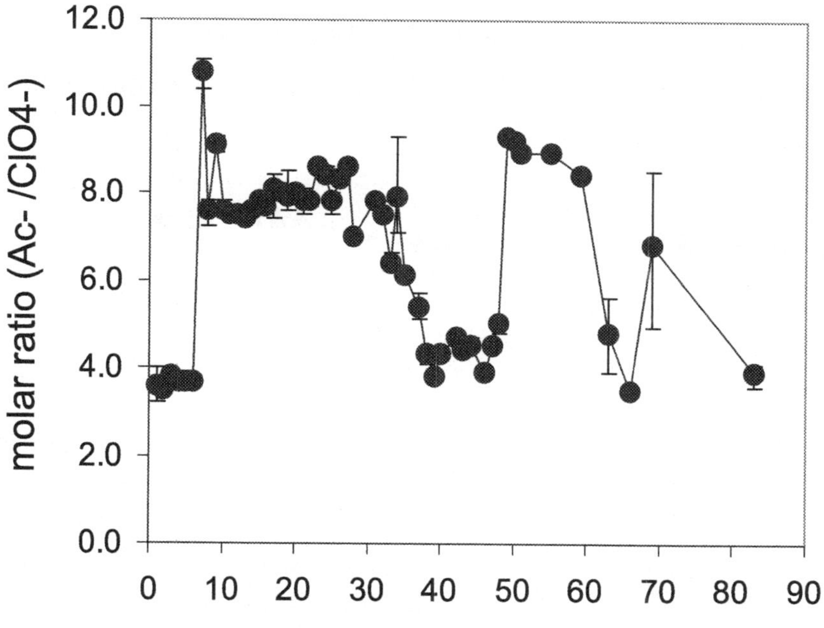


**Figure 4.6 Molar ratios of acetate-consumed to perchlorate-removed in column experiments**

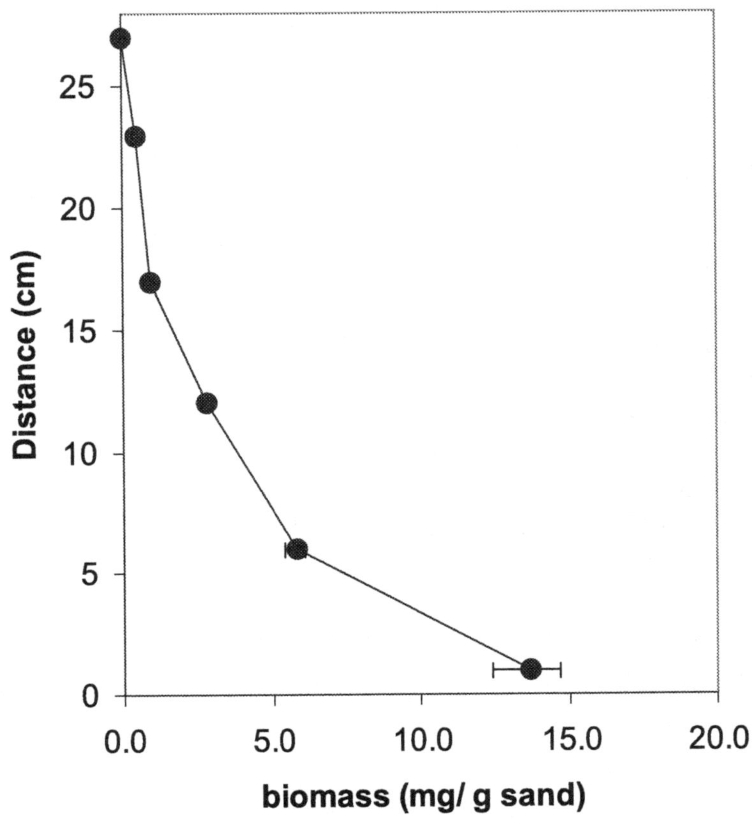


**Figure 4.7 Cell concentrations measured along the pure culture column bioreactor**

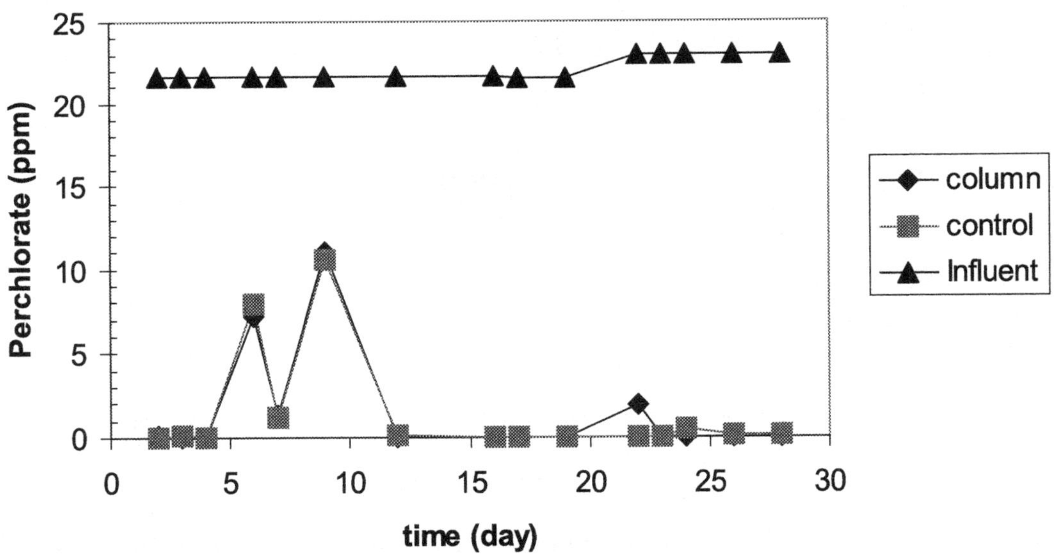


**Figure 4.8 Perchlorate concentrations in the effluent of the test and control columns during the 28 days of the column start-up period**

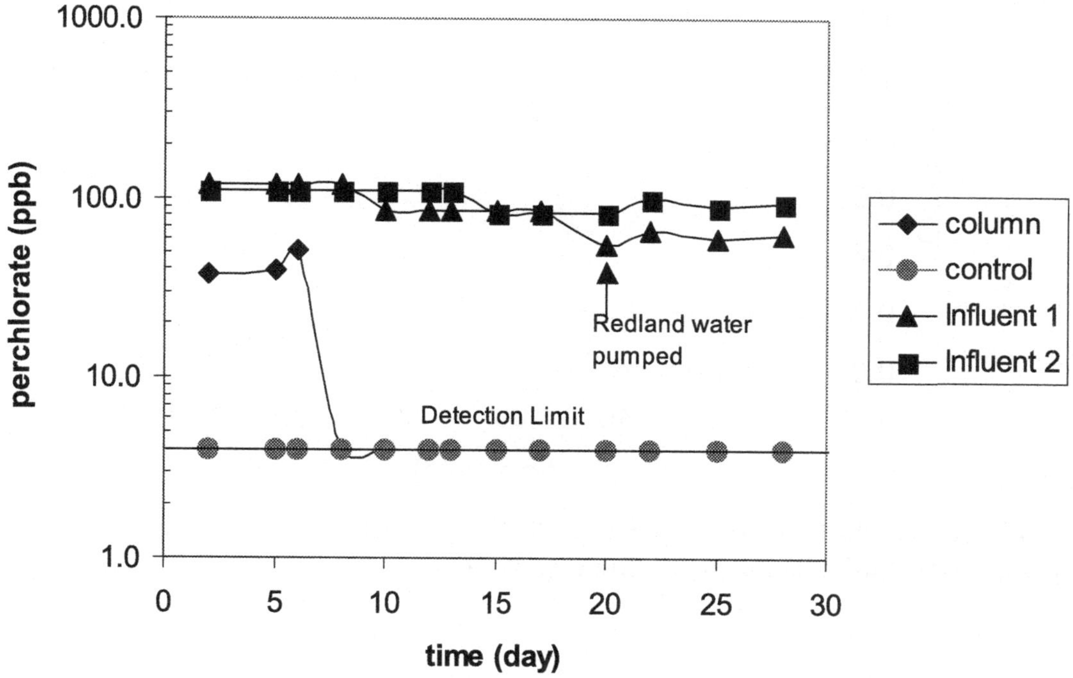


**Figure 4.9 Perchlorate concentrations in the influent and effluents of the test and the control columns**

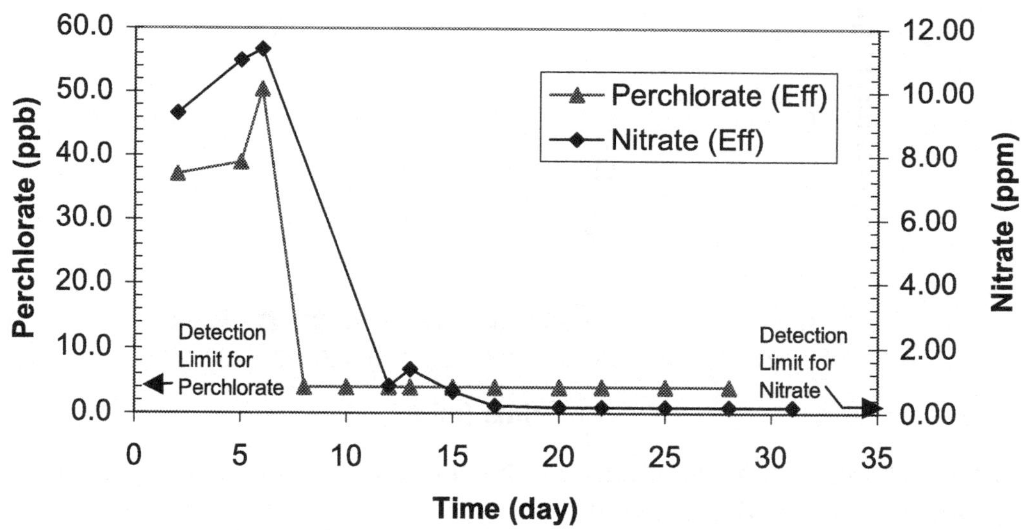


**Figure 4.10 Perchlorate and nitrate concentrations in the pure culture column. (The detection limits were 4 μg/L for perchlorate and 180 μg/L for nitrate.)**

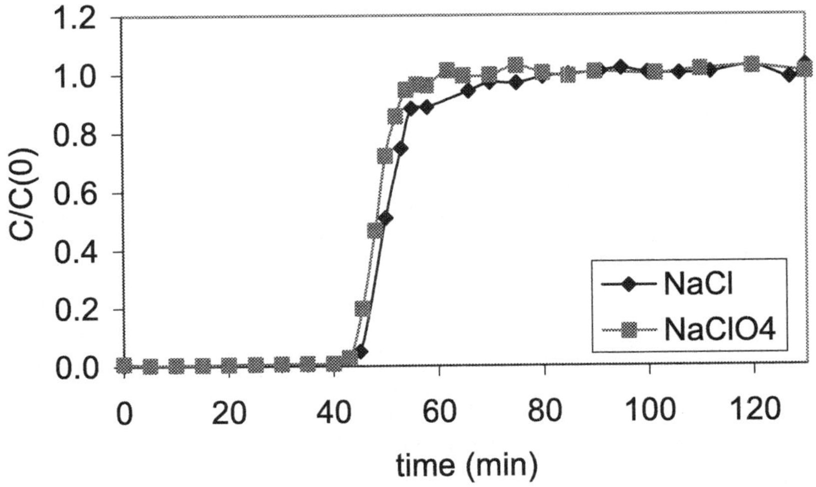


**Figure 4.11 Tracer study using NaCl and $NaClO_4$ in the control column (cleaned glass beads) prior to the biological degradation experiment**

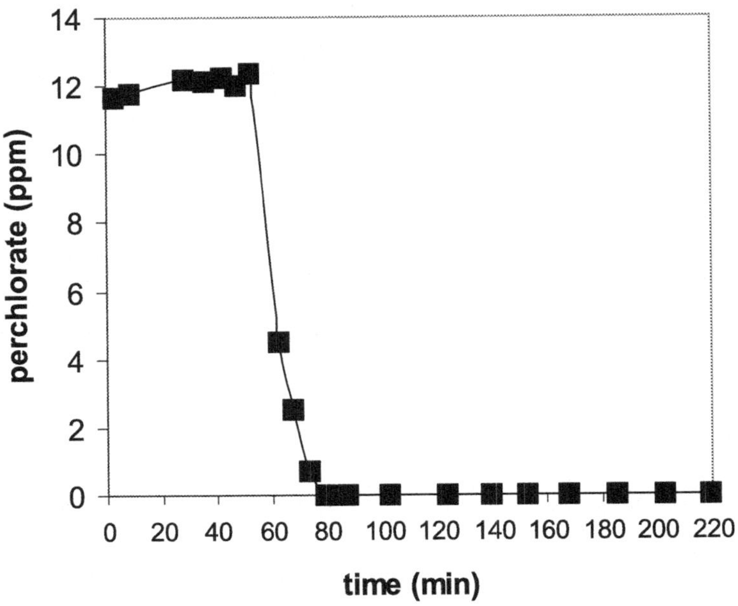


**Figure 4.12 Perchlorate concentrations in the bioreactor effluent for 23 days after the perchlorate was discontinued in the feed**

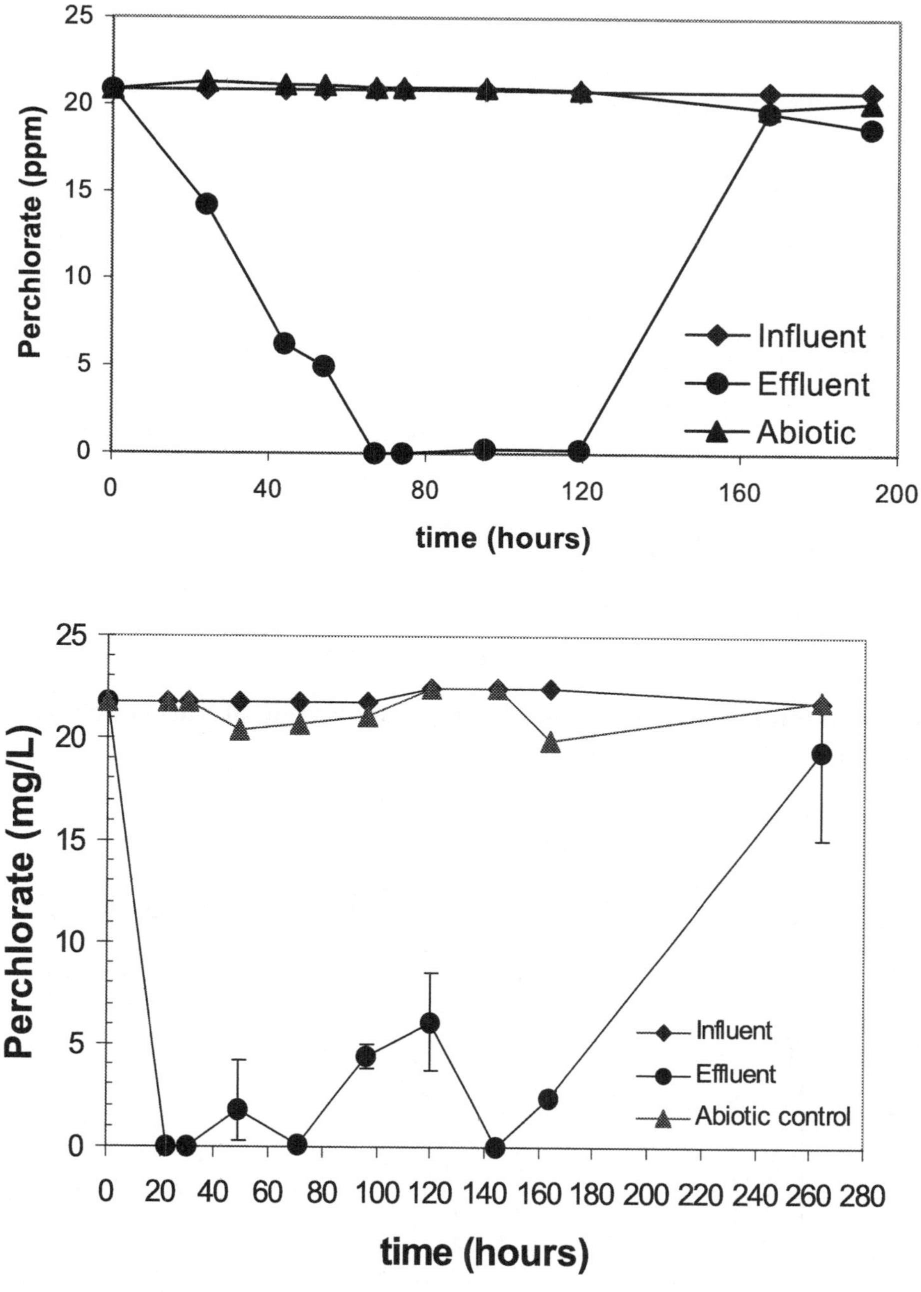


**Figure 4.13 Perchlorate concentrations in the influent and effluent samples in the bioreactor, and the effluent from the abiotic column control (cleaned glass beds): (A) first trial; (B) second trial**

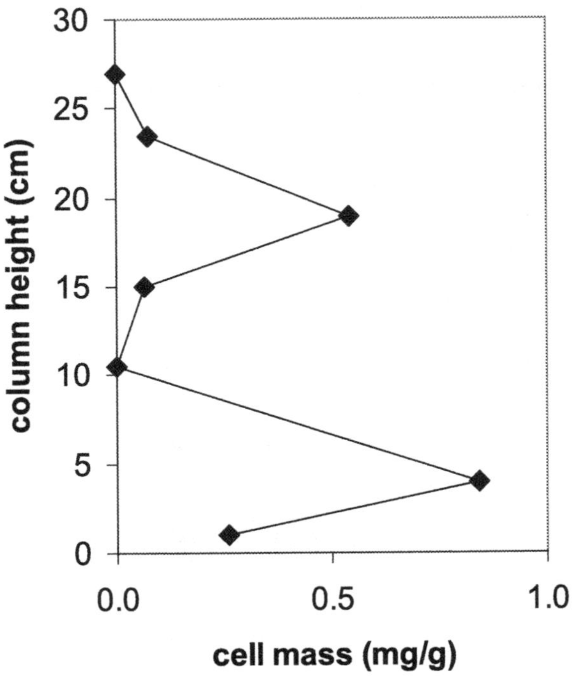


**Figure 4.14 Intracolumn cell mass distribution measured at the conclusion of the pure culture (glass bead) column experiment**

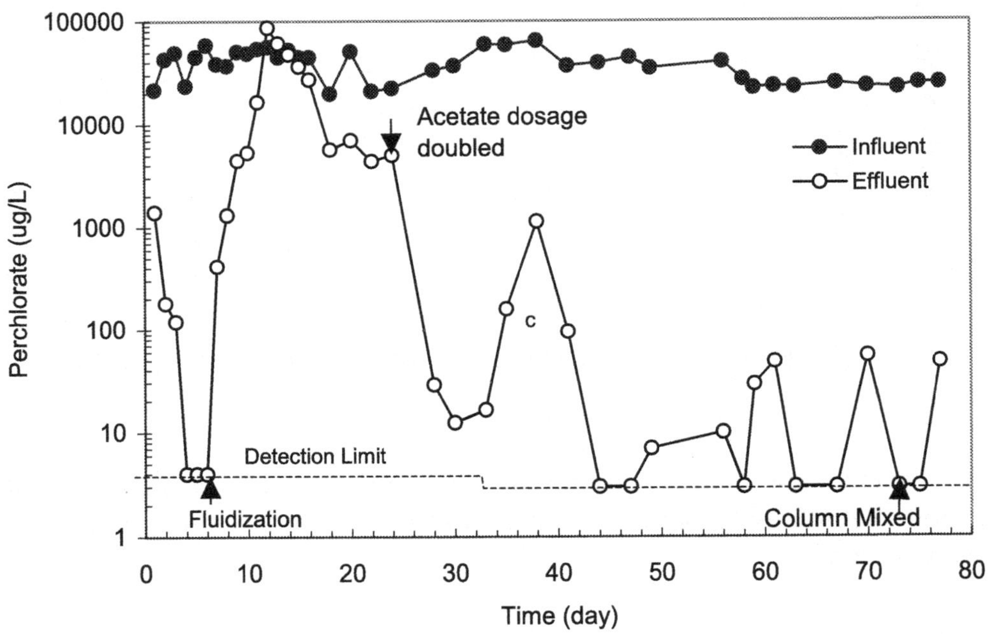


**Figure 4.15 Performance of the GAC packed column. The column was mixed on day 6 and 76 by different methods, and the acetate dosage was doubled on day 26.**

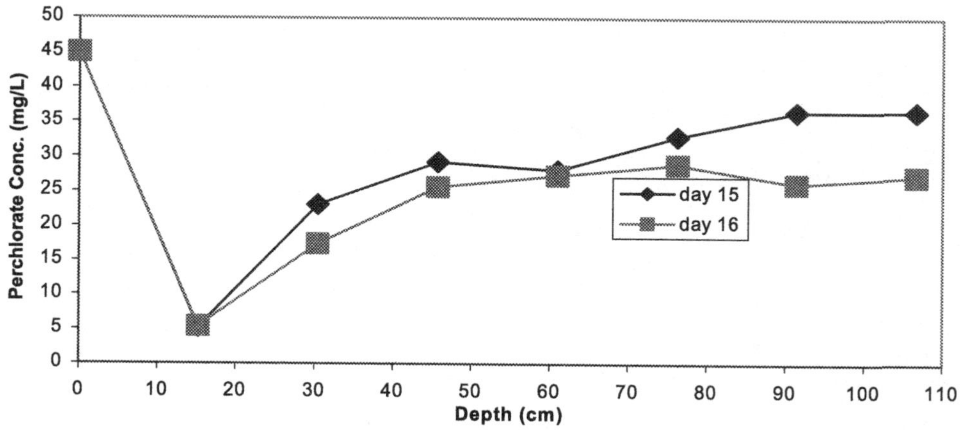


**Figure 4.16 Perchlorate concentration profiles on days 15 and 16 in the GAC column**

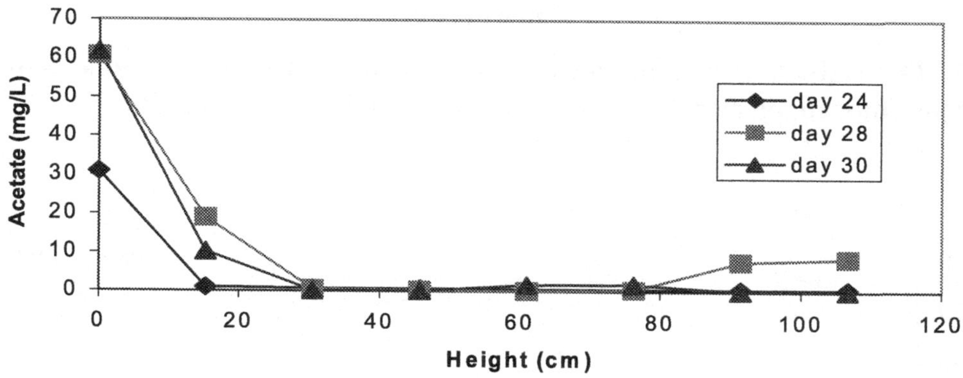


**Figure 4.17 Acetate concentration profiles on days 24, 28 and 30 in the GAC column**

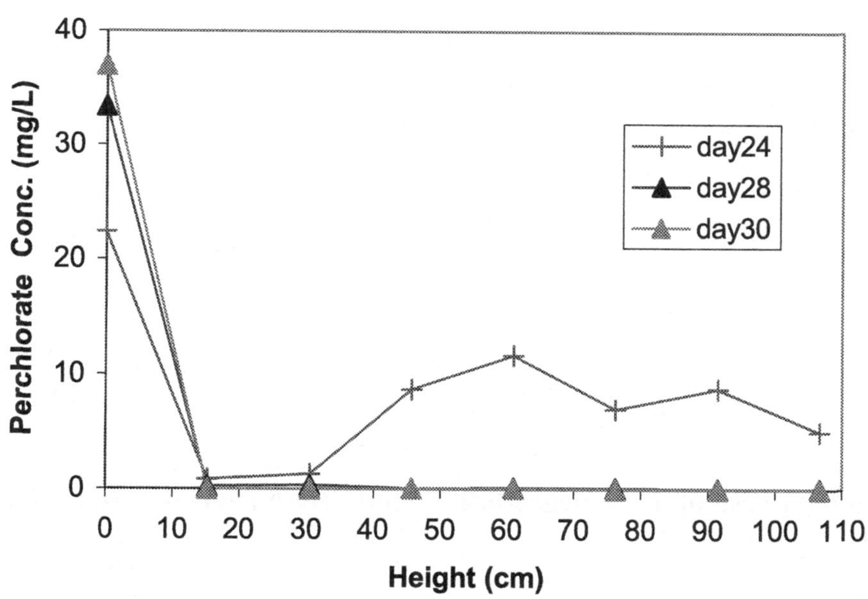


**Figure 4.18 Perchlorate concentration profiles on days 24, 28 and 30 in the GAC column experiment**

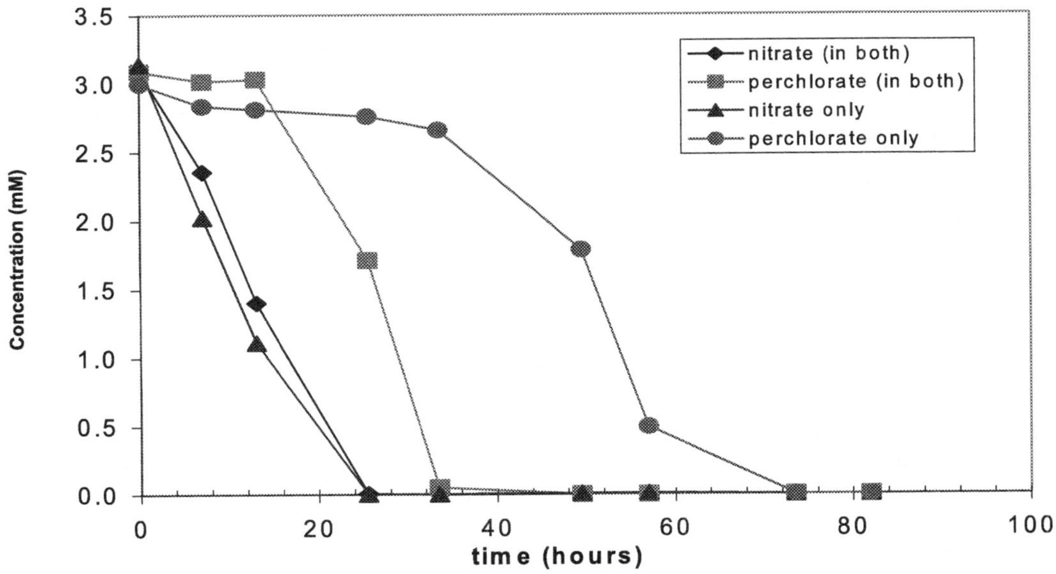


**Figure 4.19 The reduction of perchlorate and nitrate in acetate-amended batch cultures with either perchlorate or nitrate, or in the presence of both compounds**

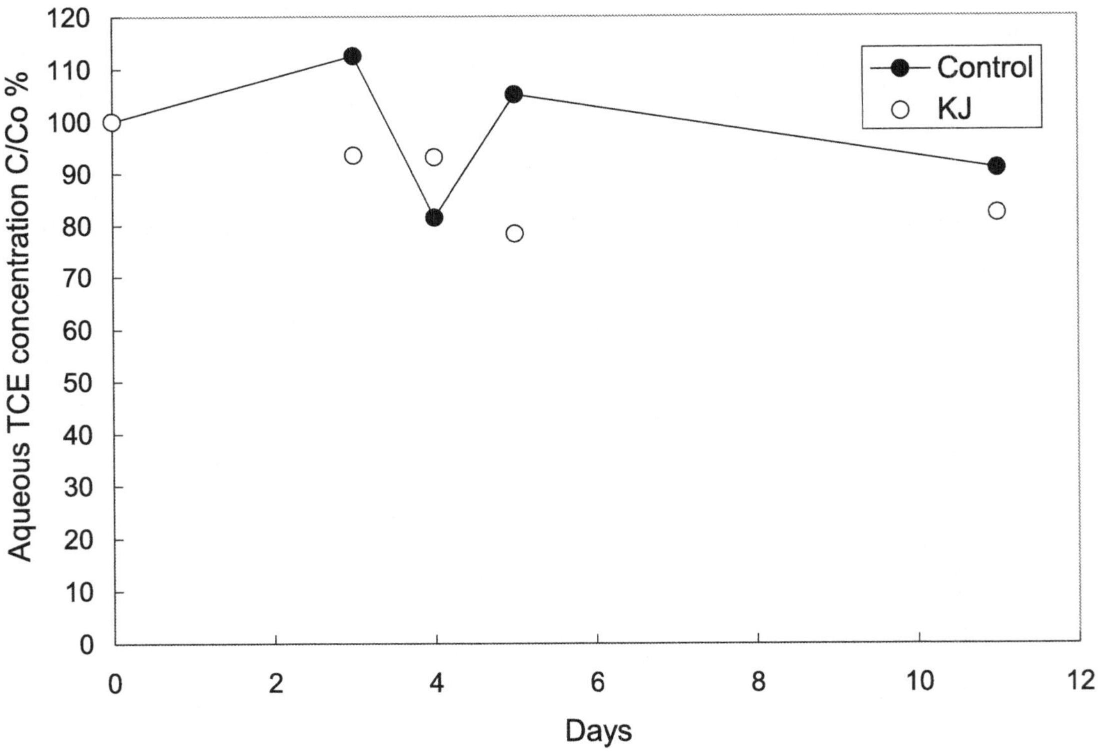


**Figure 4.20 Time course of TCE degradation study by bacterial suspension KJ versus an abiotic control**

# CHAPTER 5
# HYDROGEN-FED GAS PHASE UNSATURATED FIXED-FILM PERCHLORATE REACTORS

## INTRODUCTION

Biological treatment is considered to have the best potential for removing perchlorate from the environment (AwwaRF 1997, Urbansky 1998). Microorganisms can utilize perchlorate as the final electron acceptor during anaerobic respiration by oxidizing inorganic substrates (van Ginkle et al. 1996, Miller and Logan 2000, Giblin et al. 2000) or organic substrates (Wallace et al. 1998, Kim and Logan 2000, Giblin, Herman, and Frankenberger 2000). Bacteria that can grow on inorganic substrates such as hydrogen, however, may be preferable to those using organic substrates in water treatment processes. Organic substrates such as methanol, ethanol, and acetate—if not completely removed by downstream processes—could remain in water distribution lines and stimulate bacterial growth. Hydrogen has a low solubility in water, therefore, the potential for bacterial re-growth in the water distribution lines is low. Hydrogen is also non-toxic, and can easily be generated on site electrolytically using water. In addition, methanol, a commonly used organic substrate in water treatment reactors, is toxic to humans at doses of 0.5 mg/kg/day (USEPA 2000).

Nitrate is a common co-contaminant with perchlorate in the environment. It is currently regulated to 10 mg/L $NO_3^-$-N by the US EPA (primary drinking water standard), and 50 mg/L $NO_3^-$-N by the World Health Organization because it can cause methemoglobinemia (blue baby syndrome) in infants (Mateju et al. 1992), and because it can help stimulate algae blooms in surface water systems. Perchlorate reducing bacteria are usually capable of denitrification (Malmquist et al. 1994; Rikken, Kroon, and van Ginkel 1996; Wallace et al. 1996). A concern is that nitrate in surface waters will compete with perchlorate as the final electron acceptor thereby reducing perchlorate degradation. Herman and Frankenberger (1999) conducted experiments with a pure bacterial strain, per1ace, in batch cultures where perchlorate removal was decreased when nitrate concentrations were higher than perchlorate concentrations.

Little is known about the factors influencing perchlorate removal when hydrogen-oxidizing bacteria are used for biological perchlorate treatment. The purpose of this research project was to determine what factors affected perchlorate removal rates in hydrogen-oxidizing autotrophic unsaturated-flow bioreactors.

## MATERIALS AND METHODS

### Selection and Isolation of Hydrogen-Oxidizing Bacteria Pure and Mixed Culture

#### *Enrichment Culture*

Liquid media (60 mL) was added to five serum bottles (150 mL; Wheaton Science Products, Millville, NJ). The media (VG media) contained the following constituents: 0.240 g/L $NaClO_4^-$, 1.5 g/L $K_2HPO_4$, 0.850 g/L $NaH_2PO_4{\cdot}H_2O$, 0.5 g/L $NH_4H_2PO_4$, 0.050 g/L $MgSO_4{\cdot}7H_2O$, 3 g/L $Na_2CO_3$, and 1 mL/L of a trace metal solution (van Ginkel, Plugge, and

Stroo 1995). The pH of the medium was adjusted to approximately 7 using 6M $H_3PO_4$ following the addition of all mineral salts.

A liquid sample was obtained from a low DO activated sludge aeration tank used for nutrient removal (nitrate) at the State College Wastewater Treatment Plant. This sample served as the inoculum to enrich a culture of perchlorate reducing, hydrogen oxidizing bacteria. Activated sludge (3 mL) was added to two of the five serum bottles (control bottles), and all bottles were autoclaved for 30 minutes at 121°C. Following sterilization, 3 mL of activated sludge was added to each of the three bottles that were not inoculated previously (experimental bottles), and all bottles were sealed with butyl stoppers and crimped (Wheaton Science Products, Millville, NJ). The headspace content of each bottle was purged individually using 5%$H_2$/95%$N_2$ gas at a flow rate of 2675 $cm^3$/min for 15 minutes. All gases mentioned previously and hereafter passed through sterilized filters prior to entering the serum bottles. After a bottle was purged, it was placed horizontally on a shaker table rotating at approximately 150 rpm. It was assumed that a sufficient quantity of carbon from the initial inoculum was present in solution; therefore, bottles were not purged with carbon dioxide gas ($CO_2$). Perchlorate ($ClO_4^-$) reduction of the inoculum was confirmed using a $ClO_4^-$ probe (Orion, Boston, MA).

### *Mixed Culture Chemostat*

VG media (600 mL) was transferred to an Applikon Bioreactor (Applikon Inc., The Netherlands; Figure 5.1). All necessary connections were made to the bioreactor, and the vessel (with liquid contents) was autoclaved for 1 hour at 121°C. The fermentor was allowed to cool, and then approximately 50 mL of the contents of the enrichment bottle were added to the fermentor. The fermentor was operated in static mode (no liquid flows into or out of vessel) for approximately 10 days. During this period, the reactor was continuously purged with 30 $cm^3$/min of $CO_2$ and 60 $cm^3$/min 5%$H_2$/95%$N_2$, and was periodically sampled for $ClO_4^-$ concentration. Carbon dioxide was added to the solution as an additional carbon source, and also to preserve the weak acid concentration of the pH buffering system. Hydrogen-oxidizing bacteria use $CO_2$, which would result in a pH rise if the $CO_2$ were not replenished. This pH in a carbonate-buffered system is described by the Henderson-Hasselbach (Ray 1995) equation,

pH = pKa1 + log {(H2CO3*)/(HCO3-)}.

### *Mixed Culture Kinetics*

Mixed-culture kinetic parameters such as $V_{max}$ and $K_s$ were determined using the Lineweaver Burke and Hanes method of linearizing the equation,

$$r = \frac{V_{\max}S}{K_s + S}$$

where

$r$ = specific growth rate ($hr^{-1}$)
$V_{max}$ = maximum growth rate ($hr^{-1}$)

$S = ClO_4^-$ concentration (mg/L)
$K_s$ = Monod half-saturation constant (mg/L)

From reactor theory, the flowrate $F$ is set, and for a reactor of volume $V$ the detention time is $\theta = V/F$, and the growth rate $r = 1/\theta$.

## Hydrogen-Oxidizing Autotrophic Fixed Film Bioreactors: Design

### *Column 1 Design*

Two different reactors were used in these experiments. The first reactor (Column 1) was designed to demonstrate that it was possible to degrade perchlorate in a fixed film bioreactor using hydrogen gas (Miller 1999, Miller and Logan 2000) The Column 1 bioreactor was designed to allow unsaturated flow through the packing material (Figure 5.2). The acrylic column was 12.5 cm tall and 2.5 cm wide and packed with 3mm diameter glass beads as a support medium for microbial growth. This bead size was chosen to be large enough to permit gas to freely circulate through the bioreactor, yet small enough to produce a high ratio of surface area per unit volume (1200 $m^2/m^3$). The reactor was fed influent gases composed of a 5% hydrogen/95% nitrogen mixture and 100% carbon dioxide. These gases entered through a top port and exited with the liquid effluent through the bottom of the column. Liquid influent was pumped through plastic tubing to a pipette tip pushed through a rubber stopper. Effluent liquid was not pumped from the column, but driven by gravity and gas pressure in the column, out through the effluent tubing and into a waste bucket (Figure 5.3). The bottom ring of the column was sealed with caulking to prevent leakage of gas or liquid. Two plywood boards were clamped at the top and bottom of the column to prevent hydrogen from leaking out of the column.

### *Column 2 Design*

The Column 1 bioreactor was slightly redesigned to improve water distribution over the media, to make it easier to clean the feed lines, and to reduce hydrogen losses in the reactor (Figure 5.4). This second reactor (Column 2) was 25 cm tall and 3.5 cm wide, and was packed with the same quantity of glass beads to provide the same total surface area as the previous hydrogen column.

The Column 2 reactor was composed of three distinct chambers, with the first and second chamber separated by a thin piece of rubber, and the second and third chamber separated by a metal mesh. The first and third chamber could be removed for cleaning without disrupting the media. To more evenly distribute the flow over the glass beads, the liquid influent was pumped through a pipette tip sealed in a rubber stopper into the first chamber where the media pooled, so that the column pressure forced the water through four syringes. The second chamber contained the biofilm support medium, and the inflowing gas port. Chamber three was designed to allow the water to pool, preventing the loss of hydrogen gas through the effluent line. Because the gas did not force water out, the water was pumped out of the column at the same rate it entered the column. The entire column was clamped in place to prevent hydrogen gas from leaking out. The second chamber was clamped separately in place, so that the first and third chamber could be separately removed when necessary. The complete experimental setup for the Column 2 reactor is shown in Figure 5.5.

## Column Operation

Column 1 was initially operated at a flow rate of 2.3 ml/min (hydraulic loading rate of 0.11 gpm/ft$^2$) using a Cole-Palmer (Vernon Hills, IL) Masterflex digital console drive (Model 7523-30) equipped with a Cole-Palmer Masterflex pump head (Easy-Load Model 7518-10). The hydrogen/nitrogen gas mixture was delivered at 35 cm$^3$/min, and carbon dioxide was added at approximately 7 cm$^3$/min. The feed tubing was sterilized and/or changed for both the liquid influent and effluent to minimize biofilm growth in the lines. The liquid media added to the column was prepared using a trace metal solution (Table 5.1) from stock solutions to the final concentrations listed in Table 5.1 except as noted below.

Biofilm growth on the packing material in Column 1 was accomplished by inoculating the column with the mixed culture that was originally grown in the chemostat. The liquid suspension was recycled through the reactor until a noticeable biofilm was established on the glass beads. The influent tubing to the bioreactor was then replaced (to avoid hydrogen uptake and perchlorate degradation in the feed lines). To allow for further biofilm development, media containing approximately 50 mg/L perchlorate was fed to Column 1. Once it was demonstrated that it was possible to maintain 10 mg/L of perchlorate removal in the column, the influent perchlorate concentration was lowered to ~750 μg/L.

The column was sampled at the influent and effluent line as shown in Figure 5.4. To sample the column effluent, the sampling line was unclamped, and the effluent line clamped, resulting in the flow being pushed through the column into the sampling tube, and into a 20 ml HDPE scintillation vial. After sampling, the waste line was unclamped to return the column to normal operation. Influent samples were taken after the effluent samples to avoid disturbing the perchlorate effluent concentrations. The same procedure was used to force samples through the influent line. Samples were immediately filtered through a 0.4 μm pore diameter polycarbonate membrane filter (Osmonics, Minnetonka, MN) using a multi-port filter box (Hoeffer Scientific, San Francisco, CA) in order to remove any bacteria from the liquid, and stored in 20 mL HDPE scintillation vials. A small portion of the effluent and influent was routinely used to determine pH using pH paper with a range of 6-10 (Micro Essential Laboratory Inc., Brooklyn, NY). On occasion, results were verified using a pH meter. Samples that were not analyzed immediately were stored in a refrigerator at 8°C.

Column 2 was operated in the same manner as in Column 1, except that Column 2 was inoculated with 50% of the beads from Column 1, and 50% new beads. New influent tubing was connected to the bioreactor periodically to avoid hydrogen uptake and perchlorate degradation in the feed lines. For the first few days of operation, the column medium (Table 5.2) was set at a perchlorate concentration of 20 mg/L, and this fluid was recirculated through the column to establish a biofilm on all the packing. Once perchlorate degradation had occurred, the reactor was switched to continuous flow and maintained at a perchlorate concentration of 20 mg/L for several weeks. After the biofilm appeared well established (based on visual observations) the perchlorate concentration was reduced from 20 mg/L to 750 μg/L. The reactor was then run for an additional period of 16 days to establish steady- state operation as indicated by constant perchlorate degradation based on the influent and effluent samples (in triplicate).

### *Column Mixing*

It was noticed during the operation of Column 1 that there was a continuous thickening of the biofilm. This was shown by the column appearing almost completely red, as compared to a scattering of brownish colored biofilm at the beginning of column operation. The buildup of biofilm was a problem for the laboratory scale reactor, as this system was not designed for an in situ system to remove excess biofilm. To remove excess biofilm and to prevent reactor clogging, the reactor packing was periodically withdrawn and mixed, and placed back into the column. Prior to mixing the column, influent gas, influent media, and effluent tubing were clamped to prevent exposure of the column to oxygen, the column was separated from the tubing and transferred to the anaerobic chamber (Coy Scientific). A 3% hydrogen/97% nitrogen gas atmosphere was maintained the anaerobic chamber by a palladium catalyst that removed any oxygen (through the formation of water) that entered the chamber.

Once inside the chamber, the rubber stopper was removed from the influent end of the column and the packing material transferred to a sterile beaker. Care was taken to avoid contamination of the column during the mixing process. However, the column already contained a mixed culture and therefore potential exposure to airborne microorganisms was not a concern. The glass beads were mixed with a sterile, plastic transfer loop to evenly distribute the biofilm through the beads and also shear off large clumps of biofilm. Following the mixing process, the packing material was poured into the column and the rubber stopper was replaced. The column was removed from the anaerobic chamber, and again placed back into normal operation.

### *Column Detention Time*

The hydraulic detention time was measured using a tracer (sodium chloride) solution with a conductivity two-orders of magnitude greater than the background conductivity of the media (~200 mS/cm compared to ~1 mS/cm). A CDM230 conductivity meter (Radiometer, Copenhagen, Denmark) was used to measure the conductance of individual column effluent samples collected over time. One milliliter of tracer was injected into the influent immediately above the column inlet using a syringe. For Column 1, effluent samples were collected in borosilicate glass test tubes at 5 s intervals for a total of three minutes. The small (< 0.5 mL) sample volume was diluted with a known volume of deionized water to increase the sample size for conductivity measurements. The final volume varied between 0.5 to 1.5 mL. The conductivity of the diluted sample was plotted versus time to yield a tracer curve. The hydraulic detention time of Column 2, was measured in the same manner, except the samples were collected for 5 minutes. The tracer study on Column 2 was conducted approximately 1 month after the biofilm was established and only at a single flow rate of 2.3 mL/min.

A tracer curve was hand drawn using the conductivity data, and the hydraulic detention time was defined as the time to maximum breakthrough based on this curve. Polynomial regressions of the conductivity data were also used to find the time to maximum breakthrough (data in LaPoint 2000). However, the regression results provided a poorer fit of the conductivity data during the approach to a maximum concentration because the data were skewed by the long tails of the tracer data due to dispersion of the tracer in the column. It was decided that choosing the points for a regression analysis was as arbitrary as using the curves drawn by hand, and therefore the hand drawn curves were used.

## Column Experiments

### *Proof of Concept Study*

To prove that it was possible to remove perchlorate in the novel fixed bed, unsaturated flow reactor, after a start up period at a high perchlorate concentration (50 mg/L), Column 1 was operated for a period of approximately 125 days at a perchlorate concentration of ~750 µg/L. Perchlorate in the column influent and effluent were measured and a tracer study performed to determine the detention time in the column. Following this proof-of-concept period, additional tests were conducted to examine the perchlorate removal as a function of various operating parameters such as flow rate, perchlorate concentration and presence of competing electron acceptors such as nitrate and sulfate.

### *Perchlorate Removal as a Function of Flow Rate*

Hydraulic loading rates to Column 1 were varied to determine the effect of different flow rates, and therefore detention time, on perchlorate degradation rates. Experiments were conducted over 5 days at flow rates of 1.3, 2.3, 4.3, 7.3, and 11.3 mL/min, with a different flow rate each day. The experimental media used was the same the media listed in Table 5.2, except the perchlorate concentration was 750 µg/L.

Steady state operation at each new flow rate for Column 1 experiments was verified by monitoring effluent perchlorate concentrations. Samples of the effluent were taken in triplicate, starting at time zero, and thereafter for every two hours for typically10 hours. For one experiment (1.3 mL/min), however, samples were only collected for 8 hours. The influent was sampled once (in triplicate) at each flow rate. Each sample was filtered through a 0.2 µm pore diameter syringe filter (Acrodisc, Pall Corporation, Ann Arbor, MI). Steady state condition was tested using the ANOVA function from the Microsoft Excel program (1997). The test was whether or not the variation of the slope, analyzed by ANOVA, was significantly different than zero. A slope of effluent concentration significantly different would indicate that perchlorate concentration either decreased or increased and therefore that the column was not at steady state.

### *Effect of Perchlorate Concentration Study*

The effect of perchlorate concentration on the degradation rate of perchlorate was studied using both columns. The influent perchlorate concentration was changed each day for six days using perchlorate concentrations in Column 1 of approximately 1200, 700, 350, 120, 100, and 25 µg/L. In Column 2 tests, perchlorate concentrations were varied over 8 days with the influent perchlorate concentrations being approximately 900, 750, 650, 400, 250, 125, 80, and 40 µg/L. Effluent and influent samples were taken in triplicate each day. The samples were immediately filtered through a 25 mm syringe filter.

Perchlorate removal rates were calculated from influent and effluent perchlorate concentration and measured detention time using:

$$R = \frac{C_{in} - C_{out}}{\theta} \tag{5-1}$$

where R = removal rate (μg perchlorate/L-min), $C_{in}$ = influent perchlorate concentration (μg/L), $C_{out}$ = effluent perchlorate concentration (μg/L), and $\theta$ = hydraulic detention time. This rate was calculated as a function of the log-mean average concentration calculated as:

$$C_{lm} = \frac{C_{in} - C_{out}}{\ln(C_{in} - C_{out})} \tag{5-2}$$

where $C_{lm}$ = log mean perchlorate concentration.

### *Effect of pH on Perchlorate Removal*

To determine if low pH affected perchlorate degradation, the pH of the influent medium (Table 5.2, 1 mg/L influent perchlorate) was set at different values. To change the pH, the medium it was titrated with 6M NaOH, and 6M $H_3PO_4$ to five different pH values between a pH of 6 and 8. In each experiment, the pH of the influent medium to the reactor was changed to a new pH and held constant for one day, and the effluent then sampled (in triplicate) for perchlorate concentration and pH. Samples were tested for pH in all of these samples using a pH meter (Model 8000, VWR Scientific, Pittsburgh, PA) calibrated at a single pH of 7.

Because perchlorate removal increased with sulfate addition in Step 2 of the media alteration experiment, another experiment was conducted to determine if this effect was reproducible. Medium with a perchlorate concentration of 750 μg/L without sulfate was run in the column on day 1, 2, and the last day of the experiment. On days 3 and 4 the same medium was run in Column 2 except 32.5 mg/L of sulfate was added to the column. Samples were taken twice daily (in triplicate) for the duration of the experiment.

### *Effect of Nitrate on Perchlorate Removal Rates*

The effect of nitrate on perchlorate removal was examined at high (1000 μg/L) and low (80 μg/L) perchlorate concentrations. The column was run with the medium as listed in Table 5.2, except for the perchlorate concentration. For the first 7 days of the experiment, the feed did not contain any nitrate. During days 8-15, 25.3 mg/L of $NaNO_3^-$ was added to the media (18.5 mg/L as nitrate). Triplicate samples were taken every one or two days. Nitrate and nitrite influent and effluent concentration were analyzed using a Technicon Autoanalyzer II (Technicon Industrial Systems, Tarrytown, NY). The experiment was repeated at an influent perchlorate concentration of 80 μg/L, with nitrate added to the medium during days 8 to 14.

### *Perchlorate Removal Using Redlands Water*

Redlands water (Table 5.3) was stored in a refrigerator and used as received except it was amended with 1 g/L of carbonate and a phosphate-nitrate buffer, but not trace metals (Table 5.2) for 10 days. Samples of the influent and effluent were taken once per day. The effect of carbonate addition to the media was examined in additional experiments using Redlands water with 1 g/L of carbonate, 0.087 mg/L of phosphate, and 9.63 mg/L of nitrate added. The first and last days of the experiment were run with the modified Redlands water, while on days 2, 3, and 4, experiments were run with Redlands water that contained a phosphate buffer, but no additional carbonate.

Samples were monitored the same way as for the pH experiments previously discussed in the previous section.

### Analytical Methods

Perchlorate was measured on a Dionex DX 500 ion chromatography system (Sunnyvale, CA), with an IonPac AS 11 4 mm column, IonPac11 4mm guard column, GP50 gradient pump, ED40 electrochemical Detector, and AS40 automated sampler. Data from the DX 500 was recorded by the Dionex Peaknet™ program (version 5.01). Perchlorate concentration was measured in the range of 1-20 mg/L using a 25 µl injection loop, and in the 4-100 µg/L range using a 1000 µl loop. Samples above these perchlorate ranges were diluted with deionized water (MilliQ).

Perchlorate calibration standards were prepared with the same components as the media (Table 5.2). For the perchlorate concentrations of 1- 20 mg/L calibration standards were 1, 5, 10, and 20 mg/L. For the lower perchlorate range, four of the following concentrations were used: 5, 10, 20, 25, 50, 95, and 100 µg/L. New IC caps and vials were used for every measurement for Column 2 analysis. Column 1 caps and vials were recycled by rinsing them 5 times with MilliQ water. The DX-500 had a minimum detection limit of 1.5 µg/L in deionized water and 5 µg/L for samples made in the column influent.

Nitrate concentrations were measured on a Technicon AutoAnalyzer II as combined nitrate and nitrite, and the samples were run again as $NO_2^-$. Then the nitrite concentrations were subtracted from the nitrate and nitrite concentrations to obtain the nitrate concentrations. Three standards were run at 2, 0.2, and 0.04 mg/L of nitrate. The minimum detection level for the Technicon is 0.04 mg/L nitrate. Samples can be diluted with deionized water if the nitrate level in the sample is above the calibration range. The program used to calculate the nitrate concentrations was the Technicon GTpc V. 3.51.

## RESULTS AND DISCUSSION

### Selection and Isolation of Hydrogen-Oxidizing Bacteria Pure and Mixed Culture

Following 1 week of incubation of a wastewater sample amended with perchlorate, under a 5% hydrogen gas atmosphere, perchlorate reduction was observed. The average $ClO_4^-$ concentration of the two control bottles was 187±5 mg/L, whereas the average $ClO_4^-$ concentration of the three experimental bottles was 8.1±1.6 mg/L. All bottles were purged with the 5% $H_2$/95% $N_2$ gas two additional times between inoculation and sampling.

#### *Enrichment in a Chemostat*

The batch culture was transferred to a bioreactor, and the reactor purged of oxygen and sparged with a hydrogen-carbon dioxide gas stream. The system was initially operated in batch mode. Once a large change in $ClO_4^-$ concentration was observed (175 mg/L), the reactor was switched to chemostat mode and fed VG media with an influent $ClO_4^-$ concentration of approximately 250 mg/L. The initial influent flowrate was set at 0.54 mL/min, which corresponded to a detention time of 16.2 hrs. The reactor was periodically sampled, and from this liquid sample, absorbance (600 nm), pH, and $ClO_4^-$ concentration were monitored. When the

$ClO_4^-$ concentration had reached a steady-state condition at this particular flow rate, the influent flow rate was changed to 0.65 mL/min (detention time of 14.3 hours). An equilibration period occurred, and a steady-state reactor $ClO_4^-$ concentration was again measured using the IC. The final influent flowrate was set at 0.76 mL/min (detention time of 12.3 hours). Table 5.4 shows the steady-state $ClO_4^-$ concentrations that were measured at each of three detention times.

Because biological perchlorate degradation was maintained at a detention time of 12.3 hours, which is less than the typical detention time of 24 hours for methanogens they were considered absent from the microbial consortium. Sulfate reducers were also believed to be excluded because there was no smell, while there would have been a smell if the sulfate reducers were in the chemostat because of the production of hydrogen sulfide.

Using this data, kinetic parameters were calculated for the hydrogen oxidizing, perchlorate reducing mixed population using Lineaweaver Burke and Hanes data transformations as shown in Table 5.5. The best fit was obtained using the Hanes transformation. Thus, these results indicate a very low half saturation constant of 1.8 mg/L, and a minimum doubling time of 8.7 hours. This indicates that a suspended growth reactor would need to be operated with a cell detention time of much larger than 12.5 hours in order to maintain perchlorate-reducing conditions.

## Fixed Film Bioreactor Experiments for Column 1

### *High Influent Perchlorate Concentrations*

Perchlorate reduction was successfully achieved under hydrogen oxidizing conditions in the absence of other electron donors. The column was operated for approximately ten days at an influent perchlorate concentration of 50 mg/L (Figure 5.6). The average removal of perchlorate through the column was 10.1 ± 1.4 mg/L over the ten day period if three data points in Figure 5.6 are not included in this calculation. These three data, which indicate relatively low perchlorate removal rates, can be neglected for one of three different reasons. First, at time zero, the hydraulic loading rate was much higher than that used for the other data. The media was initially pumped at approximately 11 mL/min (equivalent to a hydraulic loading rate of 0.55 gpm/ft$^2$ for the 2.5 cm diameter column). The flowrate was changed to the 2.3 mL/min for the duration of all column experiments (0.11 gpm/ft$^2$). Second, after 50 hr, the influent media was not changed soon enough; therefore the column was operated without influent flow for a period of time prior to sampling. The third problem occurred at time prior to the sampling period at 175 hours. At some time prior to this sampling point the effluent tubing from column clogged, resulting in all liquid media leaking from the column and incomplete wetting of the media.

Stoichiometric conversion of perchlorate to chloride was measured in the column. The influent and effluent from two days was anaylzed for chloride. The average decrease in perchlorate was 0.115 mM, while the average increase in chloride was 0.1 mM. Column studies carried out by Herman and Frankenberger (1999) and Attaway and Smith (1993) also showed a stoichiometric decrease in perchlorate.

### *Low Influent Perchlorate Concentrations*

One of the main objectives of this study was to determine whether a hydrogen oxidizing bioreactor can maintain perchlorate reducing conditions at low perchlorate concentrations. The influent perchlorate concentration to the column was therefore decreased from 50 mg/L to approximately 750 µg/L. Figure 5.7 shows that the reactor successfully reduced perchlorate concentrations over an extended period of time (approximately 125 days). However, the extent of perchlorate removal measured during day-to-day sampling was somewhat erratic. Therefore, on day 43, when perchlorate removal through the column was approximately 300 µg/L, the column was mixed as described above. After mixing the perchlorate removal through the column was approximately 600 µg/L. However, the increased performance of the column did not continue, and the column once again became clogged with excess biofilm. After a break in column sampling (day 52 to 68), it can be seen that the column performance continued to decline from day 68 to 99. The column was again mixed to determine whether removal would increase, but no samples were taken in the days immediately following the mixing, so it was impossible to tell if perchlorate degradation increased due to mixing at that time. However, it can be concluded that the increase in perchlorate degradation after the first mixing was only temporary, and that overtime perchlorate degradation rates decreased significantly.

In order to better represent the efficiency of perchlorate removal during the column operation, the method of sampling was altered. Prior to day 68, the column was sampled only once each time (both influent and effluent). Due to the variability of the data, however, it was decided that more samples needed to be taken each time. Therefore, a more representative method of sampling was developed by collecting three samples over a period of approximately 12 hours, and reporting the average perchlorate concentrations in these samples. Therefore, from day 68 onward, an average decrease in perchlorate concentration through the column (triplicate samples, ± standard deviation) is shown.

An average removal rate was calculated for the column using the equation 5-2. For an average detention time of 1.21 min, and an average perchlorate decrease of 228 µg/L the average removal of perchlorate was 233 µg-perchlorate/L-min. These rates are higher than those found by others. Giblin, Herman, and Frankenberger (2000) found using an autotrophic saturated-flow hydrogen-oxidizing reactor that 740 µg/L of perchlorate was reduced to non-detectable levels (< 4 µg/L) at a residence time of 2 hours. This translates to a removal rate of 6.2 µg-perchlorate/L-min. When the flow rate was increased to 3mL/min with a residence time of 0.67 hours there was perchlorate breakthrough of 200 µg/L and the removal rate was 13.4 µg-perchlorate/L-min. Herman and Frankenberger (1999) inoculated a flow-through column bioreactor with perc1ace, and ran contaminated groundwater through the column. The groundwater had a perchlorate concentration of 738 µg/L, and a 5 hr residence time, and perchlorate was removed to non-detectable levels. The removal rate for the column was 2.5 µg/L-min.

### *Tracer Studies*

The time to maximum tracer breakthrough (43.7 mS/cm) occurred at 65 s before the column was mixed (Figure 5.8). The time at which one-half the maximum breakthrough occurred was approximately 10 seconds. After the column was mixed the maximum breakthrough occurred at approximately 82.5 seconds, with one-half the maximum breakthrough concentration occurring at 25 seconds. Thus we see that mixing the column helped to increase the detention time of the fluid in the packed bed reactor and led to greater perchlorate removal.

### *Perchlorate Removal as a Function of Flow Rate: Column 1*

The difference between influent and effluent perchlorate concentration decreased as flow rate was increased from 1.3 mL/min to 11.3 mL/min in Column 1. At 2.3 mL/min, perchlorate changed by 368 μg/L, decreasing to a change of only 164 μg/L at 7.3 mL/min (Figure 5.9). This decease in overall perchlorate removal was likely due to shorter hydraulic residence times at higher flow rates resulting in less contact of the water with the biofilm. To demonstrate that the residence times decreased with increased flow rates, tracer experiments were again performed to determine detention times in Column 1 at various flow rates. An example of the differences in the hydraulic detention times at different flow rates is shown in Figure 5.10A for flow rates of 1.3 mL/min and 11.3mL/min. The detention time was 35 seconds at 1.3 mL/min; while at 11.3 ml/min it was 15 s. Column mixing was found to increase the detention time. Examples of a tracer curve before and after column cleaning for a flow rate of 4.3 mL/min is shown in Figure 5.10B. The time until maximum breakthrough (the hydraulic detention time) was 14 s before mixing and 18 s after biofilm cleaning. The tracer curves shown in Figure 5.10 also demonstrate that there was extensive tailing of the tracer in the column due to the dispersion of the salt tracer into the biofilm and column packing. Results of the detention time experiments are summarized in Table 5.6.

From Figure 5.10 and Table 5.6 it can be concluded that mixing the column increased the detention time. Hydraulic detention times originally measured in Column 1 at 2.3 mL/min (65 s and 82.5 s before and after biofilm mixing, respectively) were about twice those measured here. The difference in residence times is due to the differences such as amount and distribution of the biofilm in the column.

The combined effects of flow rates and mixing on detention time can be seen in Figure 5.11. At lower flow rates (1.3 mL/min) there is a longer time until maximum breakthrough of the tracer producing a longer detention time. At higher flow rates of 4.3, 7.3, and 11.3 ml/min the detention times are almost identical.

A statistical test (ANOVA) was used to test the hypothesis that there was no change in the effluent concentration over time at each flow rate. The results from ANOVA showed that there was no significant change ($p<0.05$) in effluent data at all flow rates based on slopes not significantly different than zero. However the confidence intervals were relatively large at flow rates of 2.3 and 7.3 mL/min compared to those at 1.3, 4.3, and 11.3 mL/min, and the regression of the slopes for all of the flow rates had p-values greater than 0.05. The confidence interval (CI%) maintained in Table 5.7 represents the lowest confidence interval percent that the slope can fall between. Therefore, the lower the confidence interval the more likely that the reactor was operating under steady state conditions. A value of $p > 0.05$ indicates there is less than a 95% probability that the slope is different than zero. Because p-values in Table 5.7 are all above 0.05, these results support the assumption that there was no change in the effluent $ClO_4^-$ concentration, and therefore that the column was operating under steady state conditions.

## Fixed Film Bioreactor Experiments for Column 2

### *Tracer Study For Column 2*

A tracer study was conducted for Column 2, but only at the single flowrate used for this column (2.3 mL/min). The time to maximum tracer breakthrough (12 mS/cm) occurred at 90 ± 5 s

(Figure 5.12). This detention time is closer to that range originally measured for the first column (65 to 82.5 s, before and after mixing respectively).

### *Perchlorate Removal in Column 2*

During Column 2 reactor startup, influent perchlorate concentrations averaged 18 mg/L. The average decrease in perchlorate was 1.4 ± 0.3 mg/L, or a removal of 8.4% (Figure 5.12). After 9 days, the column was shifted to a lower average influent perchlorate concentration of 1200 µg/L, and the average decrease in perchlorate concentration was 340 ± 122 µg/L, or 30 ± 9% (Figure 5.13). The average removal rate for an influent perchlorate concentration of 1200 µg/L in Column 2 was 227 µg/L-min. This removal rate is very similar to the 233 µg/L-min originally measured for Column 1 at an influent perchlorate concentration of 719 µg/L.

### *Effect of Perchlorate Concentration*

The degradation rate of perchlorate increased relatively linearly ($R^2$=0.82) with the log mean perchlorate concentrations in Column 1 (Figure 5.14). The degradation rate of perchlorate in Column 2 also increased ($R^2$=0.87) in proportion to the log mean perchlorate concentration, but the overall perchlorate removal rate in Column 2 was less than that measured for Column 1. A log-log plot of perchlorate concentration vs. degradation rates shows that the slopes of Column 1 and Column 2 equal 1, which suggests perchlorate removal occurs as first order kinetics (Figure 5.15).

### *Effect of pH*

As a result of the large pH decrease in the reactor effluent when carbonate was removed from laboratory medium, a separate experiment was conducted to further examine the effect of pH on perchlorate removal. The pH of the influent water was set within five different ranges through the addition of 6M phosphoric acid to the medium. The pH did not have an effect on perchlorate degradation until the pH in the reactor dropped below 6.4 (Figure 5.16). The percent removal at an influent pH of 6.3 and effluent pH of 6.1 was only 32 ± 3%, while the average percent removal for the pH range of 8.0, 7.8, 7.3, 6.6, and 6.3 were 46 ± 6%. This is a significant decrease in perchlorate removal because the standard deviations of the average percent removal of perchlorate in the highest pH experiments do not overlap with those obtained in the low pH experiment. A pH drop from 7.1 to 5.5 in the media alteration experiment decreased the perchlorate percent removal to 8.2 ± 3.6.

### *Effect of Sulfate*

A separate experiment was also conducted to examine the effect of sulfate on perchlorate removal. Effluent perchlorate concentrations did not change in the reactor when sulfate was added to the medium (Figure 5.17). The removal rate of perchlorate when sulfate was present at a concentration of 32.5 mg/L in the feed was 201 ± 59 µg/L-min, while the removal rate with very little sulfate (63.9 µg/L) was 208 ± 37 µg/L-min. It was therefore concluded that sulfate had no effect on perchlorate removal rates in this reactor.

### *Effect of Nitrate on Perchlorate Removal Rates*

The effect of nitrate on perchlorate removal was examined at both high and low perchlorate concentrations. Prior to introducing nitrate, perchlorate was introduced at an approximate concentration of 826 μg/L although these concentrations varied during the experiment due to different media. After 6 days, 21 ± 2 mg/L of nitrate was added to the media to study its effect on perchlorate degradation. Perchlorate degradation occurred in the presence of nitrate, and was significantly decreased by nitrate according to a 95% confidence interval from a one tailed t-test using unequal variances ($p<0.02$; used throughout nitrate this section to judge statistical significance) (Figure 5.18). The average removal rate for perchlorate in the absence of nitrate was 192 ± 51 μg/L, and was 164 ± 56 μg/L in the presence of nitrate. Nitrate and perchlorate were simultaneously degraded in the column. Nitrate was significantly decreased ($p<0.016$) to a concentration of 18 ± 1 mg/L (Figure 5.19), resulting in a nitrate removal rate of 1.5 mg-$NO_3^-$/min assuming a detention time of 1.5 min.

After 5 days at the higher perchlorate concentration (826 μg/L), the influent perchlorate concentration was switched to a lower concentration (79 μg/L) that was more representative of Redlands water. After feeding the column medium for 7 days without nitrate, 25.3 mg/L of nitrate was added to the media. Perchlorate degradation still occurred in the presence of nitrate (Figure 5.20). The average removal rate of perchlorate in the absence of nitrate was 15.2 ± 3.6 μg/L, and the average removal of perchlorate in the presence of nitrate was 8.7 ± 1.6 μg/L. This is a significant decrease in perchlorate degradation after nitrate was added ($p<0.006$). Nitrate degradation was achieved at the lower perchlorate concentration at the same rate of 1.5 mg-nitrate/min (Figure 5.21) as the higher perchlorate experiment achieved in the reactor with an average influent perchlorate concentration of 73 μg/L (Figure 5.22), of 18.3 ± 4.3 μg/L. The perchlorate removal rate was 17.1 ± 2.9 μg/L-min. However the level of nitrate removal was only significant at a confidence interval of 92% instead of 95% because $p < 0.07$.

The nitrate removal rate measured for the using Redlands, CA water (1.6 mg/L-min) was nearly identical to that of the previous two nitrate experiments. Nitrate removal was significant in the reactor ($p<0.006$). However, it may be that nitrate removal was increasing over time. Initially, the nitrate removal rate was 0.32 ± 0.46 mg-nitrate/L-min for the first 3 days, but by the end of the experiment, which was after 7 days, the nitrate removal rate had increased to 3.6 mg/L-min (Figure 5.23).

An experiment was conducted using Redlands water to determine if perchlorate degradation rates changed with the addition of 10 g/L of carbonate. The perchlorate removal rate with carbonate was 26 ± 5 %, and 20 ± 6 % without the additional carbonate (Figure 5.24). The pH dropped from 6.9 to 6.6. The standard deviations overlap, therefore the rate differences with or without additional carbonate were not considered to be significant.

### *Organic Concentrations in the Effluent*

To determine how much total dissolved organic carbon was in the effluent of the reactor, DOC and biomass from the column effluent was measured. The average dissolved organic carbon from Column 2 was calculated from subtracting the influent DOC from the effluent DOC. The dissolved organic carbon from Column 2 was 0.216 ± 0.28 mg/L DOC (n=18), while the amount of biomass leaving the column was 6.67 ± 6.5 mg/L (n =3) at a perchlorate influent concentration of 750 μg/L.

**Table 5.1**
**Trace metal components in media used for perchlorate degradation experiments**

| Component | Stock concentration (mg/L) | Media concentration (μg/L) |
|---|---|---|
| $Na_2EDTA$ | 600 | 3 |
| $CaCl_2 \cdot 2H_2O$ | 200 | 1 |
| $Na_2MoO_4 \cdot 2H_2O$ | 40 | 0.2 |
| $CoCl_2 \cdot 6H_2O$ | 80 | 0.4 |
| $Na_2SeO_3 \cdot 5H_2O$ | 20 | 0.1 |
| $NiCl_2 \cdot 6H_2O$ | 20 | 0.1 |
| $ZnSO_4 \cdot 7H_2O$ | 400 | 2 |
| $CuCl_2 \cdot 2H_2O$ | 40 | 0.2 |
| $H_3BO_3$ | 120 | 0.6 |
| $FeSO_4 \cdot 7H_2O$ | 1000 | 5 |
| $MnCl_2 \cdot 4H_2O$ | 200 | 1 |

**Table 5.2**
**Media components for column perchlorate experiments. Trace metals added at 5ml/L to achieve the concentration as noted in Table 5.1.**

| Media component | Stock concentration (g/L) | Media concentration (mg/L) |
|---|---|---|
| $NaClO_4$ | 74 | 65,0.9 |
| $K_2HPO_4$ | 31 | 15.5 |
| $Na_2H_2PO_4 \cdot H_2O$ | 17 | 8.2 |
| $NH_4H_2PO_4$ | 10 | 5 |
| $MgSO_4 \cdot 7H_2O$ | 0.2 | 0.1 |
| $Na_2CO_3$ | N/A | 1000 |
| Trace metals | See Table 5.1 | See Table 5.1 |

**Table 5.3**
**Chemical composition of Redlands water as provided by CDM**

| Chemical | Concentration (mg/L) |
|---|---|
| Calcium | 56.9 |
| Magnesium | 9.3 |
| Sodium | 13.4 |
| Potassium | < 1.0 |
| Hydroxide | < 1.0 |
| Carbonate | < 1.0 |
| Bicarbonate | 181 |
| Sulfate | 32.5 |
| Chloride | 5.51 |
| Nitrate | 18.5 |
| Fluoride | 0.533 |
| Total alkalinity ($CaCO_3$) | 148 |

**Table 5.4**
**Summary of conditions during chemostat experiment**

| Influent flow rate (mL/min) | Detention time (hrs) | Steady-state $ClO_4^-$ conc. (mg/L) |
|---|---|---|
| 0.54 | 16.2 | 4.47 |
| 0.65 | 14.3 | 31.9 |
| 0.76 | 12.3 | 37.9 |

**Table 5.5**
**Summary of hydrogen-oxidizing mixed culture kinetic parameters for $ClO_4^-$**

| Method | $V_{max}$ ($hr^{-1}$) | $K_s$ (mg/L) | Doubling time (hr) |
|---|---|---|---|
| Lineweaver Burke | 0.078 | 1.15 | 8.93 |
| Hanes | 0.080 | 1.84 | 8.65 |

**Table 5.6**
**Hydraulic detention time at various flow rates before and after mixing**

| Flow rate (mL/min) | HDT (θ) (sec) | |
|---|---|---|
| | Before | After |
| 1.3 | 33 ± 5 | 35 ± 5 |
| 2.3 | 23 ± 5 | 22 ± 5 |
| 4.3 | 14 ± 5 | 19 ± 5 |
| 7.3 | 10 ± 5 | 15 ± 5 |
| 11.3 | 12 ± 5 | 15 ± 5 |

**Table 5.7**
**The statistical results from ANOVA (CI is the confidence interval)**

| Flow rate | p-value | CI % | Lower CI | Slope | Upper CI |
|---|---|---|---|---|---|
| 1.3 ml | 0.356 | 65 | -0.06 | 3.63 | 7.31 |
| 2.3 ml | 0.141 | 87 | -0.43 | 11.2 | 22.9 |
| 4.3ml | 0.978 | 5 | -0.13 | 0.09 | 0.32 |
| 7.3 ml | 0.081 | 95 | -1.74 | 8.96 | 19.7 |
| 11.3 ml | 0.399 | 65 | -0.77 | 6.26 | 13.3 |

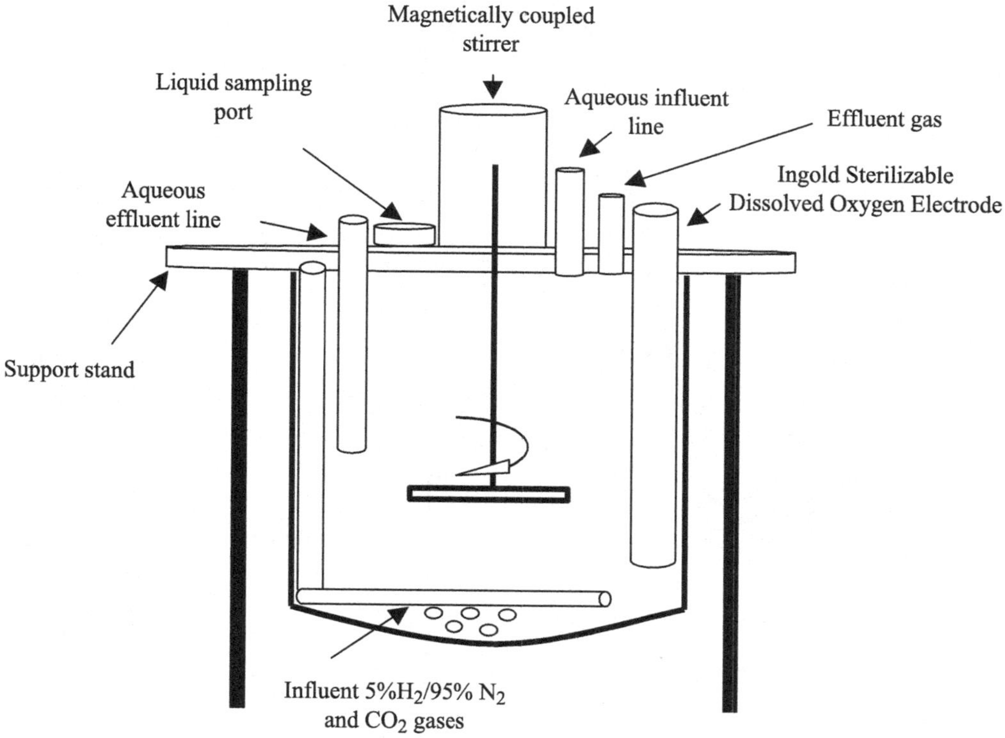


**Figure 5.1 Schematic of Applikon Bioreactor used in chemostat experiment**

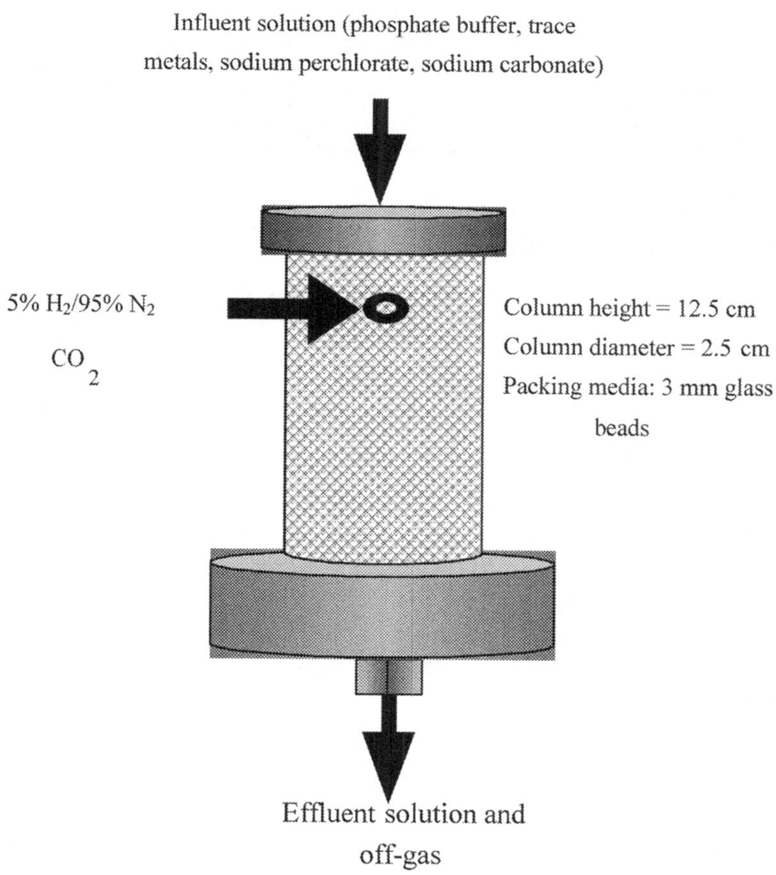


**Figure 5.2 Schematic of the first hydrogen-oxidizing reactor (column 1)**

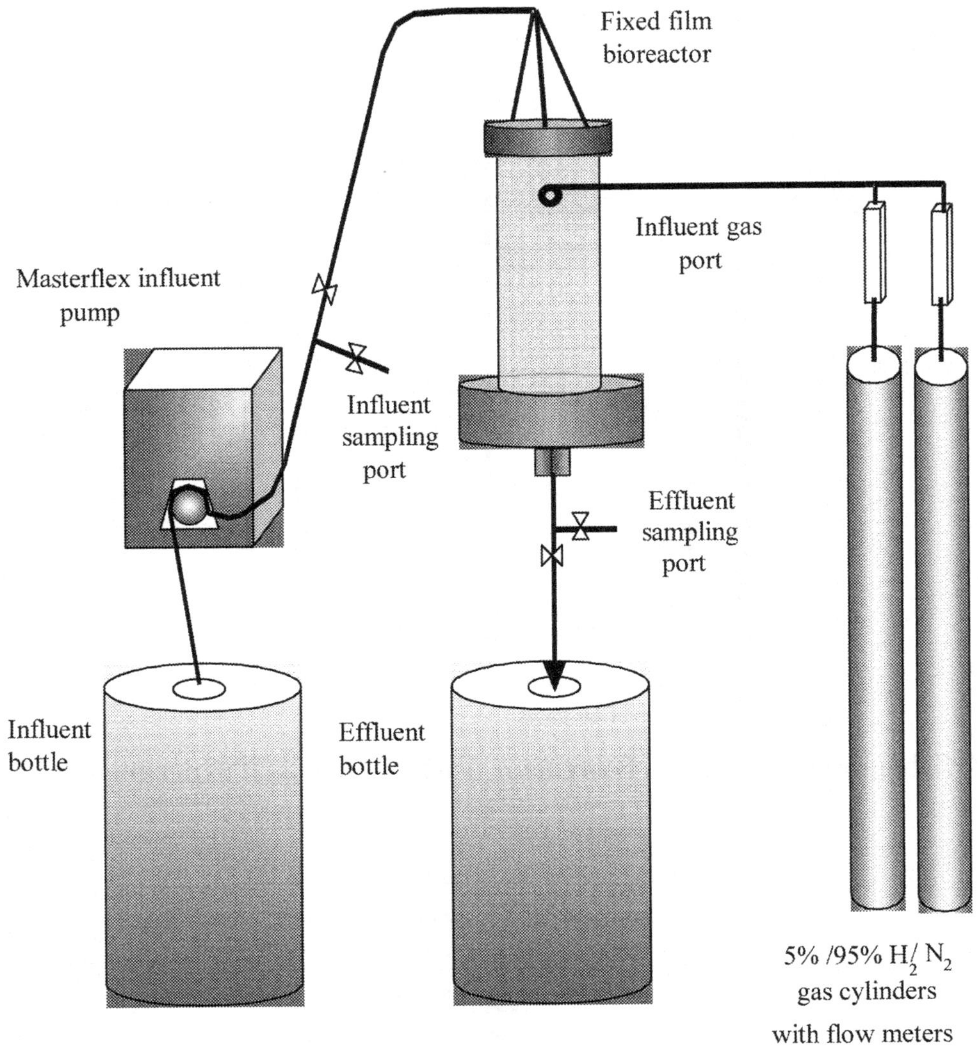


**Figure 5.3 Set up for column 1 hydrogen bioreactor**

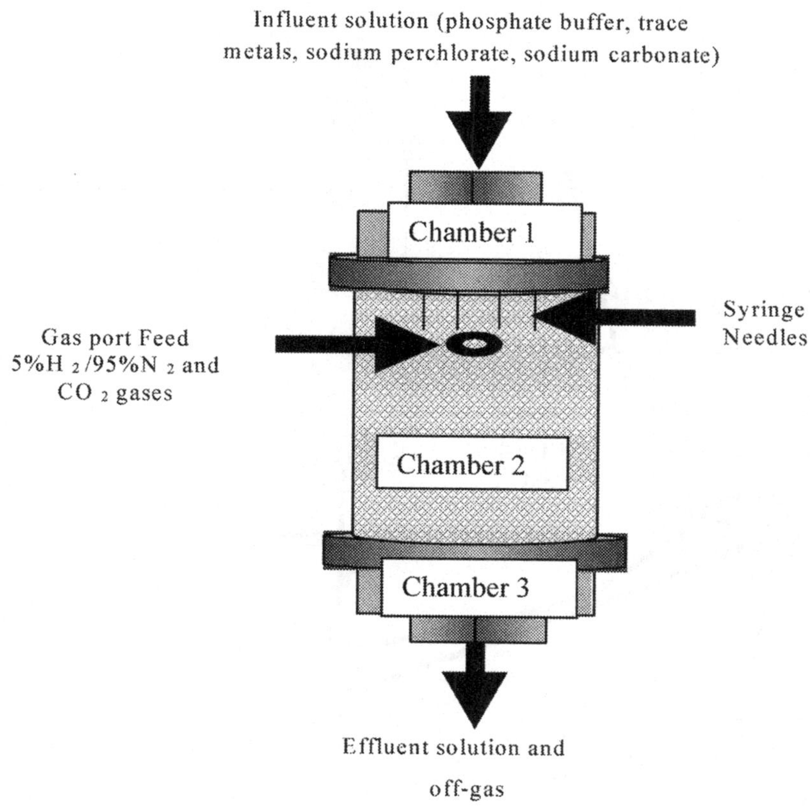


**Figure 5.4 Schematic of the column 2 reactor. The column height is 25 cm, and the column width is 3.5 cm. The column is packed with 3-mm glass beads. The gas mixture flows into the column at the rate of 30 mL/min for the 5% hydrogen/95% nitrogen mixture, and 7 mL/min for the 100% carbon dioxide.**

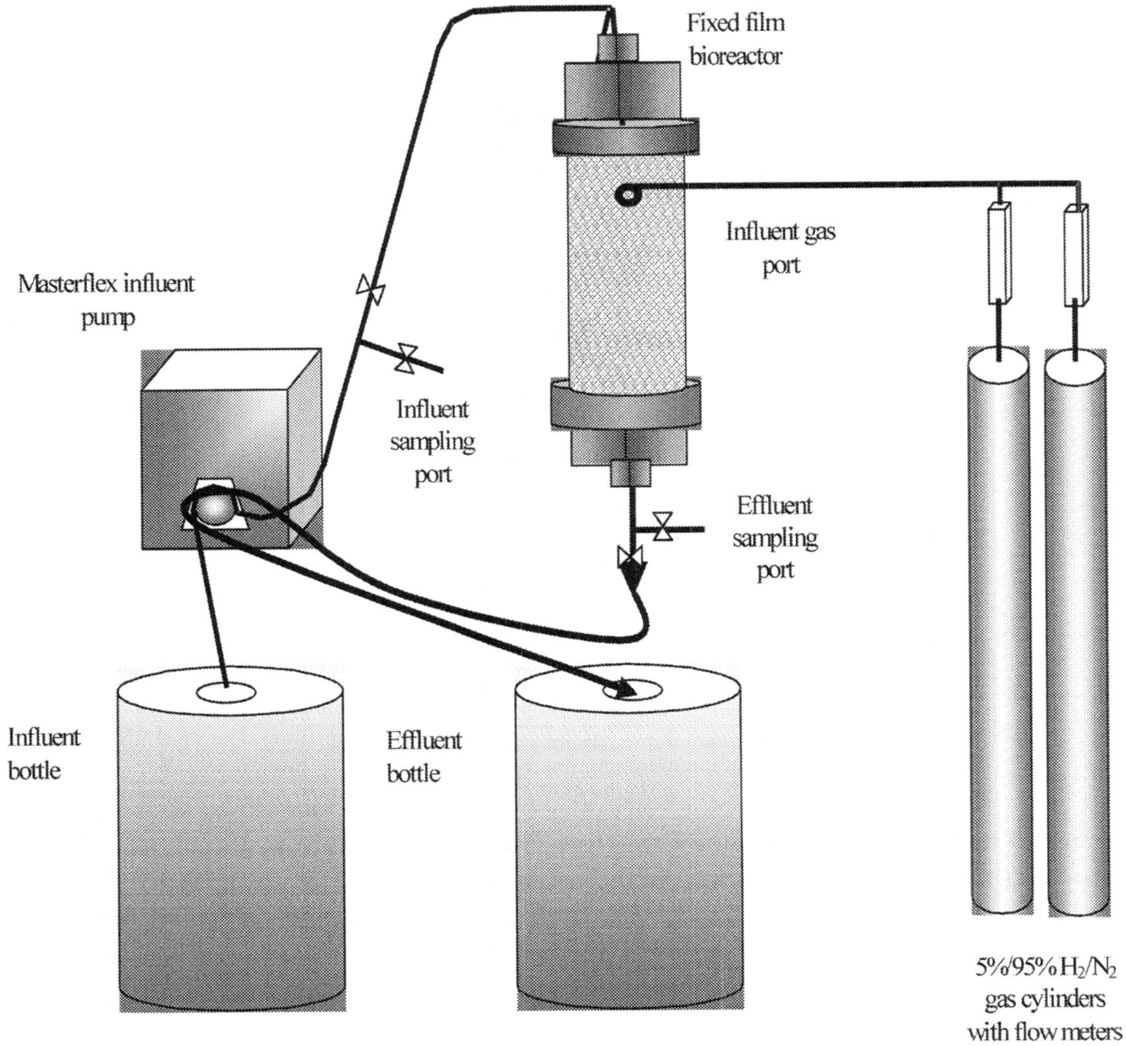


**Figure 5.5 Set up for the column 2 reactor**

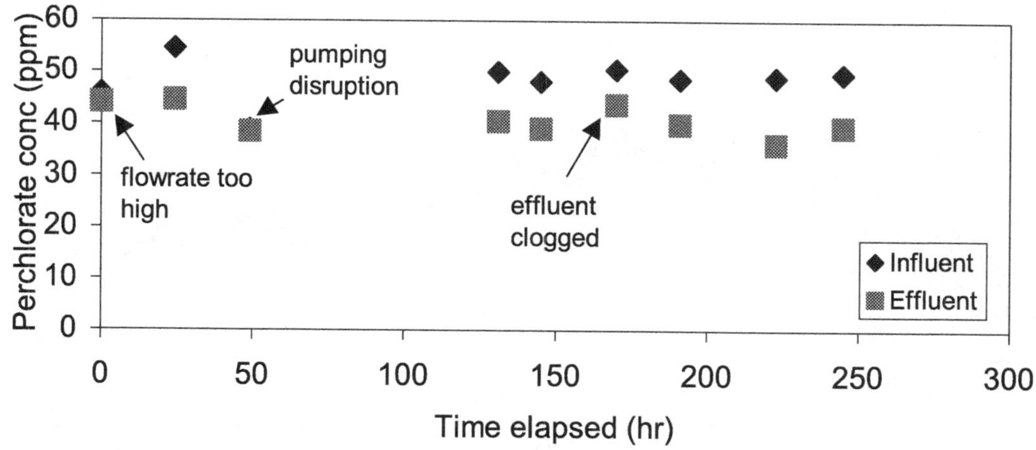


**Figure 5.6 Perchlorate reduction in hydrogen-oxidizing unsaturated column for high-perchlorate concentration feed**

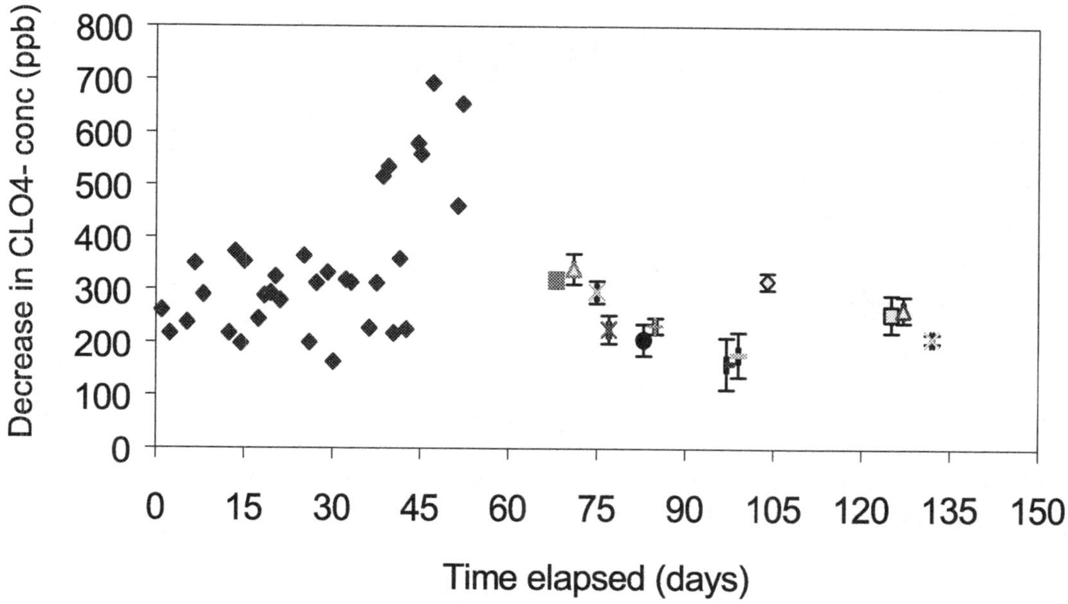


**Figure 5.7 Perchlorate removal in the hydrogen-oxidizing column at low influent perchlorate concentrations**

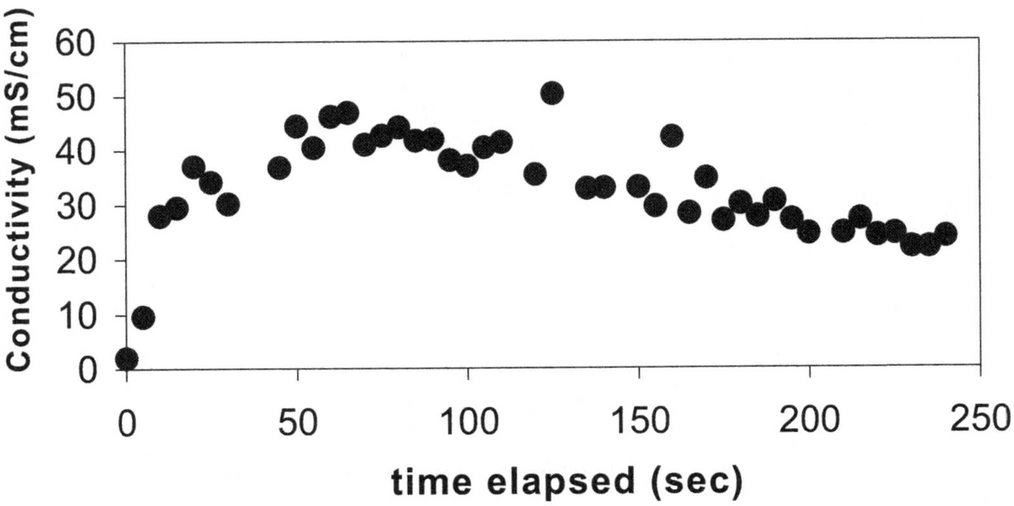


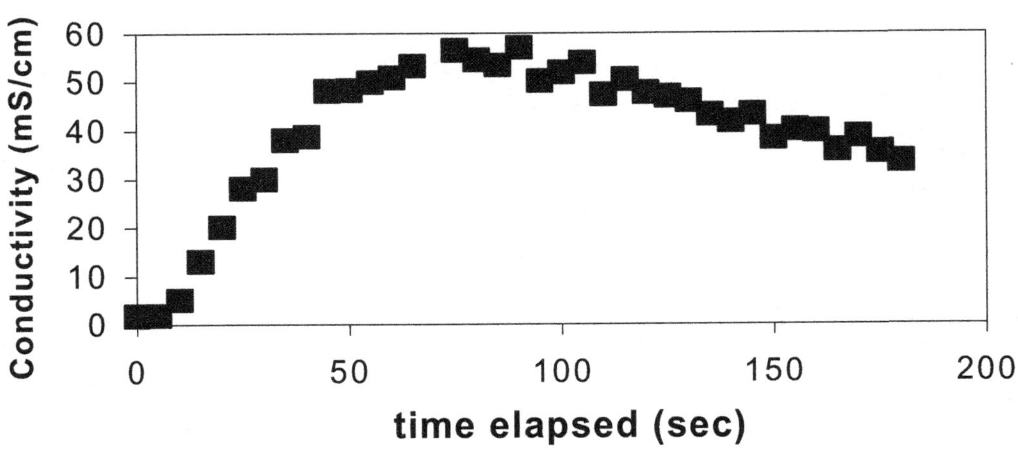


**Figure 5.8 Tracer curves taken before (upper plot) and after (lower plot) mixing the bed to redistribute the biofilm-coated glass beads**

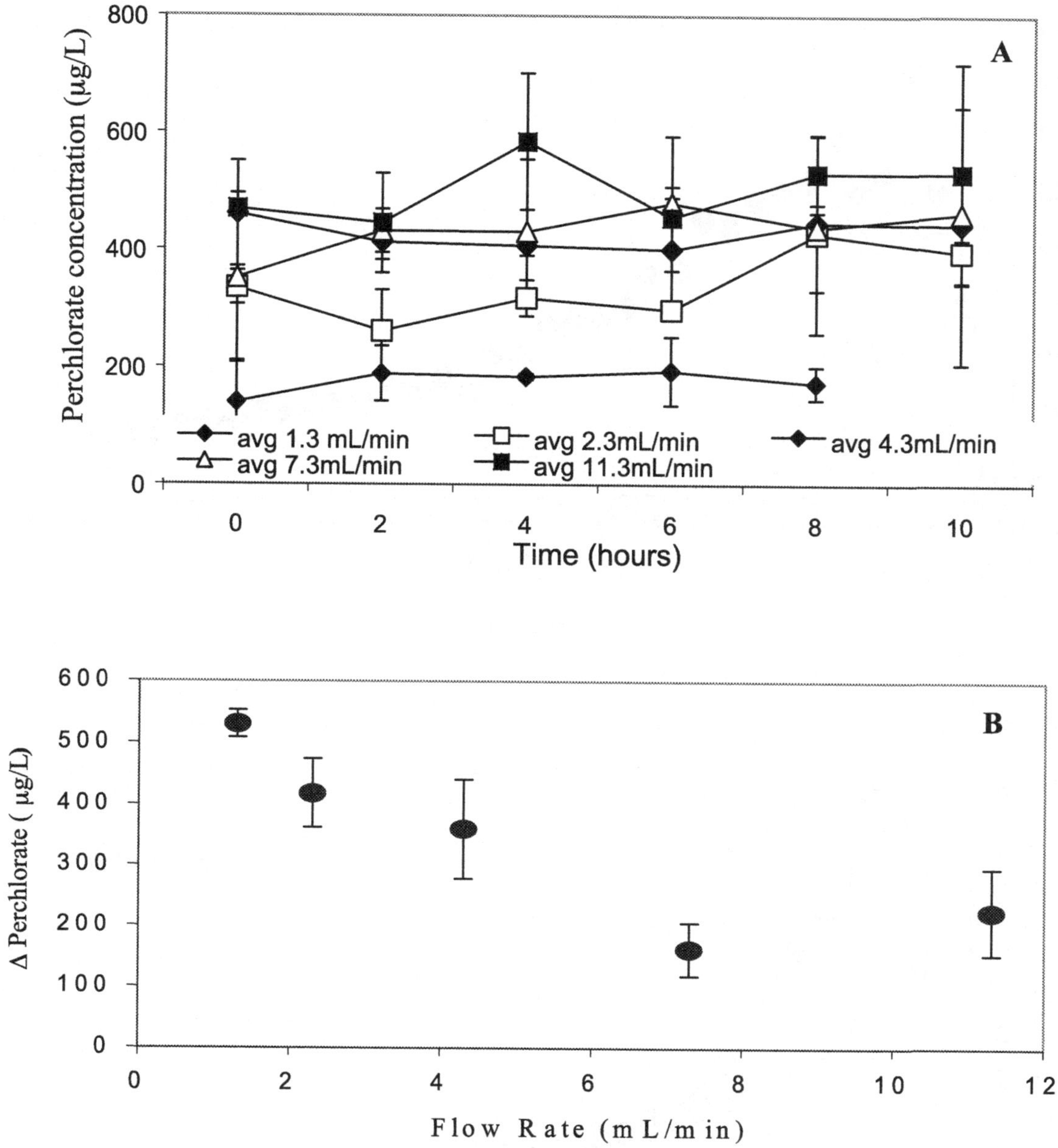


**Figure 5.9 (A) Effluent concentration of perchlorate over time for various flow rates for column 1 at an influent perchlorate concentration of 700 µg/L. (B) Decrease in perchlorate concentration from the influent media as a function of flow rate using column 1. Each point is the difference between the averages of all perchlorate influent and effluent samples at each flow rate. The error bars represent the standard deviation of the average perchlorate change.**

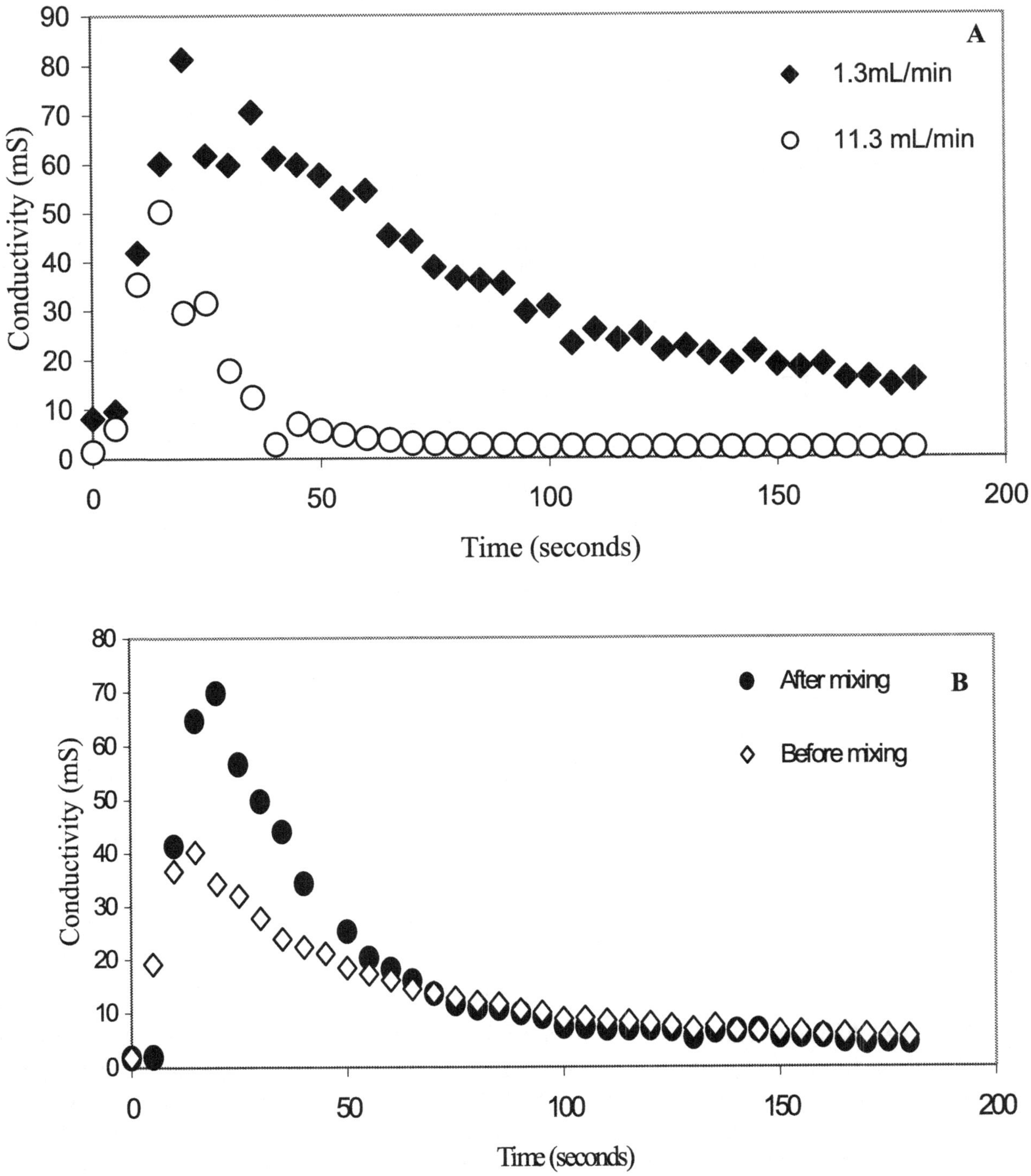


**Figure 5.10 (A) Comparison of the tracer breakthrough curves in column 1 at flow rates of 1.3 mL/min and 11.3 mL/min. (B) Comparison of the effect of mixing on tracer curves in column 1 at a flowrate of 4.3 mL/min.**

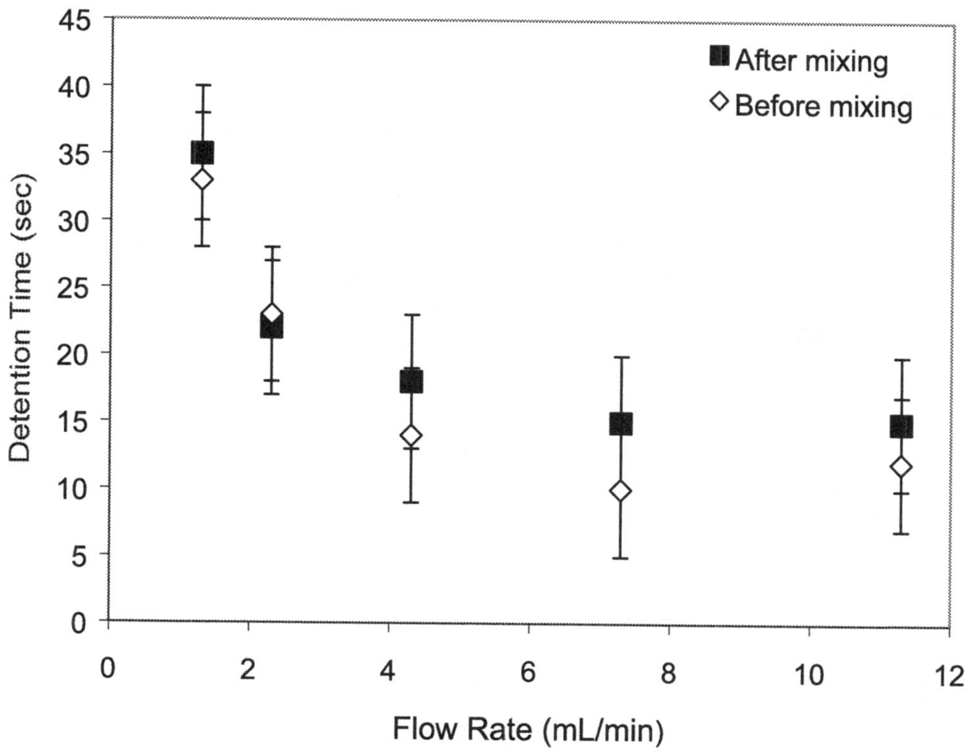


**Figure 5.11 Detention time as a function of flow rate before column mixing, and after column mixing in column 1. Error bars represent standard deviation ± 1 standard deviation.**

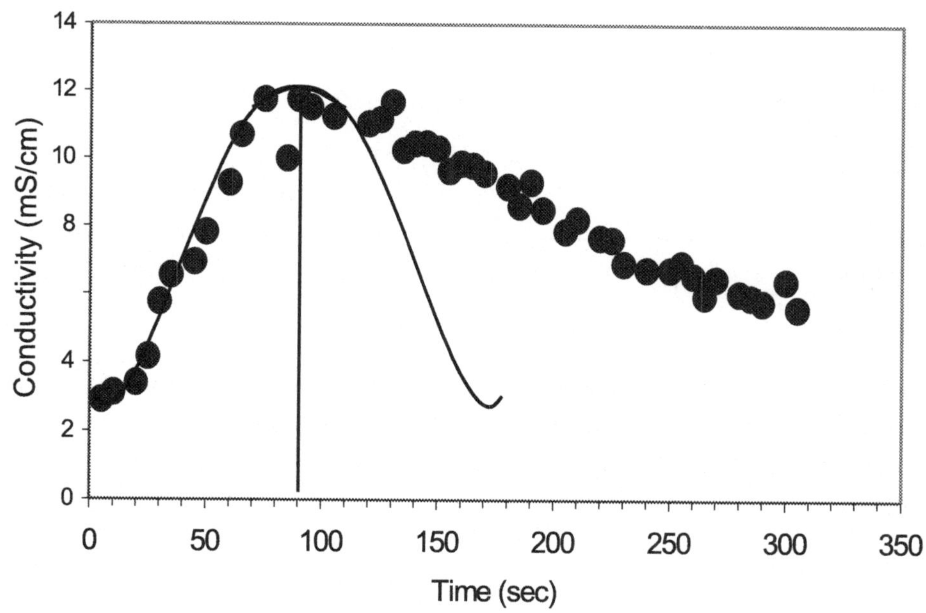


**Figure 5.12 Tracer curve for column 2 at 2.3 mL/min**

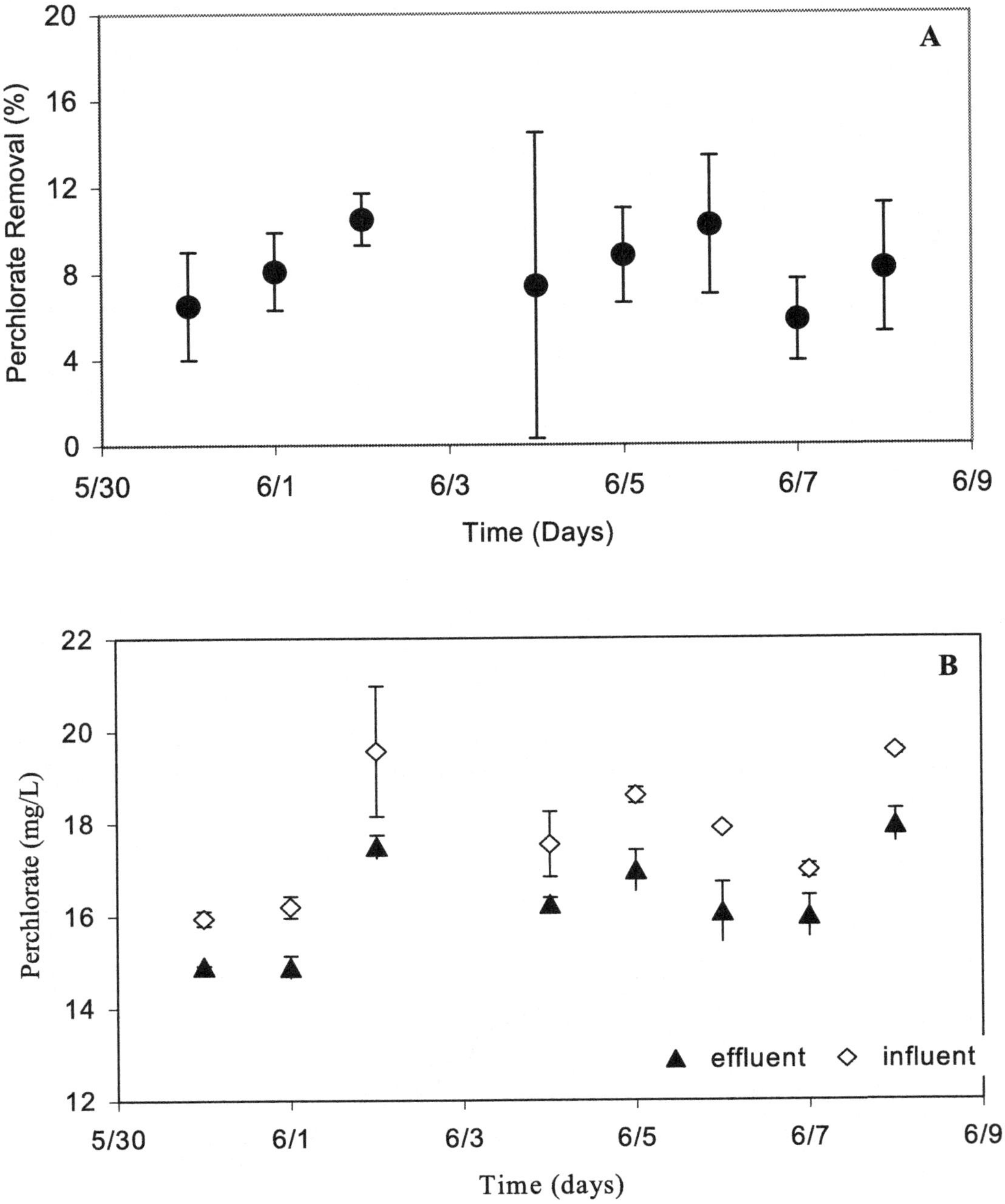


**Figure 5.13 (A) The influent, and effluent perchlorate concentration of column 2 during column startup and (B) The percent change in perchlorate at an average influent perchlorate concentration of 18 mg/L. Errors bars are ± one standard deviation based on triplicate samples.**

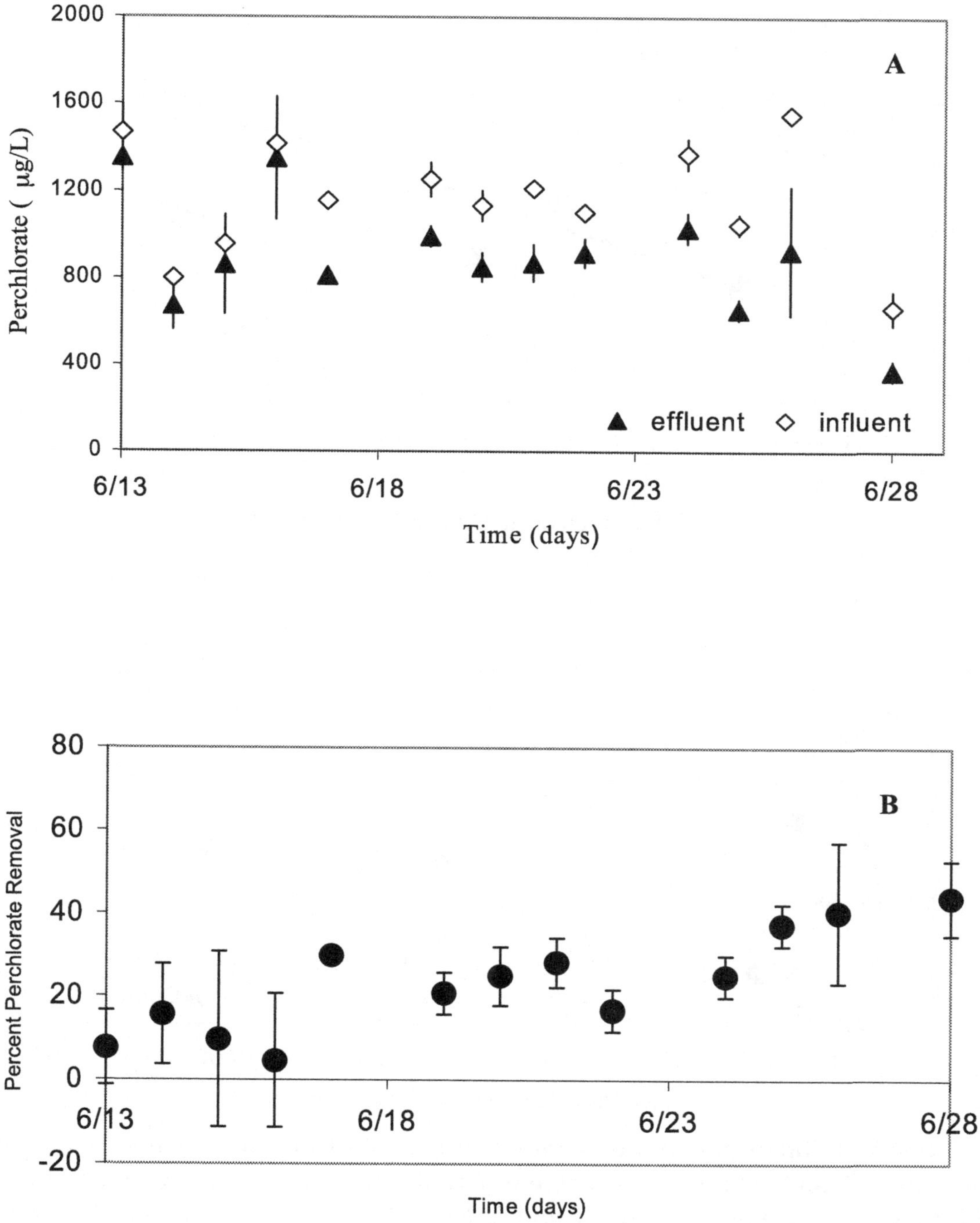


**Figure 5.14 (A) The influent and effluent perchlorate concentrations of column 2 during column startup and (B) The percent change in perchlorate at an average influent perchlorate concentration of 1200 µg/L. Errors bars are ± one standard deviation based on triplicate samples.**

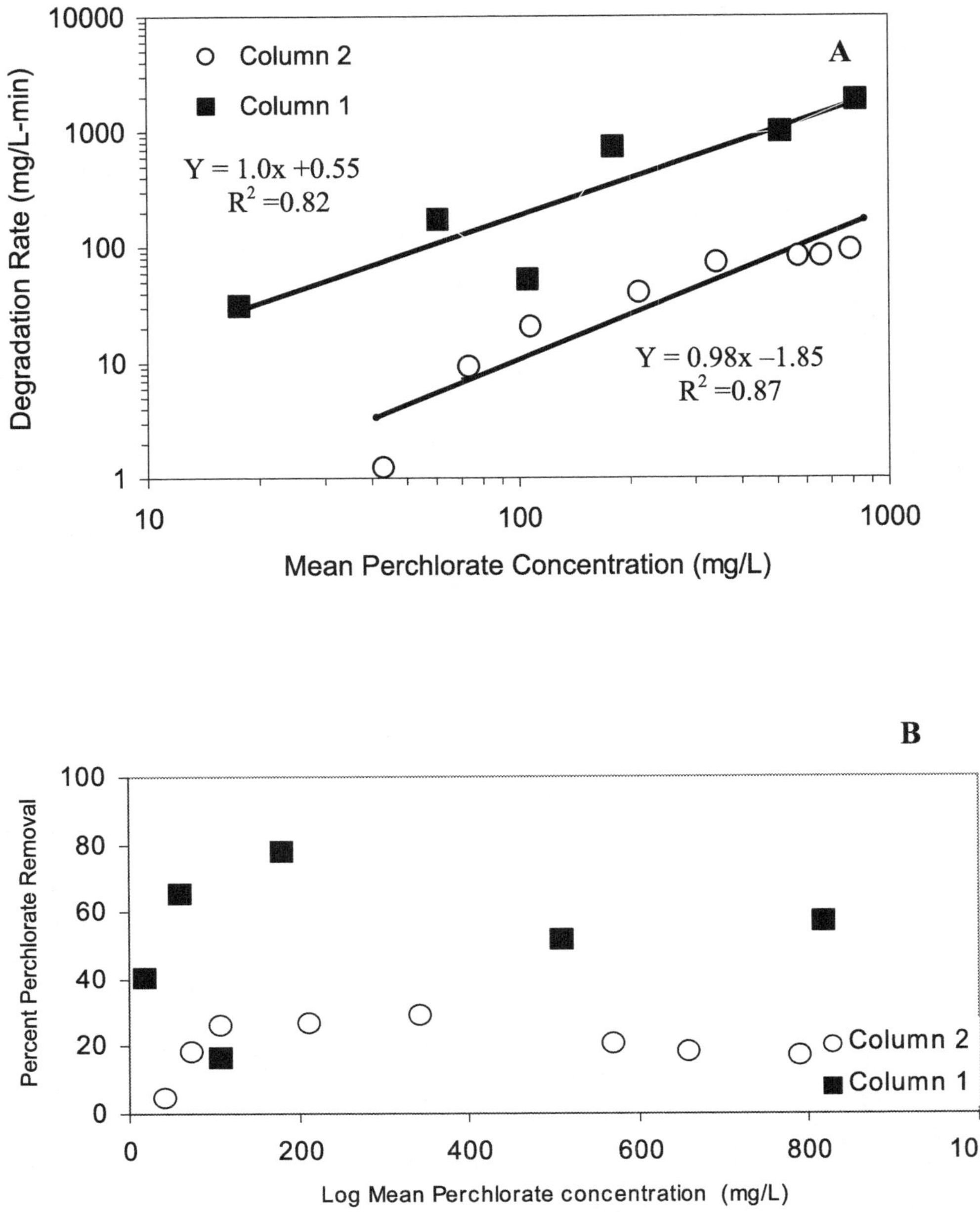


**Figure 5.15 (A) Influent perchlorate concentrations effect on the perchlorate degradation rate for column 1 and column 2. (B) Percent perchlorate removal as a function of log mean perchlorate concentration in column 1 and 2. Error bars are ± one standard deviation for the average of all of the degradation rates at each log mean perchlorate concentration.**

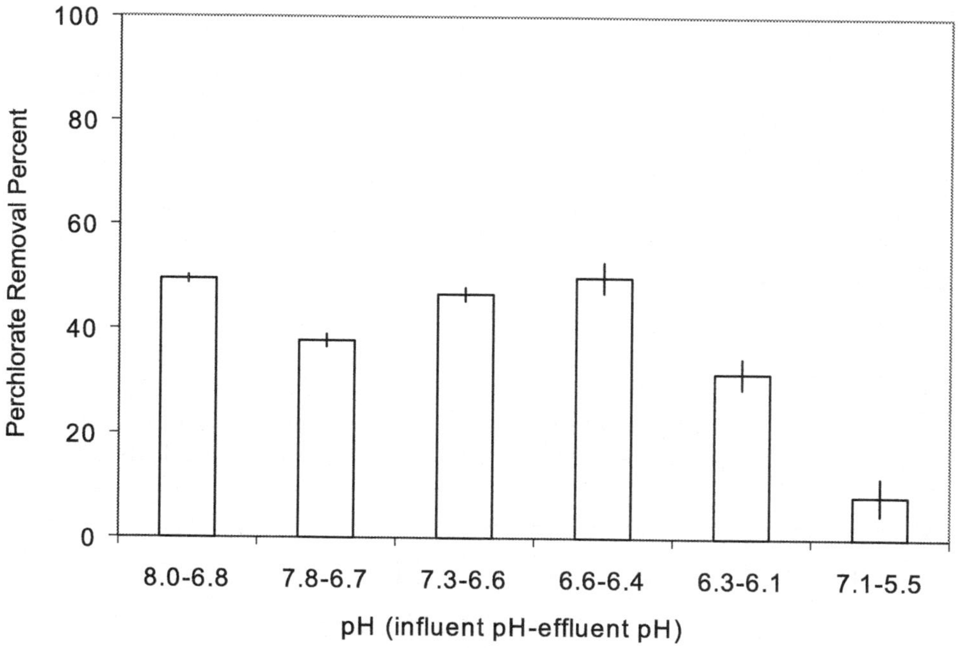


**Figure 5.16 The percent perchlorate removal in column 2 as a function of pH. The error bars represent ± one standard deviation of the mean of triplicate samples.**

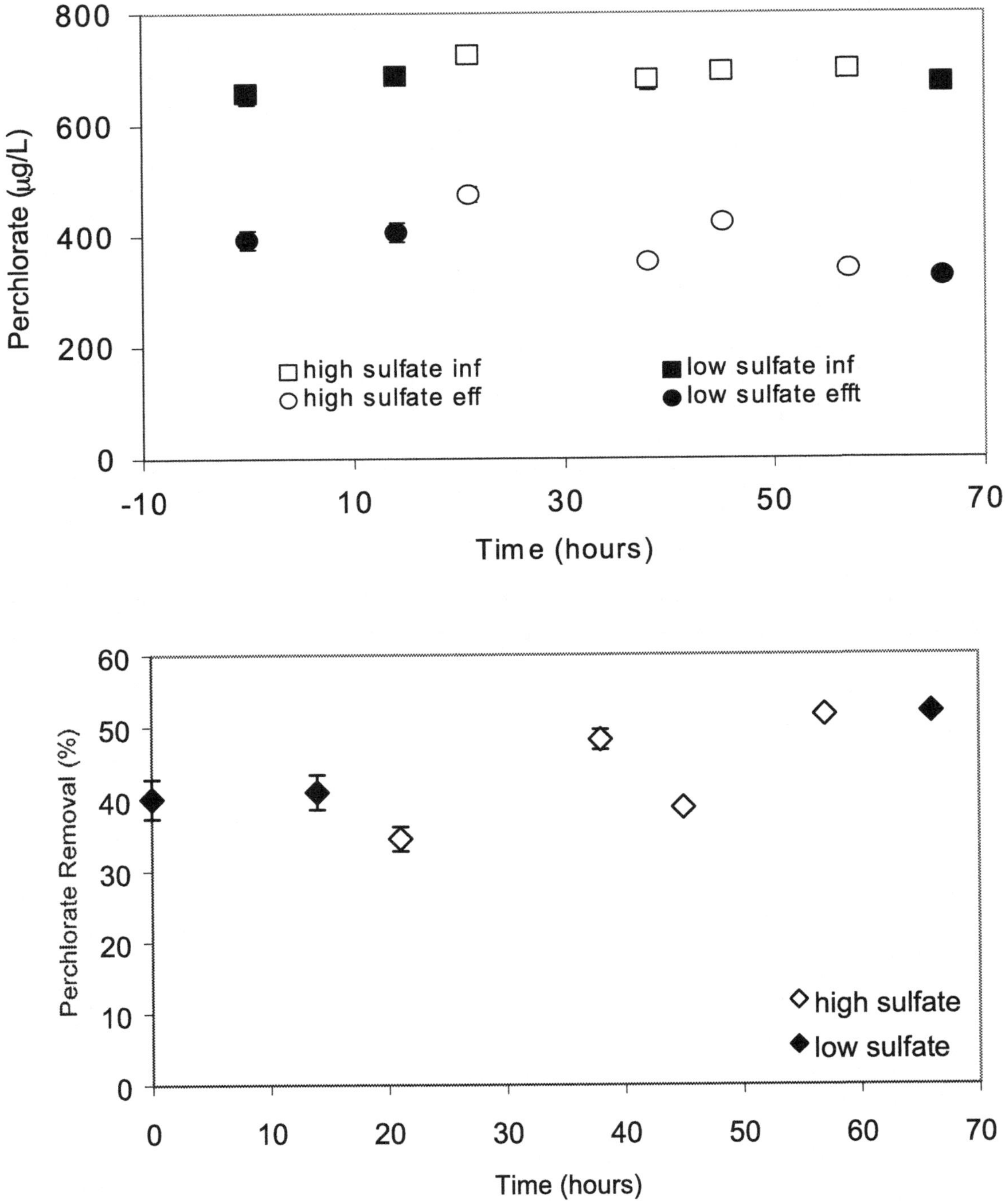


**Figure 5.17 (A) Perchlorate decrease from influent to effluent and (B) Percent perchlorate removal in the 64 mg/L sulfate (open symbols) and in the presence of 32.5 mg/L sulfate (closed symbols). Error bars represent ± one standard deviation of the mean for triplicate samples.**

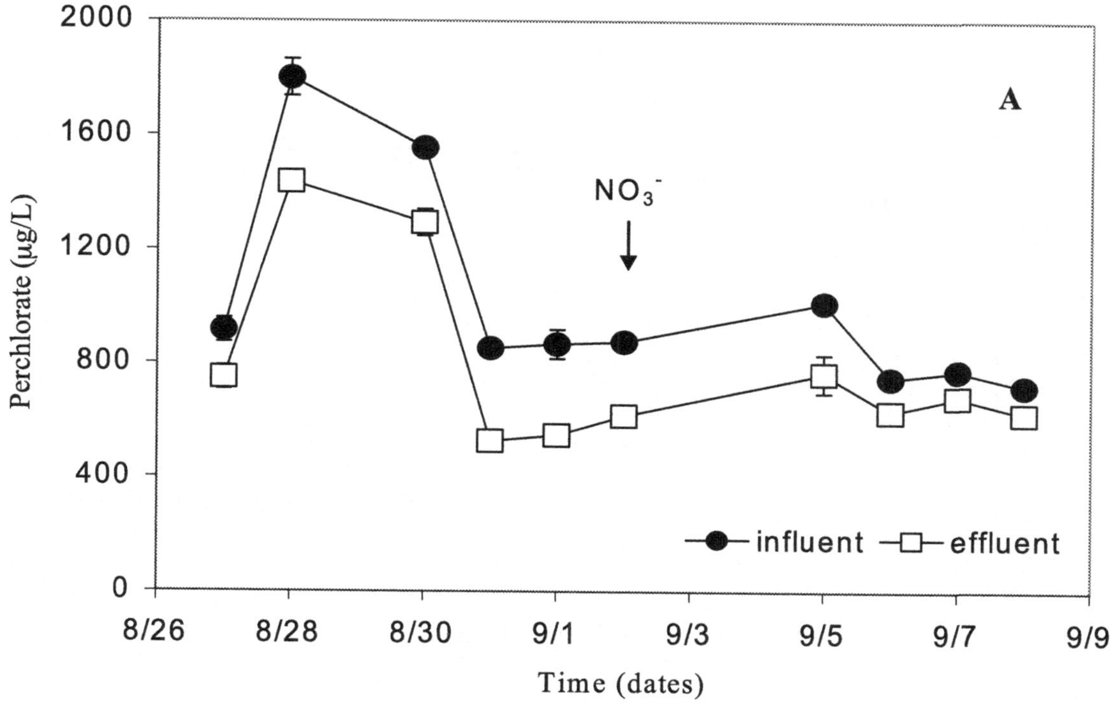


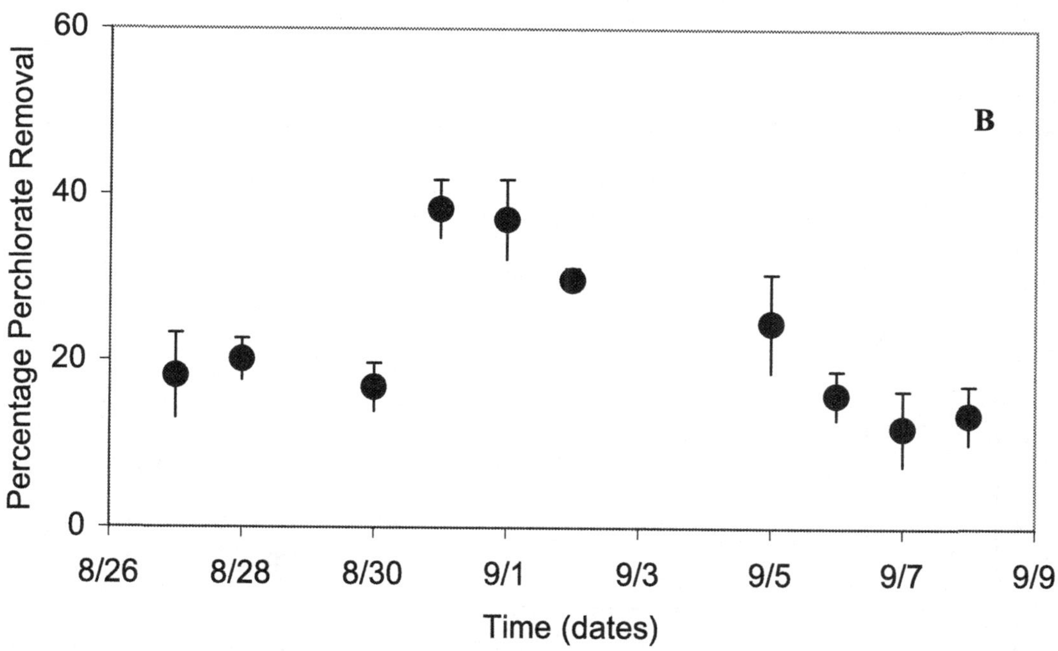


**Figure 5.18 (A) Perchlorate decrease from influent to effluent in the absence and presence (after 9/1) of nitrate for the high perchlorate experiment. (B) Percent perchlorate removal in the absence and presence (after 9/1) of nitrate. The error bars represent ± one standard deviation of triplicate samples.**

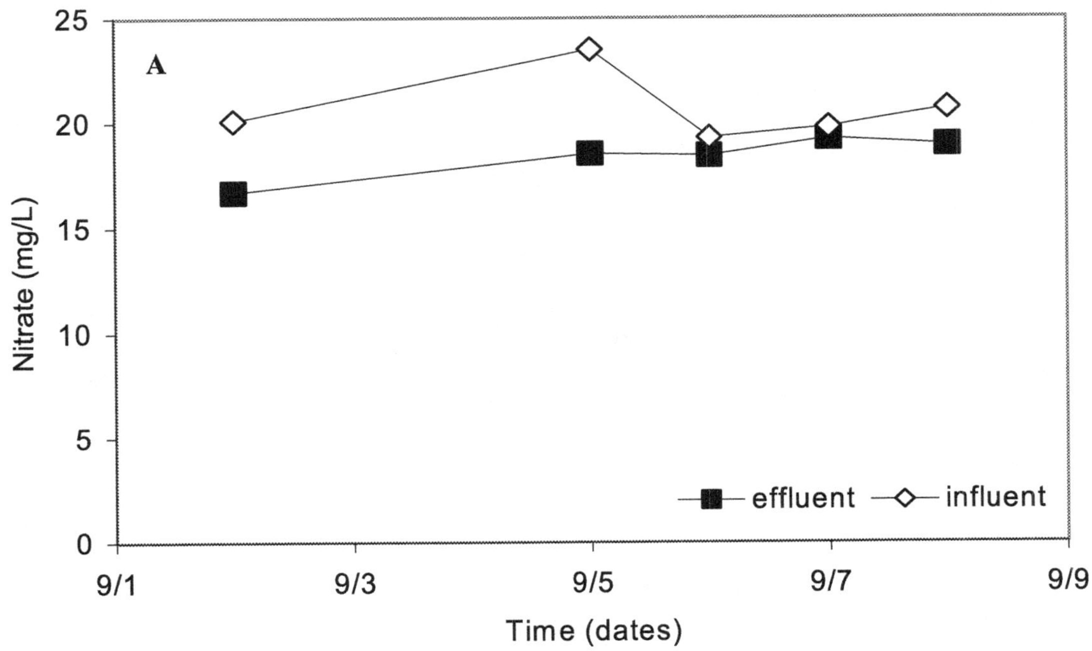


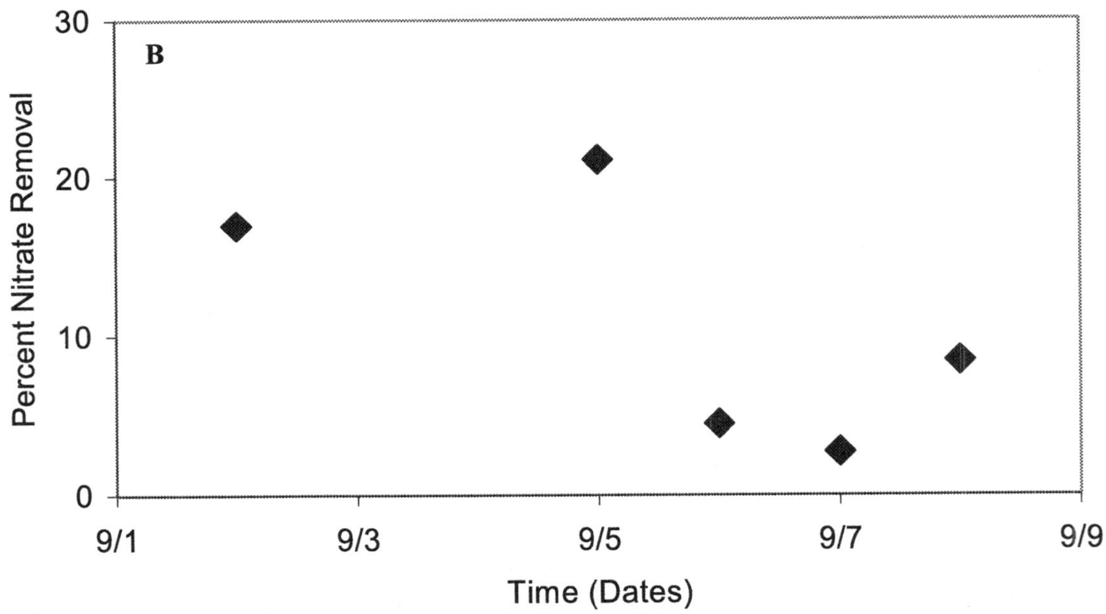


**Figure 5.19 (A) Nitrate degradation in the bioreactor at an average influent perchlorate concentration of 826 µg/L. (B) Percent nitrate removal in the bioreactor at an average influent perchlorate concentration of 826 µg/L.**

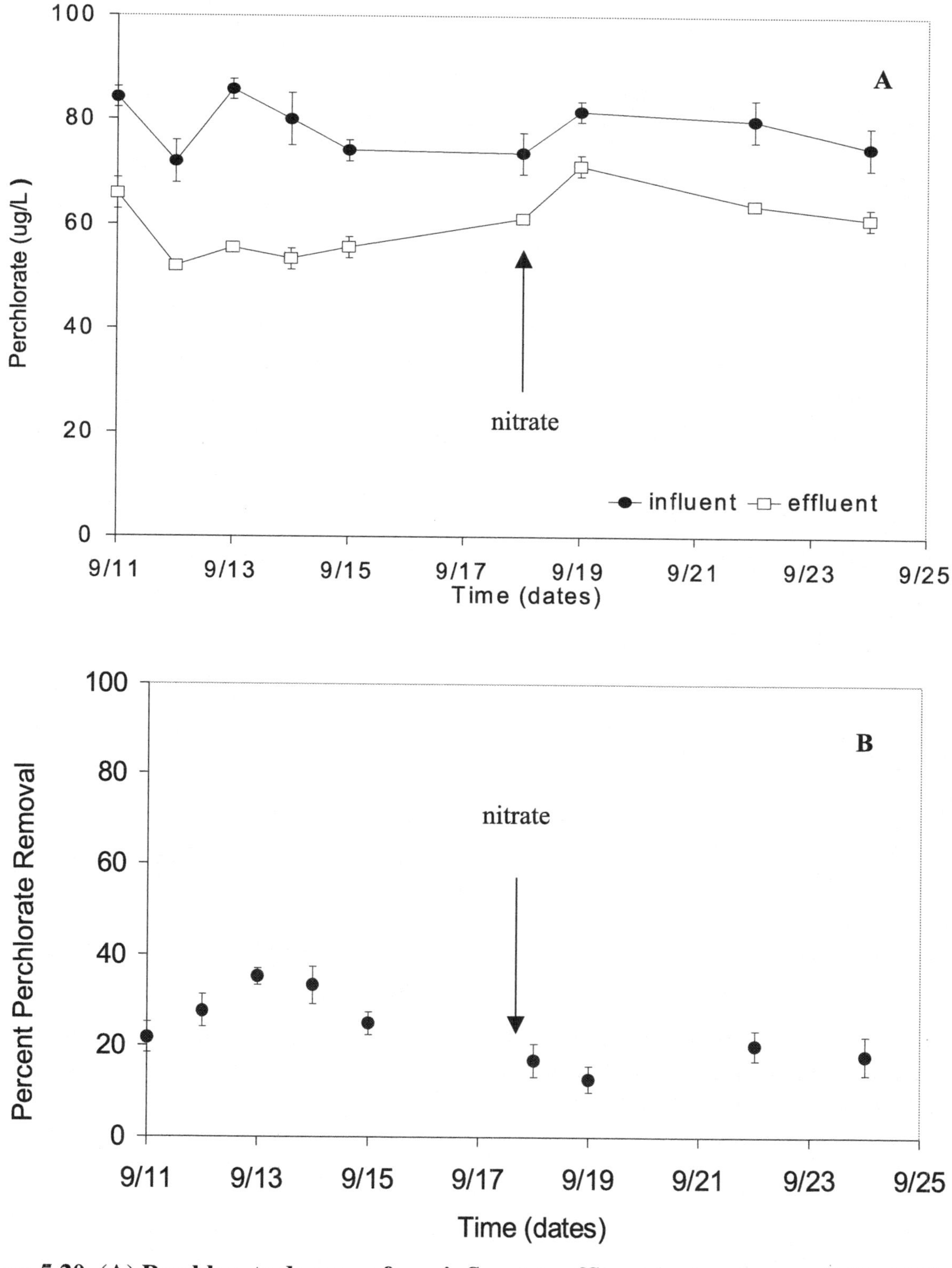


**Figure 5.20 (A) Perchlorate decrease from influent to effluent in the absence (9/11-9/17) and presence (after 9/18) of nitrate for the low perchlorate experiment. (B) Percent perchlorate removal absence and presence (after 9/18) of nitrate. Error bars represent ± one standard deviation of triplicate samples.**

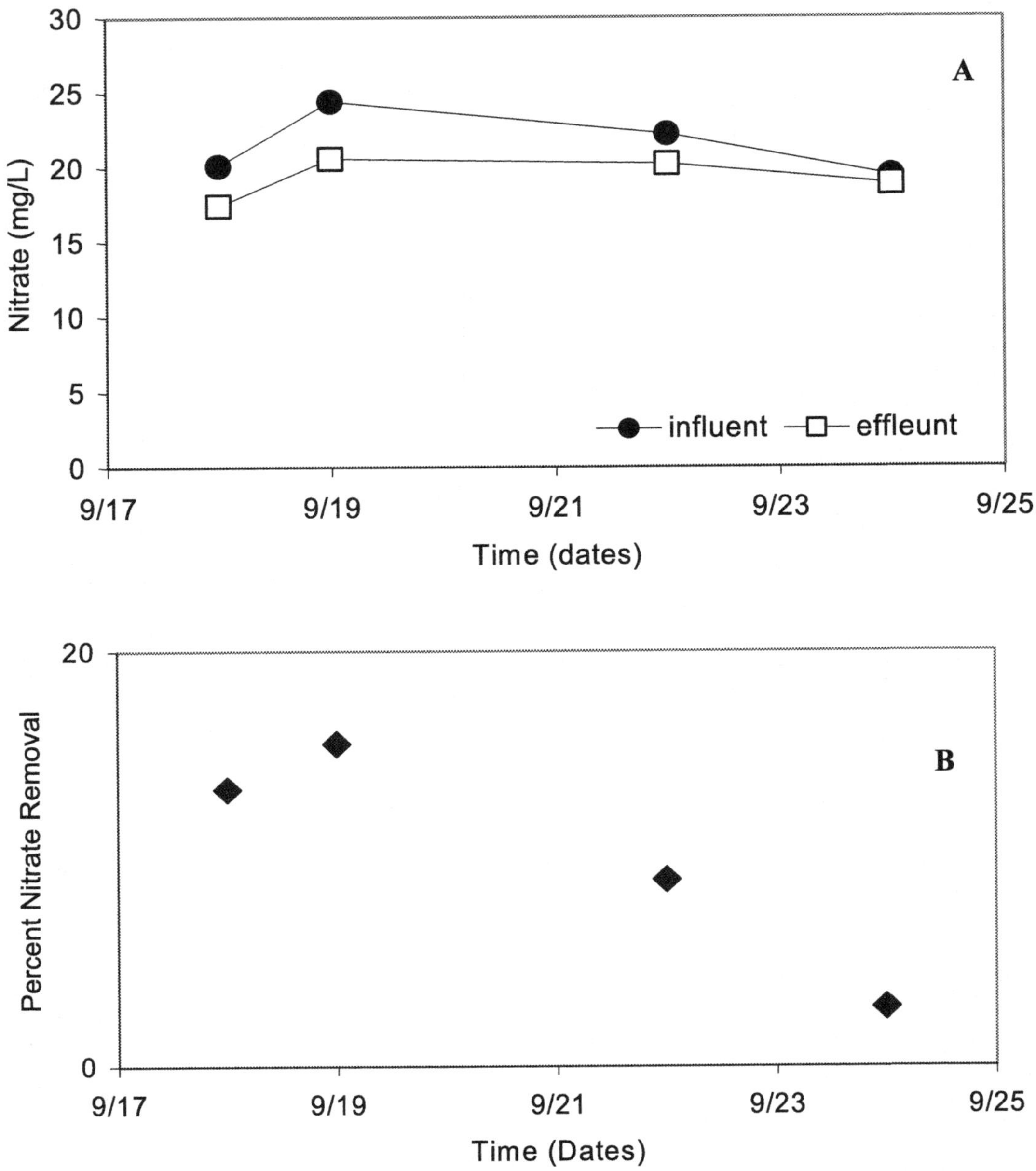


**Figure 5.21 (A) Nitrate degradation in the bioreactor at an average influent perchlorate concentration of 79 µg/L. (B) Percent nitrate removal in the bioreactor at an average influent perchlorate concentration of 79 µg/L.**

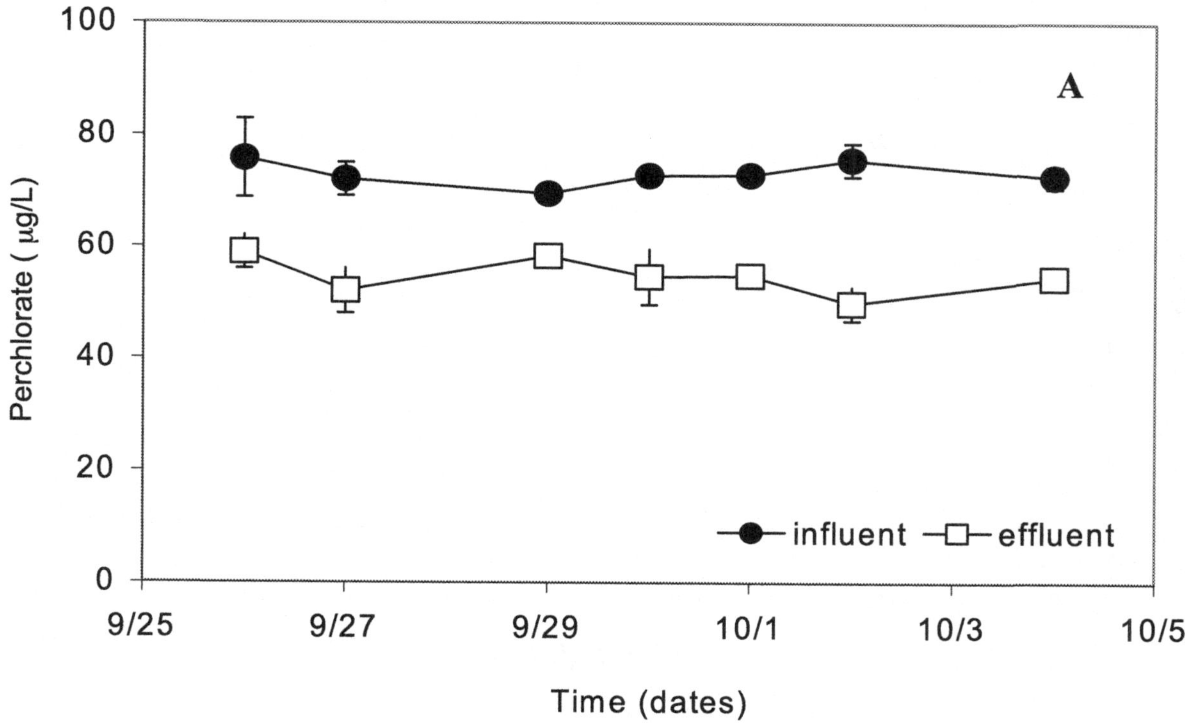


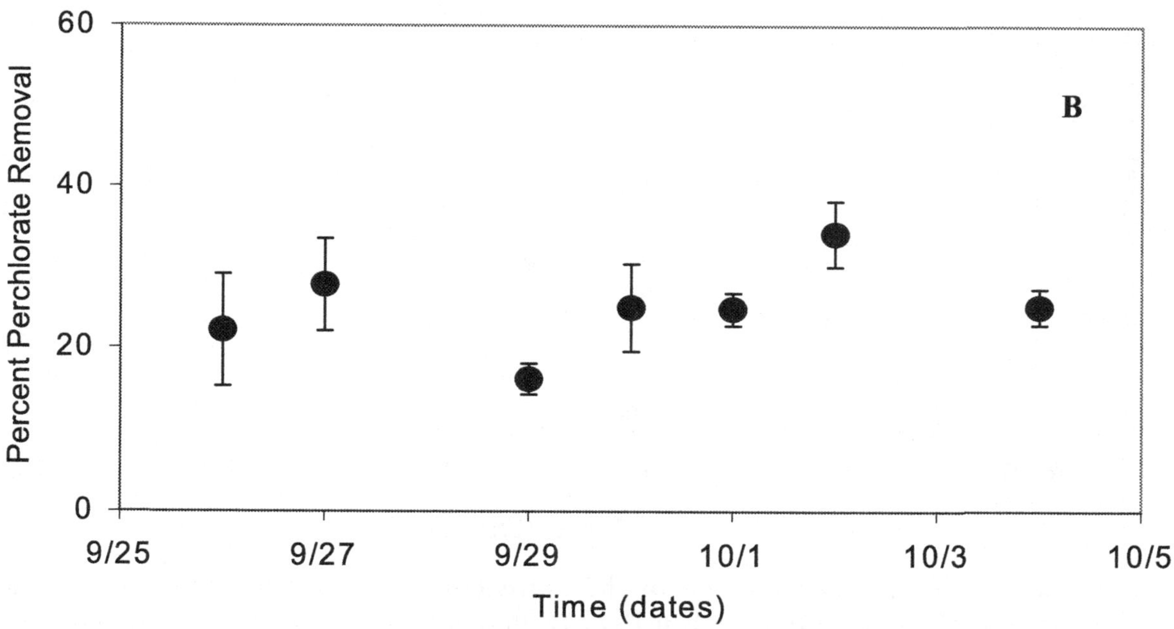


**Figure 5.22 (A) Perchlorate decrease from influent to effluent for the Redlands water experiment. (B) Percent perchlorate removal for Redlands water. Error bars represent ± one standard deviation of triplicate samples.**

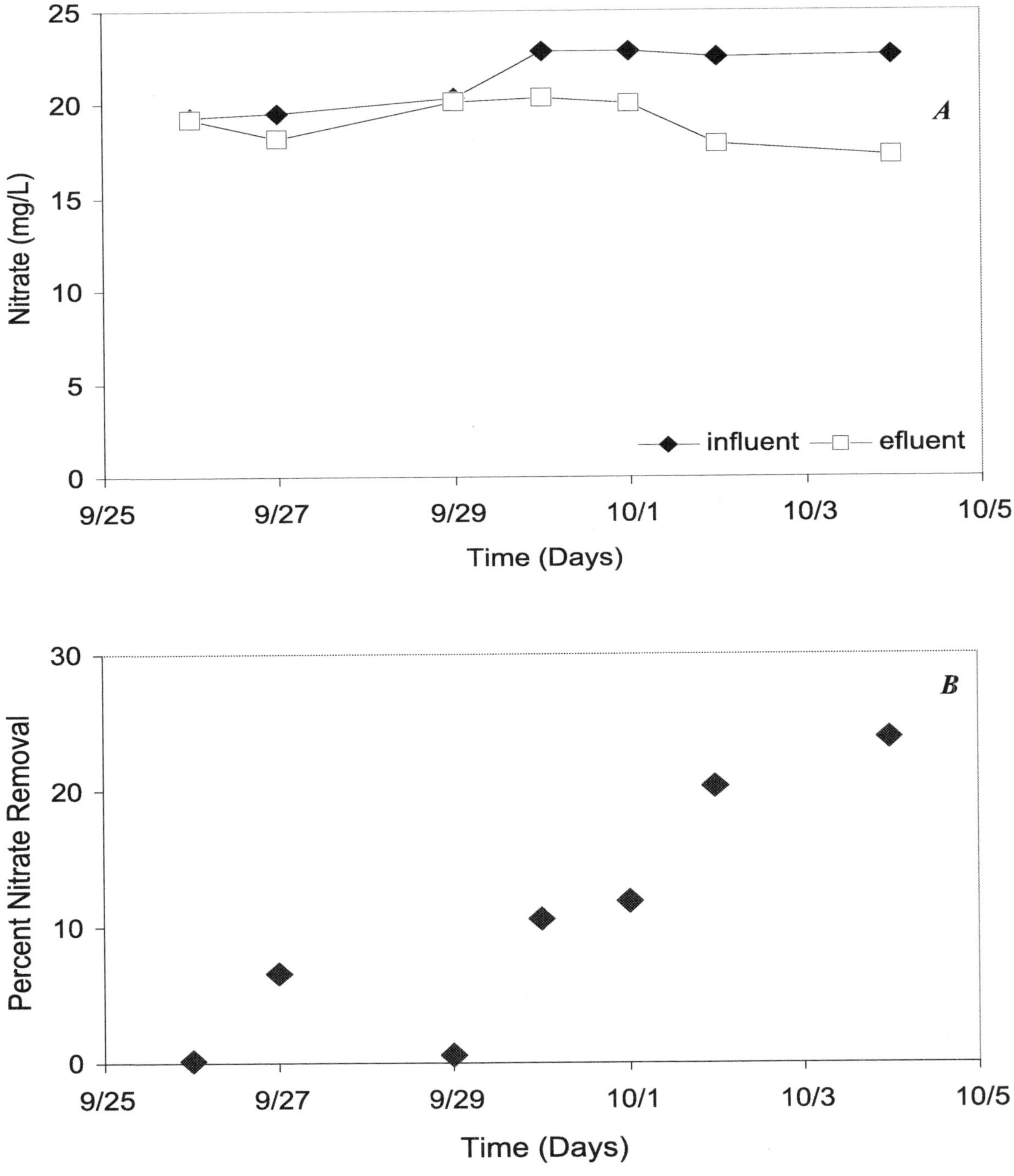


**Figure 5.23 (A) Nitrate degradation in the bioreactor at an average influent perchlorate concentration of 73 µg/L in Redlands water. (B) Percent nitrate removal in the bioreactor at an average influent perchlorate concentration of 73 µg/L in Redlands water.**

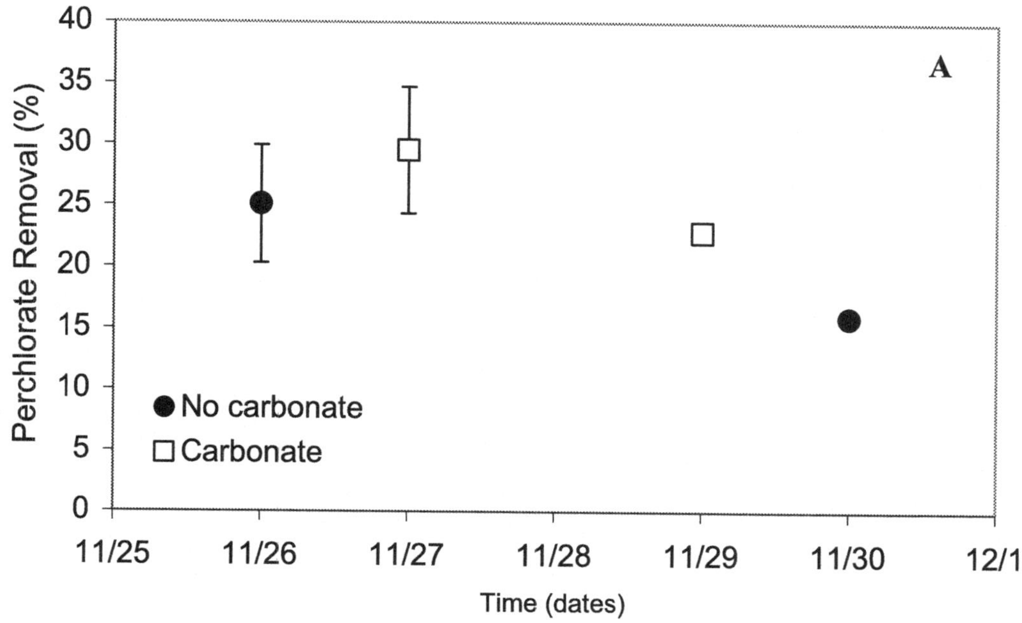


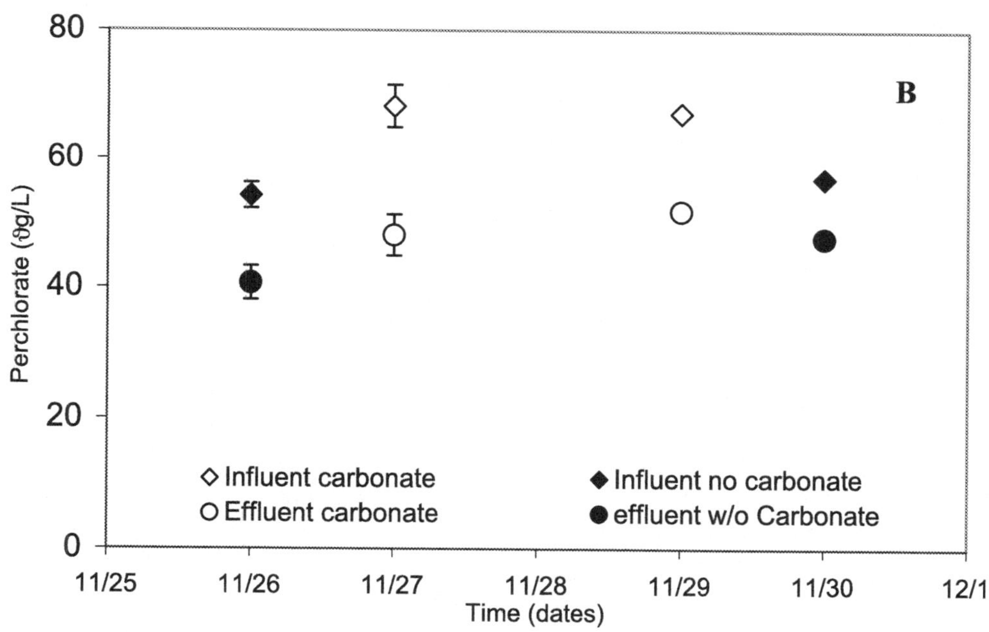


**Figure 5.24 (A) Perchlorate degradation in the column 2 for sulfate and no sulfate situations. (B) Percent perchlorate removal in the column 2 at an average influent perchlorate concentration of 73 µg/L in Redlands water. Error bars are ± one standard deviation for triplicate samples.**

# CHAPTER 6
# MEMBRANE-IMMOBILIZED BIOFILM

## EXPERIMENTAL METHODS AND MATERIALS

### Set-Up of the Membrane-Immobilized Biofilm Reactor

Figure 6.1 shows a schematic representation of the set-up of membrane-immobilized biofilm bioreactor used in all perchlorate removal experiments described in this work.

The reactor consists of two five-liter plexiglass reactors, the diffusion reactor (DR) and the biological reactor (BR), separated by a plexiglass plate holding a 10cm-diameter circular microporous membrane, to which a biofilm is attached. The two reactors and the plate are held tightly together either by four stainless steel screws or by clamps and are sealed with o-rings placed on grooves cut in the plexiglass plates. The reactors are covered by two plexiglass plates, which contain three ports used for sampling and monitoring. When these ports are not in use, they are covered by size 9 or 10 rubber stoppers. There is one waste port at the bottom of each reactor. This port is kept closed by a clamp and is used to empty the reactor. In this set-up, perchlorate diffuses from the DR reactor to the BR reactor, and it is biodegraded by the biofilm attached to the microporous membrane in the BR reactor. The final product of perchlorate biodegradation, chloride, can diffuse back into the DR reactor. There is no need to maintain a pressure difference across the membrane because the transport of ions is solely based on diffusion, and therefore the resulting process is not energy intensive.

Anaerobic conditions are maintained by covering the reactors with plexiglass covers and by purging the reactor with the compressed nitrogen gas using fine-bubble ceramic diffusing stones. Adhesive weather-strip foam is placed around the edges of the covers to completely seal the reactor. A YSI 54A dissolved oxygen meter is used to monitor the anaerobic condition in the reactor. The pH in the BR reactor is kept around neutral by phosphate buffer addition and is monitored by a Corning 450 pH/Ion meter. Complete mixing in both reactors is achieved by magnetic stirring.

Figure 6.2 shows a side view of the microporous membrane connecting the DR and the BR reactors, detailing the attachment of the membrane to a plexiglass plate which has a 10cm-diameter opening for membrane.

### Microporous Membranes Selection

Several membrane samples were obtained from Ionics, Inc., Hydranautics, Koch Membranes, Osmonics, Millipore, and US Filter/ Memcor. Some of these membranes were deemed unsuitable for the experiments due to their pore size or sensitivity to perchlorate. Therefore, it was decided to search for membranes that would be perchlorate resistant. The pore size of the microporous membrane has to be selected so that perchlorate can diffuse into the BR chamber while the back transport of microbial cells, carbon source, and macromolecules into the DR chamber is minimized.

Three membranes were chosen as potential candidates for the immobilization of a biofilm for perchlorate biodegradation. The selected membranes were Memcor BTS-55, Memcor PVDF, and Millipore FGLP membrane. All these three types of membranes were employed in the biodegradation and diffusion tests of the membrane-immobilized biofilm reactor. The characteristics of the selected membranes are shown in Table 6.1.

**Determination of Diffusion Coefficient for Perchlorate, Nitrate, and Sulfate**

The diffusion coefficient of perchlorate through these three membranes, prior to biofilm growth, was determined experimentally to evaluate the migration of perchlorate from the DR reactor (diffusion reactor) to the BR reactor (biological reactor). The testing was performed by attaching, with water-proof aluminum plumbing tape, a membrane (10 cm diameter) to a square plexiglass plate containing a 10 cm-diameter opening. The plate to which the membrane was attached was placed between the DR and BR reactors. The DR reactor was then filled with 5 liters of DI water containing 1000 mg/L perchlorate, while the BR reactor was filled with 5 liters of perchlorate-free DI water. Both reactors were purged with nitrogen gas to remove dissolved oxygen to undetectable levels before adding perchlorate to DR reactor. The dissolved oxygen (DO) level in the reactor was monitored by using a YSI 54A DO meter. Fifteen-milliliter samples were taken, at the same time, from both reactors every 15 minutes for a 4-hour testing period. Perchlorate concentration in the BR reactor was measured by a Dionex 120 ion chromatograph. One of the selected membranes, Millipore FGLP membrane, which is hydrophobic, had to be pre-treated before the testing by completely soaking in 1-propanol for 15 minutes and thoroughly rinsing with DI water.

The diffusion coefficient was calculated using Fick's law as represented in Equation 6.1 (McCleaf and Schroeder 1995). The theoretical diffusion coefficient of perchlorate in water was also calculated using the Wilke-Chang Method (LaGrega, Buckinham, and Evans 1994).

$$V_{BR}\,(\partial C_{BR}/\partial t) = [-\,(D_M A_M)/(\Delta L_M)](C_{DR} - C_{BR}) \qquad (6.1)$$

where

$V_{BR}$ = volume of BR reactor ($cm^3$)
$A_M = \pi r^2 \varepsilon$ = membrane pore area ($cm^2$)
$r$ = membrane radius (cm)
$\varepsilon$ = mean pore fraction (0.7 for all membrane tested)
$\Delta L_M$ = membrane thickness (cm)
$D_M$ = perchlorate diffusion coefficient through the membrane ($cm^2/s$)
$C_{DR0}$ = initial concentration of perchlorate in diffusion reactor (mg/L)
$C_{DR}$ = concentration of perchlorate in the diffusion chamber at time t (mg/L)
$C_{BR}$ = concentration of perchlorate in the biological reactor (mg/L)
$t$ = elapsed time (s)

Since the two reactors contained the same volume of water, one can write $C_{DR} = C_{DR0} - C_{BR}$. Integration of the above differential equation yields:

$$\ln \frac{C_{DR0} - 2C_{BR}}{C_{DR0} - 2C_{BR0}} = -\frac{2A_M D_M}{V_{BR}\Delta L_M} t \qquad (6.2)$$

Equation 6.2 is the equation of a straight line and plotting $\ln[(C_{DR0} - 2C_{BR})/(C_{DR0} - 2C_{BR0})]$ versus time (t) yields a slope of k equals $(2A_M D_M)/(V_{BR}\Delta L_M)$. Therefore, the diffusion coefficent $D_M$ can be obtained because $A_M$, $V_{BR}$, and $\Delta L_M$ are known.

The test of each membrane was performed in duplicate, and diffusion coefficients were determined by least-square regression ($R^2$) of the transformed experimental data.

In order to test the interference of nitrate and sulfate on perchlorate biodegradation, the diffusion coefficients of nitrate and sulfate through the selected membrane, in the absence of the biofilm, were also determined. It is important to perform membrane diffusivity tests for nitrate and sulfate, in the absence of a biofilm, to eliminate the possibility of diffusion controlled biodegradation. In other words, if the diffusion of sulfate and nitrate through the membrane are found to be of the same order of magnitude as that of perchlorate, all the anions will be available to the biofilm at approximately the same time.

The diffusion coefficient tests for nitrate and sulfate were performed in the same manner as that for perchlorate. The diffusion tests were performed after perchlorate biodegradation testing with membrane-immobilized biofilms deemed two of the pre-selected membranes (PVDF and FGLP) unsuitable for biofilm immobilization. Thus, only the BTS-55 membrane was used in the tests. The concentrations of nitrate and sulfate used in the diffusion tests were both 1000mg/L. Nitrate and sulfate concentrations were determined by a Dionex 120 ion chromatograph. Similar to perchlorate, the diffusion coefficients of nitrate and sulfate were calculated by Fick's law.

For comparison purposes, the diffusion coefficients of perchlorate, nitrate, and sulfate in water, in the absence of a membrane, were estimated using the Wilke-Chang method (Eq. 6.3) (Reid, Pravsnitz, and Poling 1987).

$$D = \frac{5.06 \times 10^{-7} \times T}{\mu V^{0.6}} \qquad (6.3)$$

where

T = temperature (°K)
$\mu$ = viscosity of water (centipoises, cP)
V = molar volume of anions ($cm^3$/mol)
D = diffusion coefficient ($cm^2$/sec)

By using the method of LeBas (LaGrega, Buckinham, and Evans 1994), the molar volume of perchlorate, nitrate and sulfate were calculated as 54.2 $cm^3$/mol, 40.5 $cm^3$/mol, and 58.8 $cm^3$/mol, respectively.

## Development of Perchlorate Degrading Enrichment Cultures

### *Master Culture Reactor (MCR)*

A pure perchlorate-reducing culture isolated from the Penn State wastewater treatment plant was shipped to the UNLV laboratory. In the UNLV laboratory this culture demonstrated slow kinetics and was therefore not chosen for the development of membrane-immobilized biofilms. A perchlorate-degrading mixed enrichment culture was developed from the returned activated sludge (RAS) taken from the Clark County Sanitation District (CCSD) wastewater treatment plant in Las Vegas. The CCSD treatment train contains an activated sludge system preceded by anoxic zones for biological phosphorous removal. The RAS for the enrichment culture was collected in the first anoxic zone where the RAS is mixed with the influent wastewater. The rationale was that perchlorate-reducing microorganisms would be more likely to be found in anoxic environments.

The RAS sample was used to inoculate a one-liter bottle containing buffer, nutrient/minerals, perchlorate, and a carbon source (lactate). The composition and concentration of the buffer and nutrient/minerals media used, as modified from van Ginkel, Plugge, and Stroo (1995), are shown in Table 6.2. The bottle was completely sealed with a rubber stopper and kept mixed by a magnetic stirrer at 23±2°C. The ratio of lactate to perchlorate added to the bottle was 5:1 (600 mg/L lactate and 120 mg/L perchlorate). The initial concentration of microbes in the bottle was 295 mg/L. Samples were withdrawn daily through a small opening on the rubber stopper using a 4-inch needle attached to a syringe. The opening on the rubber stopper was sealed with heavy-duty duct tape to maintain anaerobic conditions. The concentrations of perchlorate, chloride, and lactate of the sample were determined by ion chromatography. The background concentration of chloride in the bottle was about 51 mg/L. Background chloride concentrations are the result of the addition of the RAS seed which has a very high concentration of total dissolved solids (TDS). The background concentration of chloride in the mineral/nutrient solution was approximately 1 mg/L.

A second MCR was developed, with a four-liter capacity, using seed microbes from a biofilm growth reactor that was seeded with the first MCR reactor. In this reactor, the culture developed a reddish coloration with time and was named "BALI", for Batista and Liu. Initially, perchlorate (200 mg/L) and lactate (1,000 mg/L) were added to the bottle daily without wasting. Later, a portion of culture was periodically wasted to prevent accumulation of chloride in the bottle and to decrease the suspended solids concentration. A high suspended solids concentration implies the addition of large amounts of perchlorate and lactate to keep a microbial culture alive.

### SRT Studies for the Master Culture Reactor (MCR)

The objective of these studies was to determine an appropriate feed rate for the master culture, so that a desired biomass concentration could be kept. This also allowed for developing an enriched mixed culture that was more homogeneous and provided some initial data on the kinetics of perchlorate biodegradation by the enriched mixed culture. This decision was made because it was noticed that microbial composition of the master culture, when observed under the microscope, was changing with time. To keep such culture it was necessary to waste biomass, so that the cost of chemicals to maintain the reactors is minimized. An SRT of 5 days was chosen as the target SRT to be maintained in the master reactor. The reactor was operated on a semi-continuous mode. Daily, 200 ml of the mixed culture were withdrawn from the one-liter master bottle and 200 ml of solution containing an equivalent amount of perchlorate, lactate, nutrient and buffer was added to the bottle to complete the 1L volume. After the new solution was added, a 20-ml sample was taken for analysis. A portion of the solution taken out of the reactor was filtered through a 1.2 μm glass fiber filter (Whatman, GF/C) for suspended solids (SS) analysis following Standard Methods (Method #2054; APHA, AWWA, and WEF 1995). A portion of resulting filtrate was then filtered though a 0.45 μm membrane filter prior to analyzing for perchlorate, lactate, and chloride. After one SRT cycle was completed, the initial suspended solids concentration and the lactate to perchlorate ratios were varied.

## Development of Membrane Immobilized Biofilms

To investigate the biodegradation of perchlorate by an immobilized biofilm, it is necessary to first establish a biofilm on the surface of the membrane. As the research progressed, the way the biofilm was developed on the membrane was changed to obtain reliable results and for ease of operation. The purposes of the modifications were (a) to decrease the disturbance of the biofilm resulting from emptying and filling up the BR reactor prior to initiating biodegradation studies, (b) to attempt to develop a biofilm of same thickness for the different tests, (c) to minimize the amount of time needed to switch from the biofilm growth mode to the biodegradation experiments, (d) to maintain relatively stable microbial composition of the biofilm for the different tests.

In the beginning, a membrane was placed between the two reactors. Two sets of reactors (4 tanks) were available to facilitate duplication and concomitant biofilm growth and biodegradation studies. Five liters of DI water containing 1000 mg/L lactate, 200 mg/l perchlorate (lactate/perchlorate ratio of 5:1), nutrients/minerals and buffer solutions (Table 6.2) and seed microbes (0.1 g/L) from the MCR were added to the BR reactor. Five liters of DI water were also added to the DR reactor to keep the hydraulic pressure on both sides of the membrane the same. A YSI 54A oxygen meter and a Corning 450 pH/Ion meter were placed in the BR to continuously monitor the oxygen concentration and the pH. Deoxygenation in the BR reactor was obtained and kept by purging with nitrogen gas. Oxygen levels were kept undetectable at all times. The biofilm was established on the surface of a membrane inside of the BR reactor over a one week period. Biofilm growth was observed not only in the membrane, but also in the walls of the tank. However, given the hydraulics of the reactor, a thicker biofilm developed in the membrane than in the walls of the tank (Figure 6.3). After the biofilm was grown on the surface of the membrane, these reactors were cleaned and then switched to the perchlorate biodegradation testing.

After the biofilm was grown on the membrane, two plexiglass plates were placed in each reactor against the membrane to protect the biofilm. Next, the contents of the reactors were withdrawn through the waste port, the reactor walls were scrubbed with a plastic brush and thoroughly rinsed with DI water. The reactors were then dried with paper towels and used in the perchlorate biodegradation testing. Washing the reactor by the method described above was very cumbersome. Thus, a modification was made to improve the operation. The modification consisted of growing the biofilm in a separate set of reactors from that used for the biodegradation testing. The growth reactor set was used exclusively to grow the biofilm on the membrane. A clean membrane was attached to the plexiglass plate and the plate was held between the two 5-liter reactors by seven stainless steel clamps. Next, perchlorate-reducing culture, nutrients/minerals, buffer, perchlorate, and lactate were added to one side of the reactor and DI water was added to the other side. After the biofilm was developed on the membrane surface, the membrane containing the biofilm was then removed from the growth reactor and transferred to a clean biodegradation reactor set. However, sometimes, the membrane-containing biofilm was tore up when the clamps were removed to transfer the plate containing the membrane to the biodegradation testing. To overcome this problem, it was decided to grow the biofilms on a manner that would not require clamping of the membrane.

In this modification, a large, rectangular glass container fitted with a plexiglass grid to support the membrane attached to plates, was used to develop the biofilm on the surface of the membrane (Figure 6.4). The glass container was filled with 16 liters of DI water containing buffer, nutrient/minerals, perchlorate, lactate, and perchlorate-reducing microorganisms. The plates containing the membrane were placed membrane-side down on the plexiglass support so that the media containing microbes contacted the surface of the membrane (Figure 6-5). At the same time, three biofilms could be grown in this reactor assuring homogenous thickness. The reactor was purged with nitrogen gas to ensure the anaerobic condition. The DO level was monitored by a DO meter (Figure 6-6).

### Perchlorate Biodegradation by Immobilized Biofilms

The testing of perchlorate biodegradation by the membrane-immobilized biofilm was performed after the biofilm was formed on the surface of a membrane. Since modifications were made on the way the biofilm was grown on the membrane, the procedure of perchlorate biodegradation testing was also changed slightly. For the experiments in which the biofilm growth and the biodegradation testing took place in the same reactor set, the following procedure was used: after establishment of the biofilm in the BR reactor side of the membrane, a Plexiglass® plate was held against the internal surface of the reactor to minimize the contact of the biofilm with oxygen. Next, the contents of the BR and the DR reactors were discarded through the waste port and the reactors were scrubbed and rinsed well with DI water. Five liters of solution containing lactate, nutrients/minerals and buffer were then added to the BR reactor. Five liters of solution containing perchlorate were added to the DR reactor on the other side of the biofilm. Last, the Plexiglass® plate was removed and both reactors were immediately closed and kept oxygen-free with nitrogen gas. The concentrations of perchlorate, lactate, chloride, and suspended solids were monitored in both the BR and the DR reactors. The disadvantage of this procedure is that the reactors could not be thoroughly cleaned when switching from the growth to the perchlorate biodegradation mode.

For the case where the biofilm was grown on a separate reactor set from that in which the biodegradation testing took place, the following procedure was used: after a biofilm was developed on the membrane, the clamps were removed and the plate containing the membrane-immobilized biofilm was quickly transferred to the biodegradation reactors. Perchlorate was added to the diffusion reactor (DR). Lactate, nutrient and buffer were added to the BR reactor which contained the biofilm side of the membrane. As perchlorate diffused from the DR to the BR reactor, it was biodegraded by the biofilm that used lactate as a carbon source (electron donor) and perchlorate as an electron acceptor. The concentrations of perchlorate, lactate, and chloride were monitored daily. Sometimes, the membrane was damaged by the clamps when the membrane was switched from the biofilm development reactor set to the biodegradation reactor,

For the case in which the biofilm was grown on the suspended grid, the plate containing the membrane-immobilized biofilm was carefully removed from the suspended growth grid reactor and clamped to the reactors for biodegradation testing. The other steps of the biodegradation test were the same as those described above.

### Preliminary Kinetics Study of the "BALI" Culture

Preliminary kinetics studies were performed in the enrichment "BALI" culture. The objectives of these experiments were: (a) to determine the limiting carbon to perchlorate ratios needed for perchlorate biodegradation by this culture; (b) to determine the degradation rate of perchlorate by this culture. Carbon ratios are necessary for determining the amount of carbon to be added to the 5-L membrane-immobilized reactor in order to minimize excess carbon in treated water. One set of degradation tests was performed under a perchlorate-limited condition while the other set was performed under a carbon (lactate) limited condition. In the perchlorate-limited tests, the lactate concentration was 1,000 mg/L and perchlorate concentrations varied from 10 to 200 mg/L. For the lactate limited tests, the perchlorate concentration was constant and equal to 100 mg/L; lactate concentrations varied from 20 to 300 mg/L.

The tests were performed in 125-ml serum bottles in duplicate. The perchlorate limited bottles and their duplicates were labeled KC1, KC2, KC3, KC4, KC5, KC1d, KC2d, KC3d, KC4d, and KC5d, respectively. The lactate limited bottles and their duplicates were labeled KL1, KL2, KL3, KL4, KL5, KL1d, KL2d, KL3d, KL4d, and KL5d, respectively. The desired amounts of perchlorate, lactate, buffer and nutrient/minerals were added to the individual bottles. The same amount of microbes ("BALI") from the master reactor, for which the total suspended solids concentration was known, was added to each bottle so to obtain a suspended solids (SS) concentration of approximately 3.5 mg/L. Next, the bottles were sealed with butyl rubber caps and crimped with aluminum rings to assure anaerobic conditions. The bottles were then placed on a rotary shaker (30 rpm) at 20 ±2°C. Five milliliters samples were withdrawn by using a needle attached to a syringe. The samples were filtered through 0.45 μm syringe filters. The solids were determined gravimetrically after drying the filter in a drying chamber containing activated silica gel for about 15 hours. The filtrate was analyzed for perchlorate, lactate, and chloride. The concentrations of lactate and perchlorate added to each bottle are shown in Figure 6.7.

## Procedures for Characterization, Isolation and Identification of Perchlorate-Degrading "BALI" Cultures

### *Preliminary Microscopic Studies of the "BALI" Culture*

The "BALI" culture, the reddish culture enriched at the UNLV laboratory from an activated sludge wastewater treatment plant, was studied microscopically. Gram-stain and wet mount slides of the culture were prepared and observed under a Nikon Labophot-2 microscope. A Gram stain was performed by spreading a small amount of the mixed culture on a glass slide and allowing it to dry. Next, the culture was attached to the glass slide by heat fixing with a Bunsen Burner. The standard gram-staining procedure was carried out using, Crystal Violet (Primary Stain), Gram's Iodine (Mordant), 95% Ethanol (Decolorizer) and Safranin (Counterstain). Wet mounts were prepared by spreading a small amount of the culture on a glass slide and observing it with a bright field microscope.

### *Isolation and Identification of Perchlorate Degrading Bacteria From the "BALI" Mixed Culture*

Samples from the mixed enrichment culture "BALI" sealed with a butyl rubber stopper, were collected from the 1 liter master culture bottle using a sterile, 10cc syringe with a 21-gauge needle. The butyl rubber stopper was surface sterilized using 85% ethanol. The 10cc sample was transferred to a sterile, 100ml beaker. Sample (100μl) was transferred to the surface of agar plates containing perchlorate/acetate broth. The plates were inoculated in sets of five. A spread plate technique was used to distribute the sample across the agar plate surface. This involves flame sterilizing a glass rod that is bent in the shape of a hockey-stick. The flame sterilizing is accomplished using 85% ethanol and a Bunsen burner. The spread plate is then placed onto a rotating surface and the sample is spread evenly across the agar plate surface. Plates were incubated for one week at 25°C in an anaerobic environment using an anaerobe jar and BBL Gas Pak Plus system. Anaerobiosis was determined by using a methylene blue indicator. Methylene blue starts out blue and turns white when oxygen is absent.

After one week, plates were removed from the anaerobe jars and inspected for isolated bacterial colonies growing on the agar plate surface. Different colony types were transferred to fresh perchlorate/acetate agar plates using the streak plate technique in order to obtain an axenic (pure) culture of one bacterial cell type. This process was repeated three times and then phase contrast microscopic analysis (1000×) was performed on wet mounts to confirm that only one cell type was present.

### *Gram Staining Procedure*

A gram-stain was performed on each of the bacterial cultures using the following method. Cells were collected from a single isolated colony growing on the agar plate surface. The cells were suspended in a drop of sterile water that was placed on the glass slide surface. The cells were spread across the glass slide and allowed to air dry. After air drying was completed, the cells were destroyed and attached to the slide surface by gently heat fixing the bacterial smear. The standard gram-staining procedure was carried out using, crystal violet (primary stain), gram's iodine (mordant), 95% ethanol (decolorizer) and safranin (counterstain). Cells were observed using bright-field microscopy (1000×) using a Nikon Microflex SX-microscope.

### *Microscopic Characterization of the Isolates of the "BALI" Culture*

Wet microscopic slides of the isolates of the "BALI" culture were prepared and observed under a Nikon Microflex SX-Series transmission microscope. The average size of the isolates was estimated by measuring the dimensions of at least 10 cells. Micrographs of the isolates were prepared using Spot Advanced Photowear from Spot Diagnostic Instruments Inc.

### *DNA Analysis*

DNA analysis of the isolated perchlorate degraders involved cell lyasis, PCR separation, precipitation and sequencing. The cell lysate was collected using the following method. Approximately 30 mg/ml of cells (wet weight) was weighed out and resuspended in 1 ml of sterile, nanopure water in a 1.5-ml sterile microcentrifuge tube. The cell suspension was washed two times using a Tomy MC-150 High Speed Micro Centrifuge. After the washing was completed the cells were resuspended in 1 ml of sterile nanopure water.

The cells were lysed using the following method. The cell suspension (40 μl) of cell suspension, 50 μl of K-buffer and 10 μl proteinase K (1mg/ml) were placed in a sterile 1.5 ml microcentrifuge tube and incubated at 60°C for 20 minutes, 100°C for 5 minutes and then immediately placed on ice. The sample was spun at 15,000 rpm for 10 minutes. The supernatant which contains the DNA was collected and transferred to a fresh, sterile, 1.5 ml microcentrifuge tube. The DNA samples were stored at -20°C.

Polymerase chain reaction (PCR) was performed on the DNA of each sample using universal bacterial primers targeted against the 16S r-DNA region of the bacterial chromosome. A 200 μl master mix was set up using the following components. 32 μl dNTP's, 20 μl 10X buffer,

40μl Q-solution, 4μl $MgCl_2$, 10μl primer A, 10μl primer B, 2μl Taq DNA polymerase, 82 μl nanopure, nuclease free, sterile water. 40μl of master mix was transferred to sterile, thin-walled PCR tubes and 10μl DNA template was added to obtain a 50μl reaction mixture. PCR was performed using a Perkin Elmer GeneAmp PCR System 2400.

Electrophoresis (EP) of PCR product was performed with a 0.7% agarose gel; ethidium bromide was used to visualize PCR products. Tracking dye (4 μl) was added to each sample and then 50 μl of the sample was transferred to individual wells of the agarose gel. Enough TE buffer was added to the EP gel box so that the entire agarose gel was covered with TE buffer. The EP gel system was run using a 50 mV electric current until the purple portion of the tracking dye was at the middle of the gel. The gel was visually analyzed using a Gel Doc System.

PCR product bands (~1 kb in size) were removed from the gel using a surface sterilized razor blade. Bands present in the gel were visualized using an ultraviolet tray box. The bands were cut from the agarose gel and placed in a sterile 2 ml eppendorf microcentrifuge tube. A Gel Extraction Kit (Qiagen QIAquick) was used to clean up the PCR product from the agarose gel.

Sequencing buffers were prepared according to instructions provided by PE Applied Biosystems. Precipitation was performed using ethanol and a pellet was obtained by centrifuging and drying under vacuum. DNA sequencing was carried out by using ACI-PRISM™ 3700 DNA Analyzer.

**Influence of Nitrate and Sulfate on Perchlorate Biodegradation by a Membrane-Immobilized Biofilm**

Prior to testing the effects of sulfate and nitrate on perchlorate biodegradation by the membrane-immobilized biofilm, batch tests were performed to gather preliminary data on the effects of these anions on perchlorate biodegradation. In the batch testing of nitrate interference, about 100 mg/L of perchlorate, 500 mg/L of lactate, buffer, nutrient/minerals and nitrate (0, 10, 30 and 60 mg/L) were added to four 125-ml serum bottles. In the batch testing of sulfate interference, about 10 mg/L of perchlorate, 150 mg/L of lactate, buffer, nutrient/minerals, and sulfate (0, 20, 100, and 500 mg/L) were added to four 125-ml serum bottles. All the tests were performed in duplicate. Because the background sulfate concentration in the nutrient/minerals is about 41 mg/L and because it was not known at that time whether sulfate was a preferred electron acceptor to perchlorate, a larger amount of lactate was added to assure that the system was not limited by the lack of lactate.

The results of the batch tests, on the influence of nitrate or sulfate on perchlorate biodegradation, were used to guide the testing in the immobilized-biofilm reactor. The Memcor BTS-55 membrane-immobilized biofilm was used for the diffusion testing since diffusion coefficients for perchlorate, nitrate and sulfate, have been determined for this membrane. The nitrate and sulfate interference tests were performed in the same manner as those for perchlorate, and tested in duplicate.

Four tests were performed on the interference of nitrate on perchlorate biodegradation by a BTS-55 membrane-immobilized biofilm. Two concentrations of nitrate (10 and 50 mg/L) and two perchlorate concentrations (about 10 and 50 mg/L) were investigated. The perchlorate and nitrate concentrations tested are shown in Table 6.3. For the sulfate interference testing, the perchlorate and sulfate concentrations are shown in Table 6.4.

A BTS-55 membrane, containing a previously grown biofilm, was placed between the diffusion reactor (DR) and biological reactor (BR). The DR reactor was filled with 5 liters of DI water containing the desired concentration of perchlorate and either nitrate or sulfate. The BR reactor contained 5 L of medium containing lactate, nutrients/minerals and buffer. Both reactorswere flushed with nitrogen gas to remove dissolved oxygen to undetectable levels. The DO levels in the reactors were monitored using a YSI 54A DO meter. Samples were taken from each reactor at determined time intervals. Samples were then filtered through a 0.2 µm pore size membrane filter and analyzed for perchlorate and nitrate or sulfate.

### Influence of Salinity on Perchlorate Biodegradation by the "BALI" Culture

Studies were performed to determine the influence of salinity levels on the perchlorate biodegradation by the "BALI" enrichment culture. The objective of these experiments was to determine whether high TDS levels would affect the kinetics of biodegradation. Culture tube testing was used for the studies. Microbial activity was assessed by measuring, with a spectrophotometer (Spectronic 20), the changes in optical density, at 600 nm, due to microbial growth.

In the culture tube testing, the tubes, caps, nutrient/minerals, buffer, lactate, and perchlorate solutions were autoclaved prior to their use in the experiments in order to avoid contamination by other microorganisms. The buffer and nutrient/mineral media were of the same composition as those used in the membrane reactor studies throughout this project. The microbial inoculum was the "BALI" culture, which is the enrichment mixed culture used in all perchlorate biodegradation testing performed with the membrane-immobilized biofilm.

The autoclaved medium containing desired amounts of nutrients, buffer, perchlorate, and lactate were filled to the 8.5 mL culture tubes using a micro-pipette. Next, the desired volume of a suspension of the "BALI" culture was added to each tube. Microbial concentration was measured gravimetrically by filtering the samples through a 0.45 µm acetate membrane filter (Osmonics Inc.). The tubes were then capped and placed in the dark at 23 ±2°C for incubation. Triplicate tubes were prepared for each perchlorate/salinity level combination. Salinity was established by adding desired amounts of NaCl to the growth medium. The salinity levels investigated ranged from 0% to 6%.

In addition to experimental tubes, two sets of control tubes were added to the experiment; one containing no microbes (abiotic control) and another containing no lactate (electron donor control). The concentrations of perchlorate, lactate, and microbes in each tube are shown in Table 6.5. At prescribed intervals, the increase in turbidity (decrease in transmittance) was measured directly in the tubes at 600 nm using a spectrophotometer (Spectronic 20, Bausch and Lomb, Rochester, New York). Tubes were mixed well before each measurement.

## RESULTS AND DISCUSSION

### Diffusivity Tests Through Membranes Without a Biofilm

#### *Perchlorate Diffusivity Testing*

Figure 6.8 and Table 6.6 show the results of duplicate tests for perchlorate diffusion through the BT-55 membrane. The plot of the data resulted in two straight lines, both with $R^2$ values greater than 0.99 and the same slope of –0.0007. The slope of the lines, as per Equation 6.2, equals $-(2A_M D_M)/(V_{BR}\Delta L_M)$. Since $A_M$, $V_{BR}$, $\Delta L_M$ are known, the diffusion coefficient, $D_M$, was calculated to be $6.64\times10^{-6}$ cm$^2$/sec. Example calculations are shown in Appendix E.

For the diffusion coefficient data through the PVDF membrane (Figure 6.9 and Table 6.7) some variability was found in the duplicate tests. Both $R^2$ values were greater than 0.99, but the slopes were different and equal to -0.0004 and -0.0006 for tests 1 and 2, respectively. The calculated diffusion coefficients were found to be $3.0\times10^{-6}$ and $4.5\times10^{-6}$ $cm^2/sec$ for test 1 and 2, respectively. The average diffusion coefficient for this membrane is then $3.75 \times10^{-6}$ $cm^2/sec$. For the FGLP membrane (Figure 6.10 and Table 6.8), data for both tests were very similar with $R^2$ values greater than 0.97. The slope was the same (-0.0004) for both tests and a diffusion coefficient of $6.67\times10^{-6}$ $cm^2/sec$ was calculated.

The membrane characteristics as well as the calculated diffusion coefficients for all three membranes are summarized in Table 6.9. Notice in Table 6.9 that the 0.2μm membranes have similar diffusion coefficients. The FGLP has a slightly larger diffusion coefficient and it is thicker than the BTS membrane. The smallest diffusion coefficient was found for the PVDF membrane that has the largest pore sizes, but it is considerably thinner than the other membranes.

For comparison, the diffusivity of perchlorate in water, without a membrane, was calculated by using the Wilke-Chang Method, in which the molar volume of perchlorate was calculated using the Method of LeBas (Reid, Pravsnitz, and Poling 1987). The diffusion coefficient of perchlorate calculated by the Wilke-Chang method (Reid, Pravsnitz, and Poling 1987) was found to be $1.53\times10^{-5}$ $cm^2/sec$. Therefore, the diffusivity of perchlorate through the microporous membranes tested is significantly smaller than that in water. This result confirms the hypothesis that perchlorate would migrate through semi-permeable membranes by diffusion. The diffusion coefficient will allow for the calculation of the perchlorate flux through the membrane without the biofilm for different perchlorate concentrations. In addition, it will provide insight into the effects of the biofilm thickness on the transport of perchlorate to the BR reactor. All three membranes were tested for biofilm immobilization.

### *Nitrate Diffusivity Testing*

Tables 6.10 and Figure 6.11 show the results of duplicate diffusivity testing (test 1 and test 2) for nitrate through the Memcor BTS-55 membrane (pore size 0.2 μm; 125 m thick). For test 1, the initial concentration of nitrate in the DR reactor was 1066.96 ppm, while for test 2, the initial concentration of nitrate in the DR reactor was 1031.74 ppm. Figure 6.11 was generated by plotting the last columns of Tables 6.10 against time (Equation 6.2). The slopes of the generated lines were calculated and the diffusivity was determined by Fick's law. Notice that the duplicate tests provided very similar results and the same slope (-0.0005) with $R^2$ values greater than 0.996. Therefore, the calculated diffusion coefficient of nitrate through the BTS-55 membrane is $4.74\times10^{-6}$ $cm^2/sec$. McCleaf and Schroeder (1995) found diffusion coefficient for nitrate of the same order of magnitude ($3.4\times10^{-6}$ $cm^2/sec$) for a membrane with a 0.2 μm pore size and 37.5 μm thickness.

### *Sulfate Diffusivity Testing*

Tables 6.11 and Figure 6.12 show the results for duplicate diffusivity tests (test 1 and test 2) for sulfate through the Memcor BTS-55 membrane. For test 1, the initial concentration of sulfate in the DR reactor is 949.76 ppm; while for test 2, the initial concentration of sulfate in the DR reactor is 973.26 ppm. The data on the last columns of Table 6.11 were used to build Figure 6.12, according to Equation 6.2. For test 1, the slope was found to be –0.0004 with $R^2$ value of

0.988. For test 2, the slope was –0.0004 with $R^2$ value of 0.995. The calculated diffusion coefficient of sulfate through the BTS-55 membrane is therefore $3.79\times10^{-6}$ $cm^2$/sec.

Table 6.12 shows the experimentally determined diffusion coefficients for nitrate, sulfate, and perchlorate through the Memcor BTS-55 membrane. In addition, the diffusivity of these compounds in water, as estimated by the Wilke-Chang method (at 25°C) are shown in the table. The molar volume of nitrate (31.9 $cm^3$/g.mole), sulfate (58.8 $cm^3$/g.mole), and perchlorate (54.2 $cm^3$/g.mole) were calculated using the Method of LeBas. Notice that for the three compounds, as expected from the values of their molar volumes, the diffusivity in water is from the fastest to the slowest: nitrate > perchlorate > sulfate. However, the experimentally determined diffusivity values through the membrane did not show the same relationship for perchlorate. For these, the diffusivity through the BTS-55 membrane follows: perchlorate > nitrate > sulfate. Thus, through the membrane in question, the diffusivity of nitrate is higher than that of sulfate, as expected from their diffusivities in water. However, perchlorate diffusivity is larger than those for nitrate and sulfate. The implication of these findings to the performance of the membrane-immobilized biofilm reactor is that sulfate and nitrate contained in perchlorate-contaminated waters may not reach the biofilm (located on the BR side of the reactor) as fast as perchlorate and their effects on perchlorate biodegradation may be reduced.

**Development of One-Liter Master Culture Reactor (MCR)**

Table 6.13 shows the data obtained when acclimating the CCSD enriched culture to degrade perchlorate. The data presented in Table 6.13 clearly indicate that activated sludge cultures can be easily acclimated to biodegrade perchlorate. The results show that within four days about half of the perchlorate was reduced anaerobically by the biomass. After five days of incubation, perchlorate concentration decreased to undetectable levels. The concentration of chloride in the reactor was initially high (51 mg/L) due to the high TDS of the waste activated sludge sample used as the inoculum. The suspended solids concentration remained high in the presence of perchlorate and started decreasing rapidly when perchlorate was used up. The decreased suspended solids indicates the enrichment of the culture with perchlorate-reducing microbes and the decay of the microbes that cannot use perchlorate as an electron acceptor.

**Development of Immobilized Biofilms on the Membranes**

The FGLP membrane showed the largest diffusion coefficient for perchlorate (Table 6.9), and this membrane was therefore selected to be tested first for biofilm development. Table 6.14 shows the concentrations of perchlorate, lactate, and chloride in both reactors during biofilm growth on the Millipore FGLP membrane.

Notice in Table 6.14 that the initial perchlorate (203.5 mg/L), chloride (51.3 mg/L), and lactate (1122 mg/L) concentrations are higher than the concentration of these components added to the process. This occurs because these components were present in the MCR from which the seed microbes were taken.

As shown in Table 6.14 and Figure 6.13, 200 mg/L of perchlorate was rapidly and completely biodegraded by the enrichment culture (about 0.2 g/L SS). Notice that the bulk of the perchlorate biodegradation did not occur until day -three and in day –five all perchlorate had been consumed by the microbes. A lactate to perchlorate ratio of about 5.2:1 was consumed. A thick biofilm was developed in the membrane separating the two reactors and in the walls of the BR reactor as well.

The chloride concentration in the reactor increased with increasing perchlorate biodegradation, but it was not proportional to perchlorate degradation. It seems that there is a lag between perchlorate biodegradation and chloride formation. Notice in Table 6.20 that the SS concentration in the reactor quickly decreased when perchlorate was no longer available.

In the DR reactor, where only DI water was added, the concentrations of perchlorate, lactate, and chloride were monitored (Table 6.14 and Figure 6.14). Notice that only small amounts (μg/L range) of perchlorate, lactate, and chloride diffused to the DR reactor. The concentrations of perchlorate and lactate increased with time, but the chloride concentration varied somewhat during the six-day period. It was also observed that after the fifth-day that the volume in the BR reactor had increased about 0.5 L due to osmotic pressure differences between the cells. This movement of water from the DR to the BR reactor due to the difference in concentrations promoted dilution in the BR reactor. The concentration values shown in Table 6.14 are the raw numbers and have not been corrected for the dilution caused by osmotic pressure.

## Biodegradation of Perchlorate by Membrane-Immobilized Biofilms

Biofilms were developed in all three selected microporous membrane and perchlorate biodegradation by the membrane-immobilized biofilms was investigated. Table 6.15 shows the testing schedule and the number of biodegradation cycles performed for each membrane. One cycle consists of a run from an initial perchlorate concentration to a lower desired perchlorate level in the DR reactor.

### *Testing of Biofilm Immobilized on the FGLP Membrane*

In the first cycle, 223 mg/L perchlorate was added to the DR reactor separated by a biofilm immobilized on Millipore FGLP membrane (Table 6.16). In the BR reactor 1100 mg/l lactate, nutrients/minerals and buffer were added. Notice that in nine days only about 37 mg/L was biodegraded. Perchlorate diffusion was very poor through the biofilm immobilized in this membrane. This was unexpected, since this membrane showed the largest diffusion coefficient in the diffusion tests without a biofilm. It is worth mentioning that the biofilm developed in this membrane was very thick. This happened because this membrane has a plastic backing that favors microbial attachment. Therefore, transport of perchlorate to the BR reactor was limited by diffusion. Also notice that the chloride concentration in the BR and in the DR reactors increased with time, indicating perchlorate biodegradation. (Figure 6.15) The results show that chloride migrated easily from the BR to the DR reactor with time, due to its small molar volume.

Although only a small amount of perchlorate was biodegraded, a large amount of lactate was used in the BR reactor. Since the work is being performed using a mixed- biological culture, it is likely that lactate was used as a carbon source by microbes in the biofilm which are not dependent on perchlorate as an electron acceptor

In a second trial with the FGLP membrane (Table 6.17) 223.16 mg/L perchlorate was added to the DR reactor and 1099.8 mg/l lactate, nutrient and buffer were added to the BR reactor. As shown in Figures 6.16, only very small amounts of perchlorate were able to diffuse to the BR reactor. Thus, biodegradation of perchlorate was again limited by diffusion through the thick biofilm. It was concluded from the above results that the FGLP membrane was not suitable for perchlorate degradation.

### *Testing of Biofilm Immobilized on the BTS-55 Membrane*

Three testing cycles were performed with this membrane. These tests involved two steps, the growth of the biofilm on the membrane and the biodegradation experiment. To grow the biofilm, lactate, perchlorate, nutrient, buffer, and microorganisms from the enriched culture were added to the BR reactor and DI water was added to the DR reactor. After the biofilm reached a certain thickness, the biofilm was protected by a sliding plate and the tanks were emptied and rinsed with DI water. Next, lactate, nutrient, and buffer were added to the BR reactor and perchlorate was added to the DR reactor. The plate was then removed from the biofilm and the reactors were sealed and kept anaerobic by purging with nitrogen gas.

Table 6.18 and Figures 6.17 show that, after 5 days, practically all perchlorate diffused to the BR reactor where it was biodegraded. Notice that the concentration of perchlorate in the BR reactor, at all times, was very small. The lactate in the BR reactor decreased proportionally to perchlorate in the DR reactor. However, lactate kept decreasing although no more perchlorate was available. It is suspected that lactate was fermented, since the work was being performed using a mixed culture.

The small concentrations of chloride and perchlorate in the first day of the experiment in the BR reactor are the result of the previous biofilm growth experiments. When the plate is held against the biofilm to keep it anaerobic during rinsing of the reactors, a small amount of growth solution is retained in the system and will be part of the BR reactor when the plate is removed. Because of this inconvenience, the experimental set-up was been modified. In the new set-up, one set of reactors is used to grow the biofilm on a mobile plate that is removed, after the biofilm is developed, and attached to a new set of reactors by metal clamps.

An investigation of Figure 6.17 and Table 6.18 shows that, contrary to perchlorate and lactate, chloride concentrations change significantly from one reactor to another. This is due to two factors: (a) diffusion of chloride through the membrane due to its small size, (b) movement of water from the DR reactor (smaller concentration of ions) to the BR reactor (higher concentration of ions) by osmotic pressure. In all the tests performed with this membrane, migration of water from the DR to the BR reactor was observed. That means, with time, the volume of water in BR increased. As can be observed in Table 6.18 (last row), the chloride concentration stabilizes when both reactors contain about the same concentration of chloride. The fact that water is transferred by osmotic pressure to the BR reactor may have implications in the design of full-scale membrane-immobilized biofilm systems. Product water (perchlorate free) could migrate to the BR reactor decreasing the total volume of treated water produced. In the experiments, DI water was used in the DR reactor. However, natural waters contaminated with perchlorate contain several other ions and the difference of ionic gradient may not be sufficiently high to cause the water migration from the DR to the BR reactor.

Table 6.19 shows the molar ratio of perchlorate biodegraded to that of chloride formed to be 0.82 for this run. Theoretically, each mole of perchlorate would generate one mole of chloride, thus this ratio is somewhat smaller than expected. Table 6.19 also shows that about 0.25 moles of perchlorate were biodegraded by the biofilm per day per $m^2$ of membrane area. For this calculation only the first six days of the experiment were considered, since the system was perchlorate-limited after the sixth day. An example calculation for the perchlorate biodegradation rate is shown in Appendix E, example 3.

For the second and third cycles, more lactate, nutrient, and buffer were added to BR reactor and more perchlorate was added to the DR reactor. The addition was made to the resulting solutions from the previous cycles. Table 6.20, 6.22 and Figure 6.18, 6.19 show the data obtained from these tests. The high concentrations of chloride in the first day are the result of perchlorate biodegradation from the previous cycles. As can be seen in Figures 6.18 and 6.19 perchlorate was easily biodegraded by the biofilm as indicated by the decrease in perchlorate degradation and the increase in chloride concentration.

Investigation of Tables 6.20 and 6.21 indicates that perchlorate degradation in the second cycle (0.11 moles $ClO_4^-$ per day per $m^2$ of membrane area) was slower than in the first cycle (0.25 moles $ClO_4^-$ per day per $m^2$ of membrane area) (Table 6.19) and third cycles (0.22 moles $ClO_4^-$ per day per $m^2$ membrane area) (Table 6.23). Possible reasons for the differences in degradation rates in the three cycles include: (a) difference in biofilm thickness— it is not possible to measure the thickness of the biofilm between cycles, without terminating the experiment, but visually the biofilm seems thicker with time; (b) change in the microbial ecology of the reactor due to lactate degradation in the absence of perchlorate. In several runs, it was observed that when perchlorate was absent, lactate degradation still proceeded. It is possible that lactate was fermented by other microorganisms since the work was been performed with a mixed-culture. As mentioned above, in the first cycle perchlorate was biodegraded at a rate of 0.25 moles/day/$m^2$ of membrane area. After the fifth day, all perchlorate was degraded, but lactate biodegradation proceeded for ten more days. In the second cycle, perchlorate biodegradation rate was only 0.11 moles/day-$m^2$ (Table 6.21), but in cycle three, the perchlorate biodegradation was 0.22 moles/day/$m^2$ of membrane area (Table 6.23), similar to that of the first cycle. It is possible that fermentation occurred at the end of the first cycle, negatively affecting perchlorate biodegradation in the second cycle.

After the third cycle the reactors were carefully emptied and the membrane-immobilized biofilm was weighed together with the membrane. The thickness of the membrane without the biofilm was originally measured using a micrometer. However, this method was found to be inadequate to measure the thickness of the membrane and biofilm together, since the biofilm was very fragile. The amount of microbes present in the biofilm was then estimated gravimetrically. This was done by carefully removing the microbes from the membrane with DI water, filtering through a 0.45µm sterilized membrane filter and drying at 105°C for 24 h. The estimated thickness of the biofilm, based on a microbial density of 1.07 g/ml, was calculated to be 24.7 µm (Table 6.24). This value is very close to the thickness of a biofilm (27.5 µm) reported by McCleaf and Schroeder (1995), measured using a microscope with a calibrated ocular micrometer, on a denitrification study using a membrane-immobilized biofilm.

Perchlorate biodegradation rates for the BTS-55 membrane were calculated by considering a 10cm diameter biofilm with about 25 m thickness (Appendix E, example 1). The rates found varied from 110 and 251 µg $ClO_4^-$/min/L/µg microbes. These values are within the range of 1.2 to 24 g $NO_3^-$ - N per kg of biological material per day (i.e. 20 to 400 µg $NO_3^-$/min/L/µg microbes) for denitrification by microbes immobilized on agar gel (Mateju et al. 1992).

Tables 6.20 and 6.23 show the perchlorate to chloride ratios for the second and third cycle runs. Stoichiometrically, a ratio of 1.0 is expected, since each mole of perchlorate biodegraded generates one mole of chloride. For the second cycle a ratio of 1.03 (Table 6.21) was found and for the third cycle a ratio of 0.86 (Table 6.23) was found. These are considered to be satisfactory ratios, considering the size of the reactor-set (10 liters) used in the study.

*Testing of Biofilm Immobilized on the PVDF Membrane*

This was the first test performed after the experimental set-up was modified. The amount of lactate added to the BR reactor in this experiment was lower (about 700 mg/L) as compared to the other experiments performed with the BTS-55 and FGLP membranes (about 1100 mg/l lactate).

Table 6.25 and Figure 6.20 show the biodegradation of perchlorate by the biofilm immobilized on the PVDF membrane. Notice that the biodegradation rate (0.06 moles of perchlorate per day per $m^2$ of the membrane area) is much smaller than the values found for the BTS-55 membrane (0.11 to 0.25 moles of perchlorate per day per $m^2$ of the membrane area). The PVDF membrane has the smallest diffusion coefficient from all three membranes tested. Therefore, the lower biodegradation rate is the result of low diffusion of perchlorate through the membrane. Notice in Table 6.25 that perchlorate biodegradation proceeded after the lactate concentration was undetectable. The perchlorate to chloride ratio for this test was found to be 0.99 (Table 6.26), which is close to the expected value of 1.0. The perchlorate to chloride ratios and the biodegradation rates for the two memebranes are summarized in Table 6.27. This is a very good mass balance of the moles of chloride generated by the biodegradation of perchlorate.

**The "BALI" Enrichment Culture**

As reported earlier (Table 6.15), the BTS-55 membrane reactor was run for forty days. At the end of the third cycle, it was observed that the reactor's discharge tubing acquired a red coloration. Further observation indicated that the red color was due to the presence of red microbes in the system. Figure 6.21 shows the dark pink coloration of the microbes in the master reactor.

**SRT Studies**

Two cycles of SRT studies were performed in the master culture reactor. Shown below are mass balances for perchlorate, lactate, and chloride, and volatile suspended solids (VSS) for the different cycles:

*First Cycle*

Table 6.28 shows the mass balance of volatile suspended solids in the master reactor for the first cycle. The daily increase in biomass as well as the average biomass present in the reactor has been calculated from daily VSS measurement. The average VSS concentration was used to calculate the grams of perchlorate biodegraded and the grams of chloride formed by each gram of biomass present in the reactor. The daily biomass increase will be used to estimate the yield coefficient, that is, the grams of biomass formed per gram of lactate biodegraded. Table 6.29 shows the mass balance for perchlorate for the 5-day period. Notice that in the first day only 18.54 mg of perchlorate were degraded. This is the result of the acclimation period needed by the microbes to adjust to the new conditions. In the second day, the culture was already acclimated and perchlorate was biodegraded as indicated by the consumption of lactate (Table 6.30) and the increase in chloride concentration. (Table 6.31). In the fourth and fifth days the amount of perchlorate biodegraded with time increased significantly and perchlorate concentrations went to undetectable level. In the second and third days 113.82 mg and 135.82 mg of perchlorate (Table 6.30) were biodegraded and the mg $ClO_4^-$/mg VSS was calculated to be 0.38 and 0.43

respectively. For the fourth and fifth days, 54.08 and 55.12 mg of perchlorate were biodegraded and the mg $ClO_4^-$/mg VSS was calculated to be 0.19 and 0.21 respectively. These results show that each milligram of the enriched culture present was capable of biodegrading an average 0.4 mg of perchlorate per day, when the system was not perchlorate limited. In fourth and fifth days, about 55 mg perchlorate were degraded and the ratio of perchlorate to biomass for both days were practically the same (0.2 mg of perchlorate per day).

Table 6.31 shows the mass balance of chloride in the master reactor and Table 6.32 shows the daily chloride to perchlorate ratios. The data show some of the perchlorate to chloride ratios somewhat above the expected value (one). However, some of them are very close to the theoretical value. The somewhat higher values could be due to analytical accuracy. Although the calibration curves for chloride, and lactate always meet an $R^2 = 0.995$, the IC sensitivity for very low chloride and lactate levels is low and precision of measurements are within ±20%. Notice in Table 6.29, that similar to the perchlorate mass balance, the mass of chloride in the second and third days are roughly the double of those in the fourth and fifth days, reinforcing again the relationship between perchlorate biodegradation and chloride formation.

Table 6.30 shows the mass balance for lactate in the reactor. Notice that the ratio of the mass of VSS generated and the mass of lactate consumed varied considerably. Because the system was perchlorate limited in the fourth and fifth days and because the work was performed using a mixed-culture, it is believed that the cell yields do not reflect the exclusive use of perchlorate as an electron acceptor. As shown before, there is indication, from the experimental results, that lactate may have been utilized by the mixed-culture through fermentation.

### *Second Cycle*

Tables 6.33 and 6.34 show the mass balances of lactate and perchlorate in the master bottle in the second cycle. Notice that the ratio of perchlorate biodegraded to biomass present, in the second, third, fourth and fifth days are similar and in agreement with the results found in the first cycle (Table 6.29). The perchlorate to chloride ratios for the second cycle varied from 0.87 to 1.15. Again, these ratios are very reasonable considering the sensitivity of the analytical method.

Similar to the observations for cycle one, the lactate concentration varied considerably (Table 6.35) and they do not seem to correlate to perchlorate biodegradation.

In the first day, the ratio of perchlorate biodegraded to biomass present was very high (1.11 mg $ClO_4^-$/mg VSS) (Table 6.34) and the chloride formation, for that day, was correspondingly high (Table 6.36). The results show that a larger amount of perchlorate than that found in the first cycle, can be biodegraded by the culture. Table 6.37 shows that the ratio of perchlorate to chloride ranged from 0.87 to 1.15 over five days, in reasonable agreement to the expected ratio of 1.0.

## Preliminary Kinetics Study

The results of the kinetic studies clearly show that the "BALI" culture requires a minimum organic carbon to perchlorate ratio for perchlorate biodegradation to occur at acceptable rates. Notice in Figure 6.22 that under lactate limited conditions, no microbial growth was observed, except in bottles KL5 and KL5d in which the lactate to perchlorate ratio was equal to three. For all perchlorate limited bottles, microbial growth (easily identified by the red coloration of the "BALI" culture in Figure 6.22) was observed. The microbial growth was coupled with a decrease in perchlorate and lactate concentrations and an increase in chloride concentration in the bottles.

The results indicate that a lactate to perchlorate ratio of at least three is needed for perchlorate biodegradation to occur at acceptable rates. The results of these tests were used to determine the minimum amount of carbon source (lactate) to be added to the BR reactor of the membrane-immobilized biofilm reactor. This is critical, since excess carbon in the effluent of perchlorate-removing reactors may present a challenge to subsequent unit processes (e.g. disinfection).

Figure 6.23 and Table 6.38 show the biodegradation of perchlorate in the perchlorate limited bottles. Notice that a larger acclimation time (3 days) was needed for the bottles containing the highest perchlorate concentrations (100 and 200 mg/L). For the bottles containing lower perchlorate concentrations (10-60 mg/l) only about 2 days were required for complete biodegradation. After acclimation, perchlorate removal from the lower perchlorate concentration bottles was 100% (KC1, and KC2), 72.4% (KC3), and for the higher perchlorate concentration bottles, it was 52% (KC4), and 28.2% (KC5). Example calculations for the perchlorate removal after acclimation for bottles KC3 to KC5 are shown in Appendix E, example 2.

Figure 6.24 to Figure 6.28 and Table 6.39 to Table 6.43 show the concentration of perchlorate, lactate, and chloride in the perchlorate limited bottles. Notice that in all bottles, the perchlorate and lactate concentrations decreased with time while the chloride concentration increased. Notice that the suspended solids (SS) concentrations are not shown. The reason is the 5-ml samples taken daily did not contained sufficient microorganisms to be accurately weighed in a four-digit accuracy balance (0.0000 g).

Figure 6.29 and Table 6.44 show the perchlorate biodegradation for the only bottle of the lactate-limited group in which biodegradation occurred. Notice that lactate and perchlorate concentration decreased with time while the chloride concentration increased. For all the other bottles in which the lactate to perchlorate ratio was smaller than 3.0, biodegradation was not observed.

Table 6.45 shows the perchlorate biodegradation for the lactate-limited group in which perchlorate biodegradation did not occur.

### Isolation, Characterization and Identification of Perchlorate-Degrading Bacteria From the "BALI" Mixed Culture

Gram-stain photomicrographs of the "BALI" culture show that BALI is a mixed culture with Gram-positive and Gram-negative microorganisms (Figures 6.30 and 6.31). The Gram-negative microorganisms are dominant with the majority being motile rods. Also present are Gram-negative microorganisms with a helical morphology. The samples in Figures 6.30 and 6.31 were taken from the batch reactors used in the kinetics study and from the membrane-immobilized biofilm growth reactor ("BALI"). There have been at least two reports (Coates et al. 1999; Malmqvist, Welander, and Gunnarsson 1991) of perchlorate-degrading microorganisms similar in morphology to "BALI" isolated from the environment.

Three species were isolated from the "BALI" mixed culture, namely KA2-1, KL1-a and RAS-CCSD. All of them were small Gram-negative rods with average size of 3.7 μm long by 1 m wide. The characteristics of each isolates are shown in Table 6.46. Microslides of the individual isolates are shown in Figures 6.32 to Figure 6.34.

16S rRNA analyses revealed that the isolates were *Dechlorisoma sp.* with a 98% match. For confirmation, one of the isolates was sent to a commercial DNA sequencing laboratory (MIDI Labs). MIDI Labs results confirmed the isolates to be *Dechlorisoma suilla* (Appendix F).

## Influence of Nitrate on Perchlorate Biodegradation by the Membrane-Immobilized Biofilm

### *Batch Testing*

Figures 6.35 and Table 6.47 show the results for the batch testing on the interference of nitrate on perchlorate biodegradation. The lower three dashed lines show the decrease in the concentration of nitrate with time in the presence of approximately 100 mg/L perchlorate. The top four solid lines show the decrease in perchlorate concentration with time in the presence of about 10 mg/L, 30mg/L and 60 mg/L of nitrate. As shown in Figure 6.35, the rate of nitrate biodegradation is much greater than that of perchlorate. In addition, significant perchlorate biodegradation only started after the nitrate had been almost completely biodegraded. These results show that the enrichment mixed culture ("BALI") prefers nitrate to perchlorate as an electron acceptor. Thus, nitrate negatively affects perchlorate biodegradation.

### *Membrane-Immobilized Biofilm Reactor Testing*

Figures 6.36, 6.37 and Table 6.48 show the data for the biodegradation of high levels of perchlorate (about 50 mg/L) in the presence of approximately the same concentration of nitrate (50 mg/L). Both the diffusion reactor (to which perchlorate and nitrate were added) and the biological reactor (to which lactate, nutrients and buffer were added) were sampled with time and samples were analyzed for perchlorate and nitrate. Figure 6.37 shows the perchlorate and nitrate biodegradation rate expressed as percentage of nitrate or percentage of perchlorate biodegraded per hour. In the first four hours of the experiment, about the same mass of perchlorate and nitrate were biodegraded. From 4 to about 30 hours, nitrate biodegradation dominated the system and perchlorate biodegradation was hindered. After most of the nitrate had been biodegraded, perchlorate biodegradation resumed. The fast initial perchlorate biodegradation in the first hours of the experiment can probably be attributed to the faster perchlorate diffusion through the membrane. In the diffusion testing, it was reported that diffusion through the permeable membrane used decreases in the order perchlorate>nitrate>sulfate. Thus, with time, when nitrate and perchlorate concentration on the biofilm were the same, nitrate dominated the system. It may also be that the "BALI" culture needed some time to acclimate to the nitrate environment. No nitrate was present in the nutrient/mineral solution used to grow the "BALI" culture. Thus, this culture quickly adapts to the nitrate-containing environment. One can say that the bacteria contained in the "BALI" culture are capable of reducing both nitrate and perchlorate.

In Figure 6.36 the concentrations of nitrate and perchlorate in the BR reactor are shown. After about 60 hours, the perchlorate and nitrate diffusing to the BR reactor were not degraded due to carbon limitations. This figure also shows a higher concentration of perchlorate as compared to that of nitrate in the BR reactor, again confirming that the biofilm prefers nitrate to perchlorate as an electron acceptor.

Figure 6.38, 6.39 and Table 6.49 present the results of the biodegradation of about 50mg/L of perchlorate in the presence of 10mg/L nitrate by a BTS-55 membrane-immobilized biofilm reactor. This testing was performed to investigate the interference of a lower concentration of nitrate on perchlorate biodegradation. In this test, the initial concentration of perchlorate was kept the same as the test described above. Since the initial concentration of perchlorate was higher than that of nitrate and for comparison purposes, the biodegradation rate of both anions was expressed as the percentage of perchlorate or nitrate biodegradation per hour. Similar to that observed at higher concentration of nitrate (50 mg/L), in the first 4 hours of the experiment, about the same amount of perchlorate and nitrate were biodegraded. After all nitrate had been biodegraded, perchlorate biodegradation resumed until no more carbon was available.

Figure 6.40 summarizes the interference of nitrate on the biodegradation of high levels of perchlorate in two sets of experimental data. The dashed lines show the set in which the concentration of perchlorate and nitrate were approximately the same. The solid lines show the set in which the nitrate concentration was five times lower that that of perchlorate. Figure 6.40 shows that the nitrate biodegradation rates are much faster than that of perchlorate under both conditions. It also shows that at lower nitrate concentration and for the same concentration of lactate (carbon source), perchlorate biodegradation occurred to a greater extent.

Figure 6.41-6.45 and Table 6.50 and Table 6.51 show the results of interference of high nitrate concentration (about 54 mg/L and 11 mg/L respectively) on the biodegradation of low concentration perchlorate (about 1 mg/L). Notice that presence of nitrate concentration ranging from 10 to 50 mg/L significantly affected perchlorate biodegradation even when the concentration of perchlorate is very low (1 mg/L).

Table 6.52 summarizes all the results obtained from the testing of nitrate interference on the biodegradation of lower and higher perchlorate concentrations. The biodegradation rates of perchlorate and nitrate, shown in Table 6.52, were obtained from selected data highlighted in Table 6.48 to Table 6.51. The data selected considered the presence of sufficient lactate to assure that the system was not under carbon (electron donor) limiting conditions.

For higher perchlorate concentration (about 50 mg/L), the presence of 10 or 50 mg/L of nitrate equally affects perchlorate biodegradation. The biodegradation rates for perchlorate (0.51 mg/hr and 0.48 mg/hr) and for nitrate (1.04 mg/hr and 0.99 mg/hr) were quite similar despite a five-fold difference in nitrate concentration. For lower perchlorate concentrations (about 1 mg/L), a different response to the presence of nitrate is found; the higher the nitrate concentration, the greater the negative effect on perchlorate biodegradation. For the nitrate concentration of 50 mg/L, the perchlorate biodegradation rate was only 0.005 mg /hr, while for a nitrate concentration of 10 mg/L, the rate of perchlorate biodegradation (0.016 mg/hr) increased 3.2 times. At the same time, the nitrate biodegradation rate decreased from 0.38 mg/hr to 0.15 mg/hr.

Although the biofilms used in the tests described above were developed under very similar conditions, it is still not possible to assure that all the biofilms had the same thickness and the same microbial composition. However, the results described above seem to strongly suggest that the negative effect of nitrate on perchlorate biodegradation is more significant when the perchlorate concentrations are lower. Thus, for drinking waters moderately contaminated with perchlorate, the presence of nitrate will significantly hinder perchlorate biodegradation. These findings imply higher cost for the design of perchlorate removal systems because of increased carbon source (e.g.: lactate) usage and reactor volume needed for biodegradation.

## Influence of Sulfate on Perchlorate Biodegradation by the Membrane-Immobilized Biofilm

### *Batch Testing*

Table 6.53 and Figure 6.46 show the result of sulfate interference on perchlorate biodegradation in batch tests. For this test, the initial concentrations of perchlorate were about 10 mg/L, while the initial concentrations of sulfate varied from 40 mg/L to 600 mg/L. The nutrient/minerals broth used for microbial growth contained a background sulfate concentration of approximately 41 mg/L. The results showed that perchlorate was degraded from about 10 mg/L to undetectable levels within 16 hours, while the sulfate concentration remained essentially constant during the same period. Thus, sulfate concentrations ranging from 40 to 600 mg/L seem not to interfere with perchlorate biodegradation. This is confirmed by the fact that sulfate biodegradation started after all the perchlorate had been biodegraded (Table 6.53). Sulfate was very slowly biodegraded and even after 117 hours, the largest reduction of sulfate observed was about 5.2%.

Figure 6.46 shows the variations in perchlorate and sulfate concentrations with time as biodegradation occurs. In the graph, the legend "Sulfate in 40 Sul" means the concentration of sulfate in the test in which the concentration of sulfate was 40 mg/L and the concentration of perchlorate was 10 mg/L. Similarly, the "Per in 40 Sul" means the concentration of perchlorate in the test in which the concentration of sulfate was 40 mg/L and the concentration of perchlorate was 10 mg/L. The figure clearly shows that sulfate concentration remained almost constant for the experimental period, while for perchlorate, the concentrations decreased sharply to zero after 16 hours.

### *Membrane-Immobilized Biofilm Testing*

Table 6.54 to 6.57 and Figure 6.47 to 6.52 show the results of interference of sulfate on perchlorate biodegradation in BTS-55 membrane-immobilized biofilm reactors. The sulfate concentrations added to the DR reactors, in individual tests, varied from 10 mg/L to 200 mg/L while the perchlorate concentrations varied from 100 μg/L to 50 mg/L. The first test performed indicated a much higher sulfate concentration in the BR reactor than that added to the DR reactor. Investigation of the issue revealed that the nutrient/minerals broth added to the BR reactor contains about 41 mg/L sulfate resulting from the addition of $MgSO_4{\cdot}7H_2O$, $ZnSO_4{\cdot}7H_2O$, $FeSO_4{\cdot}7H_2O$, and $CuSO_4{\cdot}5H_2O$ (Table 6.2). Thus, all the perchlorate biodegradation experiments described in this report were performed in the presence of about 41 mg/L sulfate. No nitrate is present in the buffer or nutrient/minerals broth.

As seen in Figure 6.47 to Figure 6.52, the perchlorate concentration in the DR reactor decreased with time as a result of biodegradation by the biofilm as it diffused to the BR reactor, in which the perchlorate concentration remained very low. However, the sulfate concentrations in the DR and BR reactor varied considerably. This variation of sulfate concentration is the result of sulfate diffusion, in both directions, from the BR to the DR reactor. Because of this variation, it is not possible to calculate accurately the biodegradation rates for sulfate.

Table 6.58 summarizes perchlorate and sulfate biodegradation rates. The data used in this calculation were bold in Table 6.54 to Table 6.57. The data were selected so that the system was not limited by lactate. In addition, two biodegradation rates were calculated for sulfate, one used the data for which perchlorate was present, and the other used the data after all perchlorate had been biodegraded. Notice that the higher the initial perchlorate concentration, the higher the perchlorate biodegradation rate, independent of the initial sulfate concentration. The sulfate biodegradation rates, in the presence of perchlorate, were much smaller (or null) than those of perchlorate. In the absence of perchlorate, however, the sulfate biodegradation rates were larger than those in the presence of perchlorate, except for the case in which the perchlorate and sulfate concentrations were about 50 mg/L. The reason for that is not clear from the data available. However, it can be concluded that sulfate does not affect perchlorate biodegradation. Thus, the "BALI" mixed culture prefers perchlorate to sulfate as an electron acceptor.

**Influence of Salinity on Perchlorate Biodegradation by the "BALI" Culture**

Figure 6.53 shows the change in transmittance through the culture tubes with time for different salinity levels. Light transmittance, measured with a spectrophotometer at a wavelength of 240 nm, can be used as an indicator of microbial growth in liquid media. The higher the transmittance through the tubes, the clearer the solution contained in the tubes and therefore the fewer the number of microbes present. Notice that the higher the salinity level, the higher the transmittance, indicating that the microbes grow much slower at higher salinity levels. Also notice that for each individual salinity level, the transmittance first decreased with time until it reached a minimum, and then it increased slowly with time and reached a constant level. There may be two possible explanations for the increase in transmittance after it had reached a minimum. The first possibility is endogenous respiration—since the test was run for a long time and the lactate to perchlorate ratio was 3:1, it is likely that after the carbon source was depleted, endogenous respiration took place thereby decreasing the concentration of biomass in the tube. The second possibility is co-flocculation. It is possible that the increase in transmittance is the result of co-flocculation of the microbial cells. In a separate research project ongoing in the UNLV Environmental Engineering laboratory, microbial flocculation was also observed in perchlorate biodegradation batch tests at higher salinity levels. Microscopic observations of the batch tests culture used in this experiments showed flocculation. Although the test tubes used in this experiment were shaken well before each measurement, it is likely that the intensity of the shaking was not sufficient to break up all the formed flocs.

Figures 6.54 to 6.63 show the influence of different salinity levels on perchlorate biodegradation by the "BALI" enrichment mixed culture. In the figures, a, b and c represent triplicates culture tubes tested. Tube d represents a control tube to which no microbes were added, and tube e is a control tube to which no lactate was added. The raw data used to build the figures are contained in the appendicies. For each salinity level tested, three graphs were plotted. The first graph (A) shows the raw transmittance (T) data measured as microbial growth took place in the culture tubes. The second graph (B) is a plot of the absorbance (log 1/T) for the data corresponding to the exponential growth phase. For specific growth rate calculations, only the first portion of the data, those representing the exponential growth phase, are of significance. The exponential growth phase data include the data from the beginning of the experiment to the time where the minimum transmittance was obtained. The third graph (C) is a plot of the logarithmic of the absorbance with time. The slope of this curve represents the specific growth rate ($\mu$) of the culture under consideration. For exponential microbial growth, one can write that the rate of microbial growth is directly proportional to the initial number of microbes present. Thus:

$$\frac{dn}{dt} \propto n \tag{6.4}$$

The proportionality sign can be substituted by a constant, μ (the specific microbial growth rate), the equation 6.4 can be written as:

$$\frac{dn}{n} = \mu dt \tag{6.5}$$

Integrating from $n_0$ to n and from t = 0 to t, one gets

$$\ln \frac{n}{n_0} = \mu t \tag{6.6}$$

where

$n_0$ = initial number of microbes
n = number of microbes at time t
μ = specific microbial growth rate ($time^{-1}$)
t = elapsed time

The absorbance is directly proportional to the number of microbes in the tube at time t, thus one can write:

$$\ln A = \ln A_0 \pm \mu t \tag{6.7}$$

where

$A_0$ = initial absorbance of the culture tube
A = absorbance of the culture tube at time t

The transmittance data show that perchlorate biodegradation was significantly hindered with increasing salinity and at salt concentrations greater than 4% no microbial growth occurred.

Table 6.59 shows the calculated specific growth rate at different salinity levels as well as the $R^2$ obtained in fitting the data to a linear model. In the last column of Table 6.59, the fractions of the specific growth rate, for different salinity levels, as compared to the growth rate at 0% salinity are shown.

Notice that the growth rate in the absence of salinity is 0.26 $d^{-1}$. This value is well within the average growth rate for anaerobes (Rittmann and McCarty 2001). As the salinity increases, the growth rate decreases. Even at 0.5% salinity (5,000 mg/L as NaCl) there was a 27% reduction in the growth rate. At 3.5 % salt, the growth rate decreased 82 % and at a 4% salinity, no growth was observed.

Most potable waters have total dissolved solids (TDS) values between 25 and 1000 mg/L (conductivity between 50 and 2000 μmhos/cm), thus for most drinking water contaminated with perchlorate, the effects of salinity on perchlorate biodegradation may not be of great concern. However, some groundwaters contaminated with perchlorate (e.g. the Kerr McGee site in Las Vegas and the Aeroject site in California) have very high TDS and microbial growth would be affected by the salinity. The Kerr McGee groundwater contaminated with perchlorate, for example, has TDS values between 5,000 - 12,000 mg/L (NDEP 2000). However, the extent of the interference of these TDS levels with biodegradation of perchlorate contained in the Kerr McGee contaminated groundwater has to be investigated. In addition, waters with high perchlorate levels, such as the Kerr McGee contaminated groundwater, are good candidates for biodegradation, given the higher kinetics found at such high perchlorate concentrations.

The major implication of salinity on perchlorate biodegradation, however, relates to the biodegradation of regenerant brines from ion-exchange systems. These brines have salinity varying from 6 to 12% and biodegradation under such conditions would require dilution or isolation of salt tolerant microbes capable of perchlorate biodegradation.

**Table 6.1**
**Characteristics of the selected microporous membranes used in the membrane-immobilized biofilm reactor**

| Membrane name | BTS-55 | PVDF | FGLP |
|---|---|---|---|
| Manufacturer | Memcor | Memcor | Millipore |
| Pore size, μm | 0.2 | 0.45 | 0.2 |
| Thickness, μm | 125 | 99 | 220 |
| Mean pore fraction, % | 70 | 70 | 70 |
| Materials | Polysulfone | N/A | Polytetrafluoroethylene with polyethylene backing |

**Table 6.2**
**Concentration of buffer, nutrients and minerals used in the MCR media**

| Buffer | | | Nutrients and minerals | | |
|---|---|---|---|---|---|
| Component | FW (g/mole) | Concentration in the MCR (g/L) | Component | FW (g/mole) | Concentration in the MCR (mg/L) |
| $K_2HPO_4$ | 174.18 | 1.55 | $MgSO_4{\cdot}7H_2O$ | 246.47 | 100 |
| $NaH_2PO_4$ | 119.95 | 0.85 | EDTA | 372.24 | 3 |
| $NH_4H_2PO_4$ | 115.03 | 0.5 | $ZnSO_4{\cdot}7H_2O$ | 287.5 | 2 |
| | | | $CaCl_2{\cdot}2H_2O$ | 147 | 1 |
| | | | $FeSO_4{\cdot}7H_2O$ | 278.02 | 4 |
| | | | $Na_2MoO_4{\cdot}2H_2O$ | 241.95 | 0.4 |
| | | | $CuSO_4{\cdot}5H_2O$ | 249.7 | 0.2 |
| | | | $CoCl_2{\cdot}6H_2O$ | 237.9 | 0.4 |
| | | | $MnCl_2{\cdot}4H_2O$ | 197.91 | 1 |
| | | | $NiCl_2{\cdot}6H_2O$ | 237.71 | 0.1 |
| | | | $NaSeO_3$ | 149.95 | 0.1 |
| | | | $H_3BO_3$ | 61.83 | 0.6 |

*Source:* Modified from van Ginkel, Plugge, and Stroo 1995.

**Table 6.3**
**Perchlorate and nitrate concentrations tested**

| | Test 1 | Test 2 | Test 3 | Test 4 |
|---|---|---|---|---|
| Perchlorate concentration (mg/L) | 50 | 50 | 1 | 1 |
| Nitrate concentration (mg/L) | 10 | 50 | 50 | 10 |
| Lactate concentration (mg/L) | 200 | 200 | 5 (initial) added 200 later | 5 (initial) added 280 later |

**Table 6.4**
**Perchlorate and sulfate concentration tested**

| | Test 1 | Test 2 | Test 3 | Test 4 |
|---|---|---|---|---|
| Perchlorate concentration (mg/L) | 10 | 10 | 0.1 | 50 |
| Sulfate concentration (mg/L) | 10 | 200 | 50 | 50 |
| Lactate concentration (mg/L) | 220 | 1100 | 420 | 630 |

**Table 6.5**
**Contents of the culture tubes in the salinity test**

| | Salinity testing | Abiotic control | Electron donor control |
|---|---|---|---|
| Biomass (mg/L as SS) | 6.5 | No microbes | 6.5 |
| Perchlorate (mg/L) | 500 | 500 | 500 |
| Lactate (mg/L) | 1500 | 1500 | No lactate |

**Table 6.6**
**Perchlorate diffusion testing by a BTS-55 membrane**

| | Test 1 | | Test 2 (duplicate test) | |
|---|---|---|---|---|
| Time (min) | $C_A$ [conc. of BR (mg/L)] | $\ln[(C_{B0}-2C_A)/(C_{B0}-2C_{A0})]$ | $C_A$ [conc. of BR (mg/L)] | $\ln[(C_{B0}-2C_A)/(C_{B0}-2C_{A0})]$ |
| 5 | 1.88 | -0.003767087 | 1.444 | -0.002892178 |
| 15 | 5.615 | -0.011293533 | 4.927 | -0.009902872 |
| 30 | 9.212 | -0.018595836 | 9.966 | -0.020133322 |
| 45 | 15.13 | -0.030727285 | 15.104 | -0.030673663 |
| 60 | N/A | N/A | 18.3 | -0.037286585 |
| 75 | 18.69 | -0.038096545 | 24.775 | -0.050819722 |
| 90 | 30.46 | -0.062854606 | 30.345 | -0.062609716 |
| 105 | 31.15 | -0.064325211 | 34.63 | -0.07177531 |
| 120 | N/A | N/A | 40.375 | -0.084197159 |
| 135 | 45 | -0.094310679 | 45.55 | -0.095520202 |
| 150 | N/A | N/A | 51.12 | -0.107852507 |
| 165 | 52.49 | -0.110909215 | 54.51 | -0.115433298 |
| 180 | N/A | N/A | 60.99 | -0.130085907 |
| 195 | 61.51 | -0.131271092 | 63 | -0.134674903 |
| 210 | N/A | N/A | 68.28 | -0.146830868 |
| 225 | 70.84 | -0.152778289 | 81.07 | -0.176904257 |
| 240 | 74.9 | -0.162283663 | N/A | N/A |
| 255 | N/A | N/A | 82.42 | -0.180131956 |

**Table 6.7**
**Perchlorate diffusion testing by a PVDF membrane**

| | Test 1 | | Test 2 (duplicate test) | |
|---|---|---|---|---|
| Time (min) | $C_A$[conc. of BR (mg/L)] | $\ln[(C_{B0}-2C_A)/(C_{B0}-2C_{A0})]$ | $C_A$[conc. of BR (mg/L)] | $\ln[(C_{B0}-2C_A)/(C_{B0}-2C_{A0})]$ |
| 5 | 0.718 | -0.001437032 | 1.449 | -0.002902207 |
| 15 | 3.063 | -0.006144841 | 4.543 | -0.009127529 |
| 30 | 6.306 | -0.012692206 | 9.198 | -0.018567311 |
| 45 | 8.98 | -0.018123238 | 13.68 | -0.027741255 |
| 60 | 11.395 | -0.023053706 | 18.085 | -0.036840348 |
| 75 | N/A | N/A | 22.095 | -0.04519613 |
| 90 | N/A | N/A | 27.275 | -0.056094274 |
| 95 | 19.74 | -0.040280475 | N/A | N/A |
| 105 | 22.135 | -0.045279833 | 30.485 | -0.062907851 |
| 120 | 26.355 | -0.054150002 | 34.35 | -0.071173819 |
| 135 | 27.33 | -0.056210628 | 40.00 | -0.083381609 |
| 150 | 29.67 | -0.061173522 | 43.28 | -0.090537587 |
| 165 | 34.74 | -0.072011709 | 48.05 | -0.101036544 |
| 180 | 38.61 | -0.080364426 | 51.74 | -0.109234677 |
| 195 | 39.39 | -0.0820564 | 55.27 | -0.117140742 |
| 210 | 39.7 | -0.082729648 | 61.21 | -0.13058716 |
| 225 | 43.15 | -0.090252989 | 63.24 | -0.135224253 |
| 240 | 45.1 | -0.094530484 | 65.86 | -0.141241036 |
| 255 | 49.78 | -0.104871746 | N/A | N/A |

**Table 6.8**
**Perchlorate diffusion testing by a FGLP membrane**

| | Test 1 | | Test 2 (duplicate test) | |
|---|---|---|---|---|
| Time (min) | $C_A$[conc. of BR (mg/L)] | $\ln[(C_{B0}\text{-}2C_A)/(C_{B0}\text{-}2C_{A0})]$ | CA[conc. of BR (mg/L)] | $\ln[(C_{B0}\text{-}2C_A)/(C_{B0}\text{-}2C_{A0})]$ |
| 5 | 0.8828 | -0.001767161 | 0.7306 | -0.001462269 |
| 15 | 2.7045 | -0.005423682 | 2.437 | -0.004885917 |
| 30 | 4.881 | -0.009809961 | 5.467 | -0.010994216 |
| 45 | 8.296 | -0.016731189 | 5.084 | -0.010220047 |
| 60 | 10.275 | -0.020764089 | 11.51 | -0.023289098 |
| 75 | 12.83 | -0.02599496 | 12.795 | -0.025923119 |
| 90 | 16.695 | -0.033960174 | 15.64 | -0.031779667 |
| 105 | 19.53 | -0.039843307 | 19.775 | -0.040353354 |
| 120 | 22.58 | -0.046211492 | 22.9 | -0.046881986 |
| 135 | 24.33 | -0.049883762 | 25.76 | -0.052894576 |
| 150 | 29.07 | -0.059898635 | 28.01 | -0.0576503 |
| 165 | 30.04 | -0.061960514 | 27.43 | -0.056422214 |
| 180 | 32.31 | -0.066802415 | 32.77 | -0.067786457 |
| 195 | 37.34 | -0.077615655 | 29.14 | -0.060047289 |
| 210 | 36.42 | -0.075629129 | 37.6 | -0.078177781 |
| 225 | 41.96 | -0.087651582 | 44.13 | -0.092400417 |
| 240 | 44.67 | -0.093585668 | 47.3 | -0.099378444 |

**Table 6.9**
**Experimental determination of diffusion coefficient for perchlorate through three types of membranes**

| Membrane type | Memcor BTS-55 | Memcor PVDF | Millipore FGLP |
|---|---|---|---|
| Pore size, μm | 0.2 | 0.45 | 0.2 |
| Thickness, μm | 125 | 99 | 220 |
| Pore fraction, % | 70 | 70 | 70 |
| Diffusion coefficient, $cm^2$/sec | **$6.64\times10^{-6}$** | **$3.75\times10^{-6}$** | **$6.67\times10^{-6}$** |

**Table 6.10**
**Nitrate diffusivity testing by a BTS-55 membrane**

| | Test 1 | | Test 2 (duplicate test) | |
|---|---|---|---|---|
| Time (min) | $C_{BR}$ [conc. of BR reactor (mg/L)] | Ln [($C_{DR0}$-2$C_{BR}$)/($C_{DR0}$-2$C_{BR0}$)] | $C_{BR}$ [conc. of BR reactor (mg/L)] | Ln [($C_{DR0}$-2$C_{BR}$)/($C_{DR0}$-2$C_{BR0}$)] |
| 0 | 0.629 | | 0.34 | |
| 15 | 5.58 | -0.009925011 | 4.82 | -0.009057768 |
| 30 | 10.17 | -0.01865786 | 8.56 | -0.016402944 |
| 45 | 14.02 | -0.026042072 | 12.64 | -0.024477879 |
| 60 | 18.04 | -0.033810978 | 16.67 | -0.032518384 |
| 75 | 22.6 | -0.042697154 | 20.26 | -0.039735874 |
| 90 | 24.85 | -0.047111046 | 24.12 | -0.047554744 |
| 105 | 28.14 | -0.053600412 | 26.92 | -0.053264966 |
| 120 | 31.59 | -0.06045091 | 31.04 | -0.061726887 |
| 135 | 33.12 | -0.063504043 | 34.03 | -0.067913092 |
| 150 | 38.44 | -0.074193315 | 38.91 | -0.078092571 |
| 165 | 40.35 | -0.078059052 | 42.12 | -0.084845445 |
| 180 | 46.26 | -0.090116116 | 45.58 | -0.092175675 |
| 195 | 48.94 | -0.095631895 | 47.48 | -0.096223919 |
| 210 | 51.34 | -0.100597354 | 49.11 | -0.099709994 |
| 225 | 56.63 | -0.111629907 | 56.39 | -0.115429786 |
| 240 | 59.84 | -0.118384343 | 60.10 | -0.123536905 |

**Table 6.11**
**Sulfate diffusivity testing by a BTS-55 membrane**

| | Test 2 | | Test 2 (duplicate test) | |
|---|---|---|---|---|
| Time (min) | $C_{BR}$ [conc. of BR reactor (mg/L)] | Ln [($C_{DR0}$-2$C_{BR}$)/($C_{DR0}$-2$C_{BR0}$)] | $C_{BR}$ [conc. of BR reactor (mg/L)] | Ln [($C_{DR0}$-2$C_{BR}$)/($C_{DR0}$-2$C_{BR0}$)] |
| 0 | 0 | | 0 | |
| 15 | 3.77 | -0.007970528 | 2.65 | -0.005460497 |
| 30 | 4.72 | -0.009989078 | 5.08 | -0.010494013 |
| 45 | 7.45 | -0.015812536 | 8.13 | -0.01684787 |
| 60 | 10.69 | -0.022768189 | 10.93 | -0.022716677 |
| 75 | 13.1 | -0.027973553 | 12.92 | -0.026908761 |
| 90 | 12.15 | -0.02591841 | 15.96 | -0.033346869 |
| 105 | 17.75 | -0.038094326 | 18.59 | -0.038950318 |
| 120 | 22.42 | -0.048362779 | 20.72 | -0.043511598 |
| 135 | 24.58 | -0.053148113 | 26.35 | -0.055669081 |
| 150 | 27.98 | -0.060727285 | 25.93 | -0.054757009 |
| 165 | 29.22 | -0.063505811 | 30.96 | -0.065735217 |
| 180 | 31.35 | -0.068296697 | 31.91 | -0.067822236 |
| 195 | 34.20 | -0.074743152 | 37.18 | -0.079479468 |
| 210 | 39.94 | -0.08785405 | 37.73 | -0.080703935 |
| 225 | 40.06 | -0.088129988 | 40.52 | -0.086938522 |
| 240 | 41.15 | -0.09063992 | 43.17 | -0.092896475 |

**Table 6.12**
**The diffusion coefficients of perchlorate, nitrate, and sulfate in water and through Memcor BTS-55 membrane**

| | Perchlorate | Nitrate | Sulfate |
|---|---|---|---|
| With BTS-55, $cm^2/sec$ (Testing Data) | $6.64 \times 10^{-6}$ | $4.74 \times 10^{-6}$ | $3.79 \times 10^{-6}$ |
| Without BTS-55, $cm^2/sec$ (Calculated by Wilke-Chang Method) | $1.53 \times 10^{-5}$ | $2.12 \times 10^{-5}$ | $1.47 \times 10^{-5}$ |

**Table 6.13**
**Perchlorate biodegradation in the first master culture reactor**

| Day | $ClO_4^-$ (mg/L) | $Cl^-$ (mg/L) | Lactate (mg/L) | SS (g/L) |
|---|---|---|---|---|
| 1 | 120 | 51.0 | 600 | 0.295 |
| 2 | * | * | * | * |
| 3 | * | * | * | * |
| 4 | 66.04 | 110.4 | 153.7 | 0.39 |
| 5 | < 0.005 | 122.7 | 194.3 | 0.26 |
| 6 | < 0.005 | 139.4 | 204.6 | 0.29 |
| 7 | < 0.005 | 143.2 | 168.0 | 0.19 |

* Not sampled

**Table 6.14**
**Biofilm development on a Millipore FGLP membrane**

| | BR reactor | | | | | DR reactor | | |
|---|---|---|---|---|---|---|---|---|
| Day | pH | SS (g/L) | $ClO_4^-$ (mg/L) | Lactate (mg/L) | $Cl^-$ (mg/L) | $ClO_4^-$ (mg/L) | Lactate (mg/L) | $Cl^-$ (mg/L) |
| 1 | 6.576 | 0.1 | 203.5 | 1122 | 51.3 | 0 | 0.196 | 0.275 |
| 2 | 6.673 | 0.11 | 203.3 | 825.9 | 59.1 | 0.022 | 0.238 | 0.191 |
| 3 | 6.957 | 0.13 | 208.3 | 676.8 | 65.5 | 0.070 | 0.253 | 0.226 |
| 4 | 6.916 | 0.21 | 136.9 | 232.7 | 120.0 | 0.129 | 0.295 | 0.218 |
| 5 | 7.105 | 0.07 | < 5 ppb | 195.2 | 169.4 | 0.181 | 0.141 | 0.228 |
| 6 | 7.277 | 0.09 | < 5 ppb | 64.6 | 193.2 | 0.198 | 0.470 | 0.265 |

**Table 6.15**
**The time sheet of perchlorate biodegradation testing**

| | Testing period for different membrane types | | |
|---|---|---|---|
| Testing cycle | FGLP | BTS-55 | PVDF |
| First cycle | 6/7/99–6/13/99 | 6/28/99–7/12/99 | 9/7/99–9/24/99 |
| Second cycle | 6/17/99–7/10/99 | 7/13/9–7/23/99 | |
| Third cycle | | 7/23/99–8/6/99 | |

**Table 6.16**
**Perchlorate biodegradation by a biofilm immobilized on a Millipore FGLP membrane (1st cycle)**

| | DR reactor ($ClO_4^-$ only) | | | BR reactor (lactate, nutrients/minerals, buffer) | | |
|---|---|---|---|---|---|---|
| Day | $ClO_4^-$, (mg/L) | $Cl^-$, (mg/L) | Lactate (mg/L) | $ClO_4^-$, (µg/L) | $Cl^-$, (mg/L) | Lactate (mg/L) |
| 1 | 223.2 | 0.346 | <30 ppb | <5 | 1.32 | 1076 |
| 2 | 217.3 | 0.65 | <30 ppb | <5 | 1.69 | 1030. |
| 3 | 214.8 | 1.78 | <30 ppb | <5 ppb | 2.01 | 266.1 |
| 4 | 205.2 | 3.22 | <30 ppb | <5 ppb | 2.07 | 249.7 |
| 5 | 199.4 | 3.77 | <30 ppb | <5 ppb | 2.73 | 191.9 |
| 6 | 190.8 | 5.42 | <30 ppb | <5 ppb | 2.99 | 89.28 |
| 7 | 202.6 | 6.28 | <30 ppb | <5 ppb | 3.45 | 151.7 |
| 8 | 193.2 | 6.56 | <30 ppb | <5 ppb | 3.68 | 164.2 |
| 9 | 186.5 | 7.13 | <30 ppb | <5 ppb | 4.02 | 161.7 |

**Table 6.17**
**Perchlorate biodegradation by a biofilm immobilized on a FGLP membrane (2nd cycle)**

| | DR reactor ($ClO_4$- only) | | | BR reactor (lactate, nutrients/minerals, and buffer) | | |
|---|---|---|---|---|---|---|
| Day | $ClO_4^-$ (mg/L) | Lactate (mg/L) | $Cl^-$ (mg/L) | $ClO_4^-$ (mg/L) | Lactate (mg/L) | $Cl^-$ (mg/L) |
| 1 | 223.16 | 0 | 0.32 | 0 | 1099.8 | 1.32 |
| 2 | 217.32 | 0 | 0.69 | 0 | 929.6 | 1.69 |
| 3 | 214.76 | 0 | 1.78 | 0 | 287.11 | 2.01 |
| 4 | 205.24 | 0 | 3.38 | 0 | 249.73 | 2.07 |
| 5 | 199.48 | 0 | 4.00 | 0 | 191.9 | 2.73 |
| 6 | 190.8 | 0.013 | 5.42 | 0 | 89.28 | 2.99 |
| 7 | 202.65 | 0.016 | 6.28 | 0 | 151.7 | 3.45 |
| 8 | 192.96 | 0.012 | 6.56 | 0 | 164.2 | 3.68 |
| 10 | 188.36 | 0.021 | 8.18 | 0.036 | 148.38 | 2.14 |
| 11 | 197.44 | 0 | 9.47 | 0.024 | No data | No data |
| 12 | 186.24 | 5.67 | 8.88 | 0 | 35.77 | 4.35 |
| 13 | 178.08 | 0 | 8.99 | 0 | 38.12 | 6.79 |
| 14 | 193.84 | 0 | 10.02 | 0.04 | 20.81 | 6.81 |
| 15 | 187.44 | 0 | 10.72 | 0.25 | 1.75 | 4.67 |
| 16 | 185.08 | 0 | 10.78 | 0.83 | 0 | 8.79 |
| 17 | 187.88 | 0 | 7.42 | 0 | 0 | 9.36 |
| 18 | 183.4 | 0 | 10.3 | 0.34 | 0 | 10.3 |
| 19 | 182.08 | 13.32 | 9 | 0.65 | 0 | 11.6 |
| 20 | 156.56 | 7.91 | 10.3 | 0.93 | 0 | 12.78 |
| 21 | 178.22 | 13.92 | 11.59 | 0.78 | 0 | 15.22 |
| 22 | 176.56 | 0 | 13.21 | 0.13 | 0 | 13.83 |
| 23 | 169.76 | 0 | 11.46 | 0.82 | 0 | 15.37 |
| 24 | 169.96 | 0 | 11.75 | 1.77 | 0 | 16.4 |

**Table 6.18**
**Perchlorate biodegradation by a biofilm-immobilized on a BTS-55 membrane (1st cycle)**

| | DR reactor ($ClO_4$- only) | | | BR reactor (lactate, nutrient/mineral, and buffer) | | |
|---|---|---|---|---|---|---|
| Day | $ClO_4^-$ (mg/L) | Lactate (mg/L) | $Cl^-$ (mg/L) | $ClO_4^-$ (mg/L) | Lactate (mg/L) | $Cl^-$ (mg/L) |
| 1 | 231.12 | 0 | 0.58 | 0.194 | 1101.13 | 12.09 |
| 2 | 171.4 | 14.52 | 10.19 | 26.39 | 240.18 | 2.63 |
| 3 | 109.45 | 74 | 33.29 | 4.19 | 215.03 | 25.2 |
| 4 | 3.21 | 0.744 | 93.36 | 0 | 208.5 | 23.99 |
| 5 | 0.025 | 0 | 90.42 | 0 | 115.19 | 32.97 |
| 6 | 0 | 14.91 | 72.13 | 0.078 | 114.1 | 41.92 |
| 7 | 0 | 24.74 | 74.94 | 0 | 108.7 | 36.87 |
| 8 | 0 | 12.45 | 90.84 | 0 | 35.97 | 36.77 |
| 9 | 0 | 15.01 | 65.84 | 0 | 29.41 | 38.54 |
| 10 | 0 | 0 | 65.08 | 0 | 5.2 | 37.05 |
| 11 | 0.028 | 0 | 64.43 | 0.034 | 5.55 | 37.4 |
| 12 | 0.019 | 15.17 | 61.15 | 0 | 0.27 | 53.87 |
| 13 | 0 | 13.69 | 60.55 | 0.088 | 0 | 49.22 |
| 15 | 0 | 0 | 56.05 | 0 | 0 | 57.31 |

**Table 6.19**
**BTS-55 membrane: molar ratio of perchlorate to chloride and biodegradation rate (1st cycle)**

| | |
|---|---|
| Moles $ClO_4^-$ / Moles $Cl^-$ | 0.82 |
| Moles $ClO_4^-$ biodegraded/day-m$^2$ of membrane area (for first 6 days) | 0.25 |

**Table 6.20**
**Perchlorate biodegradation by a biofilm immobilized on a BTS-55 membrane (2nd cycle)**

| | DR reactor ($ClO_4$- only) | | | BR reactor (lactate, nutrients/minerals, and buffer) | | |
|---|---|---|---|---|---|---|
| Day | $ClO_4^-$ (mg/L) | Lactate (mg/L) | $Cl^-$ (mg/L) | $ClO_4^-$ (mg/L) | Lactate (mg/L) | $Cl^-$ (mg/L) |
| 1 | 246.96 | 0.94 | 49.89 | 0.19 | 1058.9 | 44.71 |
| 2 | 187.08 | 12.73 | 61.6 | 0.35 | 266.54 | 64.86 |
| 3 | 134.21 | 0 | 93.22 | 0.055 | 204.93 | 72.14 |
| 4 | 117.74 | 0 | 93.25 | 0.024 | 172.05 | 78.91 |
| 5 | 105.58 | 0 | 86.69 | 0.003 | 99.22 | 80.00 |
| 6 | 96.42 | 0 | 95.37 | 0 | 33.66 | 84.65 |
| 7 | 89.24 | 0 | 89.58 | 0 | 0 | 88.47 |
| 8 | 79.82 | 0 | 80.99 | 3.09 | 0 | 69.02 |
| 9 | 67.6 | 0 | 80.3 | 1.43 | 5.44 | 74.7 |
| 10 | 57.88 | 0 | 80.81 | 2.26 | 7.06 | 75.16 |
| 11 | 50.07 | 0 | 82.73 | 3.25 | 9.28 | 79.12 |

**Table 6.21**
**BTS-55 membrane: molar ratio of perchlorate to chloride and biodegradation rate (2nd cycle)**

| | |
|---|---|
| $ClO_4^-$ / $Cl^-$ | 1.03 |
| Moles $ClO_4^-$ biodegraded/day-$m^2$ of membrane area | 0.11 |

**Table 6.22**
**Perchlorate biodegradation by a biofilm immobilized on a BTS-55 membrane (3rd cycle)**

| | DR reactor ($ClO_4$- only) | | | BR reactor (lactate, nutrients/minerals, and buffer) | | |
|---|---|---|---|---|---|---|
| Day | $ClO_4^-$ (mg/L) | Lactate (mg/L) | $Cl^-$ (mg/L) | $ClO_4^-$ (mg/L) | Lactate (mg/L) | $Cl^-$ (mg/L) |
| 1 | 323.72 | 0 | 82.47 | 3.42 | 1037.84 | 71.08 |
| 2 | 248.6 | 0 | 89.58 | 0 | 373.64 | 80.48 |
| 3 | 212.42 | 0 | 100.84 | 0 | 286.16 | 83.39 |
| 4 | 188.42 | 0 | 106.47 | 0 | 247.34 | 92.35 |
| 5 | 159.04 | 0 | 115.48 | 0 | 253.31 | 90.01 |
| 6 | 124.38 | 0 | 139.62 | 0 | 255.1 | 97.72 |
| 7 | 89.94 | 0 | 148.77 | 0 | 260.39 | 98.5 |
| 8 | 49.57 | 0 | 164.75 | 0 | 243.88 | 106.18 |
| 9 | 12.64 | 0 | 177.84 | 0 | 223.41 | 106.92 |
| 10 | 0 | 8.8 | 175.97 | 0 | 240.09 | 114.08 |
| 11 | 0 | 16.6 | 170.58 | 0 | 212.95 | 115.85 |
| 12 | 0 | 0 | 161.42 | 0 | 211.07 | 125.32 |
| 13 | 0 | 11.27 | 157.83 | 0 | 184.19 | 127.91 |
| 14 | 0 | 11.27 | 151.09 | 0 | 192.1 | 128.3 |
| 15 | 0 | 11.27 | 155.12 | 0 | 148.09 | 134.32 |

**Table 6.23**
**BTS-55 membrane: molar ratio of perchlorate to chloride and biodegradation rate (3rd cycle)**

| | |
|---|---|
| $ClO_4^-$ / $Cl^-$ | 0.86 |
| Moles $ClO_4^-$ biodegraded/day-m$^2$ of membrane area (from the first to the 9th days) | 0.22 |

**Table 6.24**
**Estimation of thickness of biofilm immobilized on BTS-55 membrane**

| | |
|---|---|
| Wet weight of membrane +biofilm+o-ring, g | 12.7258 |
| Wet weight of membrane ± o-ring, g | 8.2616 |
| Wet weight of microbes, g | 4.4642 |
| Dry weight of microbes, g | 0.2072 |
| Membrane diameter, cm | 10.00 |
| Membrane surface area, $cm^2$ | 78.50 |
| Density of microbes, $g/cm^3$ | 1.07 |
| Biofilm thickness, μm | 24.7 |

**Table 6.25**
**Perchlorate biodegradation by a biofilm immobilized on a PVDF membrane**

| | DR reactor ($ClO_4$- only) | | | BR reactor (lactate, nutrients/minerals, and buffer) | | |
|---|---|---|---|---|---|---|
| Day | $ClO_4^-$ (mg/L) | Lactate (mg/L) | $Cl^-$ (mg/L) | $ClO_4^-$ (mg/L) | Lactate (mg/L) | $Cl^-$ (mg/L) |
| 1 | 207.68 | 0 | 0 | 0.23 | 761.95 | 5.85 |
| 2 | 188 | 6.4 | 6.63 | 16.57 | 702.06 | 6.53 |
| 3 | 143.84 | 21.58 | 19.55 | 13.57 | 656.7 | 12.18 |
| 4 | 106.68 | 0 | 30.24 | 0.17 | 375.2 | 19.55 |
| 5 | 93.6 | 0 | 34.03 | 0.16 | 105.53 | 21.71 |
| 6 | 83.04 | 0 | 37.74 | 0 | 7.82 | 25.34 |
| 7 | 73.7 | 0 | 36.94 | 3.9 | 0 | 26.04 |
| 8 | 67.88 | 0 | 37.35 | 6.56 | 0 | 27.18 |
| 9 | 59.46 | 0 | 37.66 | 7.65 | 0 | 29.95 |
| 10 | 54.44 | 0 | 39.19 | 9.19 | 0 | 30.15 |
| 11 | 49.62 | 0 | 39.51 | 9.97 | 0 | 32.57 |
| 12 | 46.65 | 0 | 40.43 | 11.77 | 0.27 | 33.53 |
| 13 | 31.86 | 0 | 40.63 | 13.74 | 0 | 34.43 |
| 14 | 40.21 | 0 | 40.45 | 14.37 | 0 | 34.78 |
| 15 | 39.14 | 0 | 36.85 | 15.74 | 0 | 30.96 |
| 16 | 32.15 | 0 | 31.95 | 16.46 | 0 | 28.78 |
| 17 | 31.56 | 0 | 32.21 | 17.55 | 0 | 29.33 |
| 18 | 30.04 | 0 | 33.73 | 18.35 | 0 | 29.7 |

**Table 6.26**
**PVDF membrane: molar ratio of perchlorate to chloride and biodegradation rate (1st cycle)**

| | |
|---|---|
| $ClO_4^-$ / $Cl^-$ | 0.99 |
| Moles $ClO_4^-$ biodegraded/day-m$^2$ of membrane area | 0.06 |

**Table 6.27**
**Summary of relevant parameters calculated for the biofilms immobilized on the BTS-55 and PVDF membranes**

| Membrane | Parameters | 1st cycle | 2nd cycle | 3rd cycle |
|---|---|---|---|---|
| BTS-55 | $ClO_4^-$ / $Cl^-$ | 0.82 | 1.03 | 0.86 |
| | Moles $ClO_4^-$ biodegraded/day-m$^2$ | 0.25 | 0.11 | 0.22 |
| PVDF | $ClO_4^-$ / $Cl^-$ | 0.99 | | |
| | Moles $ClO_4^-$ biodegraded/day-m$^2$ | 0.06 | | |

**Table 6.28**
**Mass balance of VSS in the reactor (1st cycle)**

| Day | VSS measured (mg)[a] | VSS left (mg)[b] | Average VSS (mg) [c] | VSS increase (mg) [d] |
|---|---|---|---|---|
| 0 | 265 | | | |
| 1 | 312 | 249.6 | 288.5 | 47 |
| 2 | 352 | 281.6 | 300.8 | 102.4 |
| 3 | 352 | 281.6 | 316.8 | 70.4 |
| 4 | 284 | 227.2 | 282.8 | 2.4 |
| 5 | 292 | 233.6 | 259.6 | 64.8 |

NOTES:
a. VSS measured = the VSS mass present in the 1L bottle each day.
b. VSS left = mass of biomass present in the bottle after daily withdrawal of 200 ml from the master reactor.
c. Average VSS = the average microbial mass present in the reactor.
d. VSS increase = the daily increase in biomass in the master reactor.

**Table 6.29**
**Mass balance for perchlorate (1st cycle)**

| Initial [a] (mg) | Final [b] (mg) | Biodeg. [c] (mg) | Out [d] (mg) | In [e] (mg) | Th [f] (mg) | MS [g] (mg) | VSS [h] avg. (mg) | (mg) $ClO_4^-$/ (mg) VSS [i] |
|---|---|---|---|---|---|---|---|---|
| 233.12 | 214.58 | 18.54 | 42.916 | 40 | 211.66 | 210.84 | 288.5 | 0.06 |
| 210.84 | 97.02 | 113.82 | 19.404 | 40 | 117.62 | 135.12 | 300.8 | 0.38 |
| 135.12 | 0 | 135.12 | 0 | 40 | 40.00 | 54.08 | 316.8 | 0.43 |
| 54.08 | 0 | 54.08 | 0 | 40 | 40.00 | 55.12 | 282.8 | 0.19 |
| 55.12 | 0 | 55.12 | 0 | 40 | 40.00 | 51.88 | 259.6 | 0.21 |

NOTES:
The initial suspended solids concentration was 265 mg/l
a. Initial = initial mass of perchlorate in the reactor
b. Final = final mass of perchlorate in the reactor after 1 day
c. Biodeg. = mass of perchlorate biodegraded after 1 day
d. Out = mass of perchlorate taken out in the 200 ml volume
e. In = theoretical increase in perchlorate mass when 200 ml of fresh solution was added
f. Th = theoretical total mass of perchlorate in the reactor
g. MS = measured perchlorate mass in the reactor
h. VSS = average biomass in the reactor
i. $mgClO_4^-/mgVSS$ = mg of perchlorate is biodegraded per mg of biomass present in the reactor

**Table 6.30**
**Mass balance lactate (1st cycle)**

| Initial [a] (mg) | Final [b] (mg) | Biodeg. [c] mg | Out [d] (mg) | In [e] (mg) | Th [f] (mg) | MS [g] (mg) | VSS [h] increase (mg) | (mg) VSS/ (mg) lactate [i] |
|---|---|---|---|---|---|---|---|---|
| 1763.52 | 1621.12 | 142.4 | 324.224 | 200 | 1496.90 | 1543.01 | 47 | 0.33 |
| 1543.01 | 1230.64 | 312.37 | 246.128 | 200 | 1184.51 | 1256.48 | 102.4 | 0.33 |
| 1256.48 | 764.7 | 491.78 | 152.94 | 200 | 811.76 | 789.09 | 70.4 | 0.14 |
| 789.09 | 656 | 133.09 | 131.2 | 200 | 724.80 | 692.83 | 2.4 | 0.02 |
| 692.83 | 482 | 210.83 | 96.4 | 200 | 585.60 | 573.82 | 64.8 | 0.31 |

NOTES:
a. Initial = initial mass of lactate in the reactor
b. Final = final mass of lactate in the reactor after 1 day
c. Biodeg. = mass of lactate biodegraded after 1 day
d. Out = mass of lactate taken out in the 200 ml volume
e. In = theoretical increase in perchlorate mass when 200 ml of fresh solution is added
f. Th = theoretical total mass of perchlorate in the reactor
g. MS = measured lactate mass in the reactor
h. VSS increase = the increase of biomass in the reactor after one day
i. mgVSS/mg lac. = mg of biomass generated per mg of lactate consumed

**Table 6.31**
**Mass balance for chloride (1st cycle)**

| Initial [a] (mg) | Final [b] (mg) | Produced [c] (mg) | Out [d] (mg) | In [e] (mg) | Th [f] (mg) | MS [g] (mg) | VSS [h] avg. (mg) | (mg) $Cl^-$/ (mg) VSS [i] |
|---|---|---|---|---|---|---|---|---|
| 204.43 | 186.32 | -18.11 | 37.26 | 0 | 149.06 | 155.97 | 288.5 | -0.06 |
| 155.97 | 194.73 | 38.76 | 38.95 | 0 | 155.78 | 158.71 | 300.8 | 0.13 |
| 158.71 | 203.04 | 44.33 | 40.61 | 0 | 162.43 | 161.66 | 316.8 | 0.14 |
| 161.66 | 176.55 | 14.89 | 35.31 | 0 | 141.24 | 135.68 | 282.8 | 0.05 |
| 135.68 | 151.09 | 15.41 | 30.22 | 0 | 120.87 | 121.12 | 259.6 | 0.06 |

NOTES:
a. Initial = initial mass of chloride in the reactor
b. Final = final mass of chloride in the reactor after 1 day
c. Produced = mass of chloride produced after 1 day
d. Out = mass of chloride taken out in the 200 ml volume
e. In = theoretical increase in chloride mass when 200 ml of fresh solution is added
f. Th = theoretical total mass of chloride in the reactor
g. MS = measured chloride mass in the reactor
h. Avg. VSS = the average biomass present in the reactor
i. mg$Cl^-$/mgVSS = mg of chloride generated per mg of VSS present in the reactor

**Table 6.32**
**The daily molar ratio of perchlorate to chloride (1st cycle)**

| Day | 1 | 2 | 3 | 4 | 5 |
|---|---|---|---|---|---|
| $ClO_4^-$ / $Cl^-$ | -0.37 | 1.05 | 1.09 | 1.30 | 1.28 |

**Table 6.33**
**Mass balance of VSS in the reactor (2nd cycle)**

| Day | VSS measure (mg) | VSS left | Average VSS | VSS increase |
|---|---|---|---|---|
| 0 | 155 | 124 | | |
| 1 | 232 | 185.6 | 193.5 | 77 |
| 2 | 236 | 188.8 | 210.8 | 50.4 |
| 3 | 224 | 179.2 | 206.4 | 35.2 |
| 4 | 244 | 195.2 | 211.6 | 64.8 |
| 5 | 208 | 166.4 | 201.6 | 12.8 |

**Table 6.34**
**Mass balance for perchlorate (2nd cycle)**

| Initial (mg) | Final (mg) | Biodeg. (mg) | Out (mg) | In (mg) | Th (mg) | MS (mg) | VSS avg. (mg) | (mg) $ClO_4^-$/ (mg) VSS |
|---|---|---|---|---|---|---|---|---|
| 214.72 | 0 | 214.72 | 0 | 40 | 40.00 | 36.66 | 193.5 | 1.11 |
| 36.66 | 0 | 36.66 | 0 | 40 | 40.00 | 43.59 | 210.8 | 0.17 |
| 43.59 | 0 | 43.59 | 0 | 40 | 40.00 | 43.76 | 206.4 | 0.21 |
| 43.76 | 0 | 43.76 | 0 | 40 | 40.00 | 46.29 | 211.6 | 0.21 |
| 46.29 | 0 | 46.29 | 0 | 40 | 40.00 | 46.59 | 201.6 | 0.23 |

**Table 6.35**
**Mass balance for lactate (2nd cycle)**

| Initial (mg) | Final (mg) | Biodeg. (mg) | Out (mg) | In (mg) | Th (mg) | MS (mg) | VSS increase (mg) | (mg) VSS/ (mg) lac. |
|---|---|---|---|---|---|---|---|---|
| 1051.4 | 289.42 | 761.98 | 57.884 | 200 | 431.54 | 409.16 | 77 | 0.10 |
| 409.16 | 111.08 | 298.08 | 22.216 | 200 | 288.86 | 214.56 | 50.4 | 0.17 |
| 214.56 | 63.53 | 151.03 | 12.706 | 200 | 250.82 | 221.07 | 35.2 | 0.23 |
| 221.07 | 26.46 | 194.61 | 5.292 | 200 | 221.17 | 198.12 | 64.8 | 0.33 |
| 198.12 | 0 | 198.12 | 0 | 200 | 200.00 | 221.77 | 12.8 | 0.07 |

**Table 6.36**
**Mass balance for chloride (2nd cycle)**

| Initial (mg) | Final (mg) | Produced (mg) | Out (mg) | In (mg) | Th (mg) | MS (mg) | VSS avg. (mg) | (mg) $Cl^-$/ (mg) VSS/ |
|---|---|---|---|---|---|---|---|---|
| 45.22 | 115.87 | 70.65 | 23.17 | 0 | 92.70 | 91.53 | 193.5 | 0.37 |
| 91.53 | 108.32 | 16.79 | 21.66 | 0 | 86.66 | 80.14 | 210.8 | 0.08 |
| 80.14 | 97.96 | 17.82 | 19.59 | 0 | 78.37 | 78.82 | 206.4 | 0.09 |
| 78.82 | 92.38 | 13.56 | 18.48 | 0 | 73.90 | 72.46 | 211.6 | 0.06 |
| 72.46 | 90.2 | 17.74 | 18.04 | 0 | 72.16 | 69.43 | 201.6 | 0.09 |

**Table 6.37**
**The daily molar ratio of perchlorate to chloride (2nd cycle)**

| Day | 1 | 2 | 3 | 4 | 5 |
|---|---|---|---|---|---|
| $ClO_4^-$ / $Cl^-$ | 1.08 | 0.78 | 0.87 | 1.15 | 0.93 |

**Table 6.38**
**Perchlorate biodegradation in perchlorate limited bottles**

| Time (day) | KC1 (10ppm) | KC2 (20ppm) | KC3 (60ppm) | KC4 (100ppm) | KC5 (200ppm) |
|---|---|---|---|---|---|
| | $ClO_4^-$ | $ClO_4^-$ | $ClO_4^-$ | $ClO_4^-$ | $ClO_4^-$ |
| 0 | 13.09 | 21.9 | 63.65 | 108.72 | 232.2 |
| 1 | 11.42 | 20.65 | 63.09 | 108.14 | 239.78 |
| 3 | 0 | 0 | 17.54 | 104.14 | 222.85 |
| 5 | 0 | 0 | 0 | 52.17 | 166.68 |
| 6 | 0 | 0 | 0 | 0 | 93.34 |
| 7 | 0 | 0 | 0 | 0 | 0 |

**Table 6.39**
**Perchlorate, lactate, and chloride concentration of preliminary kinetics study (KC1)**

| Time (day) | Perchlorate (mg/L) | Lactate (mg/L) | Chloride (mg/L) |
|---|---|---|---|
| 0 | 13.09 | 897.07 | 0.38 |
| 1 | 11.42 | 816.62 | 0.63 |
| 3 | 0 | 95.57 | 3.74 |
| 5 | 0 | 104.39 | 3.48 |
| 6 | 0 | 95 | 5.43 |
| 7 | 0 | 81.78 | 5.64 |
| 8 | 0 | 87.55 | 5.53 |
| 9 | 0 | 87.03 | 6.64 |

**Table 6.40**
**Perchlorate, lactate, and chloride concentration in preliminary kinetics study (KC2)**

| Time (day) | Perchlorate (mg/L) | Lactate (mg/L) | Chloride (mg/L) |
|---|---|---|---|
| 0 | 21.9 | 870.18 | 0.39 |
| 1 | 20.65 | N/A | 0.2 |
| 3 | 0 | 93.43 | 6.7 |
| 5 | 0 | 77.22 | 6.46 |
| 6 | 0 | 72.7 | 8.01 |
| 7 | 0 | 67.48 | 7.26 |
| 8 | 0 | 41.64 | 8.11 |
| 9 | 0 | 43.89 | 8.62 |

**Table 6.41**
**Perchlorate, lactate, and chloride concentration in preliminary kinetics study (KC3)**

| Time (day) | Perchlorate (mg/L) | Lactate (mg/L) | Chloride (mg/L) |
|---|---|---|---|
| 0 | 63.65 | 972.58 | 0.35 |
| 1 | 63.09 | 784.61 | 0.45 |
| 3 | 17.54 | 103.38 | 11.46 |
| 5 | 0 | 110.83 | 17.11 |
| 6 | 0 | 105.94 | 19.01 |
| 7 | 0 | 95.23 | 17.52 |
| 8 | 0 | N/A | 17.62 |
| 9 | 0 | 74.59 | 15.96 |

**Table 6.42**
**Perchlorate, lactate, and chloride concentration in preliminary kinetics study (KC4)**

| Time (day) | Perchlorate (mg/L) | Lactate (mg/L) | Chloride (mg/L) |
|---|---|---|---|
| 0 | 108.72 | 1000 | 0 |
| 1 | 108.14 | N/A | 0 |
| 3 | 104.14 | 690.11 | 3.1 |
| 5 | 52.17 | N/A | 20.77 |
| 6 | 0 | 91.02 | 36.45 |
| 7 | 0 | N/A | 34.73 |
| 8 | 0 | 71.58 | 25.29 |
| 9 | 0 | 73.28 | 33.42 |

**Table 6.43**
**Perchlorate, lactate, and chloride concentration in preliminary kinetics study (KC5)**

| Time (day) | Perchlorate (mg/L) | Lactate (mg/L) | Chloride (mg/L) |
|---|---|---|---|
| 0 | 232.2 | 711.29 | 0.33 |
| 1 | 239.78 | N/A | 0 |
| 3 | 222.85 | 664.01 | 1.24 |
| 5 | 166.68 | 94.91 | 15.93 |
| 6 | 93.34 | 90.56 | 38.05 |
| 7 | 0 | N/A | 72.47 |
| 8 | 0 | N/A | 50.48 |
| 9 | 0 | 65.88 | 71.22 |

**Table 6.44**
**Perchlorate, lactate, and chloride concentration in preliminary kinetics study (KL5)**

| Time (day) | Perchlorate (mg/L) | Lactate (mg/L) | Chloride (mg/L) |
|---|---|---|---|
| 0 | 103.34 | 314.51 | 0.054 |
| 1 | 102.72 | 0 | 0.39 |
| 3 | 81.52 | 0 | 9.78 |
| 5 | 28.73 | 37.29 | 23.3 |
| 7 | 0 | 0 | 36.13 |
| 9 | 0 | 0 | 32.9 |
| 17 | 0 | 0 | 38.11 |

**Table 6.45**
**Perchlorate concentration of preliminary kinetics study in lactate-limited bottles (KL1-KL4)**

| Time | Perchlorate concentration (mg/L) | | | |
|---|---|---|---|---|
| (day) | KL1 | KL2 | KL3 | KL4 |
| 0 | 89.31 | 84.50 | 87.50 | 84.51 |
| 9 | 87.80 | 87.60 | 86.44 | 83.76 |
| 12 | 90.76 | 85.56 | 86.62 | 85.45 |
| 17 | 88.4 | 85.20 | 84.38 | 85.00 |

**Table 6.46**
**The characteristics of three isolates of "BALI" mixed culture**

| Isolate name | Characteristics |
|---|---|
| KA2-1 | • Colony characteristics: Small, translucent, shiny, entire, convex<br>• Microscopy: Gram-negative rod shaped bacterium<br>• Size (average of 10 cells): 3.7 μm long × 1 μm wide<br>• 16S ribosomal RNA sequencing: *Dechlorisoma* sp (SDGM); 97% match |
| KL1-A | • Colony characteristics: Small, translucent, shiny, entire, convex<br>• Microscopy: Gram-negative rod shaped bacterium<br>• Size (average of 10 cells): 3.8 μm long × 1 μm wide<br>• 16S ribosomal RNA sequencing: *Dechlorisoma* sp (SDGM); 92% match<br>• 16S ribosomal RNA sequencing from Midi labs: *Dechlorisoma suilla*; 98% match |
| RAS-CCSD | • Colony characteristics: Small, white, opaque<br>• Microscopy: Gram-negative rod shaped bacterium<br>• Size: (average of 10 cells): 3.8 μm long × 1 μm wide<br>• 16S ribosomal RNA sequencing: *Dechlorisoma* sp (Iso2); 98% match |

**Table 6.47**
**The interference of nitrate on perchlorate biodegradation (batch testing)**

| | 0 NIT | 10 NIT | | 30 NIT | | 60 NIT | |
|---|---|---|---|---|---|---|---|
| Time (hr) | $ClO_4^-$ (ppm) | $ClO_4^-$ (ppm) | $NO_3^-$ (ppm) | $ClO_4^-$ (ppm) | $NO_3^-$ (ppm) | $ClO_4^-$ (ppm) | $NO_3^-$ (ppm) |
| 0 | 108.36 | 100.23 | 10.2 | 103 | 31.95 | 105.34 | 64.9 |
| 2.5 | 104.7 | 99.12 | 9.82 | 101.42 | 32.06 | 101.72 | 68.16 |
| 22 | 96.02 | 92.07 | 0.61 | 95.61 | 0.61 | 99 | 25.23 |
| 34 | 0 | 0 | 0 | 0 | 0 | 0 | 0 |

**Table 6.48**
**The interference of nitrate on perchlorate biodegradation (test 1)**

| | DR reactor ($ClO_4^-$ and $NO_3^-$) | | | | BR reactor (lactate, nutrient/minerals, and buffer) | |
|---|---|---|---|---|---|---|
| Time (hour) | $ClO_4^-$ (mg/L) | $NO_3^-$ (mg/L) | mg $ClO_4^-$ biodegraded/hr | mg $NO_3^-$ biodegraded/hr | $ClO_4^-$ (mg/L) | $NO_3^-$ (mg/L) |
| **0** | **50.65** | **58.50** | 0.00 | 0.00 | 0.00 | 0.00 |
| 4 | 46.70 | 55.50 | 4.94 | 3.75 | 0.00 | 0.83 |
| 17 | 42.39 | 27.34 | 1.66 | 10.83 | N/A | N/A |
| 20 | 42.30 | 21.03 | 0.15 | 10.52 | N/A | N/A |
| 24 | 39.96 | 16.90 | 2.93 | 5.16 | N/A | N/A |
| 26 | 38.32 | 14.58 | 4.10 | 5.80 | N/A | N/A |
| 29 | 35.19 | 12.51 | 5.22 | 3.45 | N/A | N/A |
| 31 | 34.26 | 11.50 | 2.33 | 2.53 | N/A | N/A |
| 40.5 | 28.70 | 9.31 | 2.93 | 1.15 | N/A | N/A |
| **48** | **26.17** | **8.71** | 1.69 | 0.40 | N/A | N/A |
| 54 | 24.33 | 7.90 | 1.53 | 0.68 | 0.00 | 0.00 |
| 66 | 21.26 | 6.67 | 1.28 | 0.51 | 2.03 | 0.65 |
| 73 | 19.93 | 6.00 | 0.95 | 0.48 | N/A | 0.73 |
| 78.5 | 19.17 | 5.72 | 0.69 | 0.25 | 3.59 | 1.06 |
| 89.5 | 18.37 | 4.99 | 0.36 | 0.33 | 5.20 | 1.39 |
| 97 | 17.34 | 4.29 | 0.69 | 0.47 | N/A | 1.34 |
| 102.25 | 15.57 | 3.54 | 1.69 | 0.71 | 6.45 | 1.19 |
| 113.5 | 13.24 | 2.88 | 1.04 | 0.29 | 7.15 | 0.78 |
| 126 | N/A | 2.14 | N/A | 0.30 | N/A | 0.83 |

**Table 6.49**
**The interference of nitrate on perchlorate biodegradation (test 2)**

| Time (hour) | DR reactor ($ClO_4^-$ and $NO_3^-$) $ClO_4^-$ (mg/L) | $NO_3^-$ (mg/L) | % of $ClO_4^-$ biodegraded/hr | % of $NO_3^-$ biodegraded/hr | BR reactor (lactate, nutrient/minerals, and buffer) $ClO_4^-$ (mg/L) | $NO_3^-$ (mg/L) |
|---|---|---|---|---|---|---|
| **0** | **51.58** | **10.38** | 0.00 | 0.00 | 0.00 | 0.00 |
| **4** | **49.08** | **9.80** | 1.21 | 1.40 | 0.00 | 0.00 |
| 17 | 43.45 | 0.00 | 0.88 | 7.69 | 0.00 | 0.00 |
| 20 | 40.29 | 0.00 | 2.42 | 0 | 0.00 | 0.00 |
| 24 | 39.00 | 0.00 | 0.80 | | 0.00 | 0.00 |
| 26 | 38.02 | 0.00 | 1.26 | | 0.00 | 0.00 |
| 29 | 35.35 | 0.00 | 2.34 | | 0.00 | 0.00 |
| 31 | 34.08 | 0.00 | 1.80 | | 0.00 | 0.00 |
| 40.5 | 26.51 | 0.00 | 2.34 | | 0.00 | 0.00 |
| 48 | 21.68 | 0.00 | 2.43 | | 0.00 | 0.00 |
| 54 | 18.45 | 0.00 | 2.48 | | 0.00 | 0.00 |
| 66 | 14.73 | 0.00 | 1.68 | | 0.00 | 0.00 |
| 73 | 13.40 | 0.00 | 1.29 | | 0.00 | 0.00 |
| 78.5 | 12.50 | 0.00 | 1.22 | | 0.45 | 0.00 |
| 89.5 | 11.24 | 0.00 | 0.92 | | 1.63 | 0.00 |
| 97 | 10.38 | 0.00 | 1.02 | | 2.03 | 0.00 |
| 102.25 | 9.19 | 0.00 | 2.18 | | 2.46 | 0.00 |
| 113.5 | 7.62 | 0.00 | 1.52 | | 2.67 | 0.00 |

**Table 6.50**
**The interference of nitrate on perchlorate biodegradation (test 3)**

| | DR reactor ($ClO_4^-$ and $NO_3^-$) | | | | BR reactor (lactate, nutrient/minerals, and buffer) | | |
|---|---|---|---|---|---|---|---|
| Time (hour) | $ClO_4^-$ (mg/L) | $NO_3$- (mg/L) | % of $ClO_4^-$ Biodegraded/hr | % of $NO_3^-$ Biodegraded/hr | $ClO_4^-$ (mg/L) | $NO_3^-$ (mg/L) | Lactate (mg/L) |
| 0 | 1.03 | 54.00 | 0.00 | 0.00 | 0.00 | 0.00 | 4.99 |
| 2 | 0.99 | 52.92 | 1.94 | 1.00 | 0.00 | 0.00 | |
| 4 | 1.00 | 51.03 | -0.51 | 1.79 | 0.00 | 0.00 | |
| 6 | 0.98 | 50.50 | 1.00 | 0.52 | 0.00 | 0.00 | |
| 8 | 0.94 | 49.88 | 2.04 | 0.61 | 0.00 | 0.00 | |
| 9.5 | 0.92 | 47.07 | 1.42 | 3.76 | 0.00 | 0.00 | |
| 22.5 | 0.88 | 43.18 | 0.33 | 0.64 | 0.00 | 0.00 | |
| **25** | **0.84** | **43.37** | 1.82 | -0.18 | 0.00 | 0.00 | **0,281.93** |
| 27 | 0.79 | 40.20 | 2.98 | 3.65 | 0.00 | 0.00 | |
| 29 | 0.80 | 39.92 | -0.63 | 0.35 | 0.00 | 0.00 | |
| 31 | 0.76 | 37.90 | 2.50 | 2.53 | 0.00 | 0.00 | |
| **33** | **0.73** | **35.13** | 1.97 | 3.65 | 0.00 | 0.00 | **143.99** |
| 46.5 | 0.35 | 19.21 | 3.86 | 3.36 | 0.00 | 0.00 | |
| 49 | 0.35 | 17.46 | 0.00 | 3.64 | 0.00 | 0.00 | |
| 51 | 0.35 | 14.18 | 0.00 | 9.39 | 0.00 | 0.00 | |
| 53 | 0.31 | 13.24 | 5.71 | 3.31 | 0.00 | 0.00 | |
| 55 | 0.31 | 11.44 | 0.00 | 6.80 | 0.00 | 0.00 | |
| 56.5 | 0.25 | 10.59 | 12.90 | 4.95 | 0.00 | 0.00 | 101.89 |
| 70.5 | 0.12 | 3.81 | 3.71 | 4.57 | 0.00 | 0.00 | 53.69 |
| 72 | 0.1 | 3.56 | 11.11 | 4.37 | 0.00 | 0.00 | |
| 74 | 0.093 | 3.04 | 3.50 | 7.30 | 0.00 | 0.00 | |
| 75 | 0.086 | 2.88 | 7.53 | 5.26 | 0.00 | 0.00 | |
| 76 | 0.082 | 2.49 | 4.65 | 13.54 | 0.00 | 0.00 | |
| 77 | 0.075 | 2.44 | 8.54 | 2.01 | 0.00 | 0.00 | |
| 78.5 | 0.068 | 2.24 | 6.22 | 5.46 | 0.00 | 0.00 | |
| 80 | 0.064 | 2.00 | 3.92 | 7.14 | 0.00 | 0.00 | 43.83 |
| 81 | 0.057 | 1.83 | 10.94 | 8.50 | 0.00 | 0.00 | |
| 94.5 | 0.027 | 0.91 | 3.90 | 3.72 | 0.00 | 0.00 | 15.07 |
| 97 | 0.025 | 0.72 | 2.96 | 8.35 | 0.00 | 0.00 | 0 |

**Table 6.51**
**The interference of nitrate on perchlorate biodegradation (test 4)**

| | DR reactor ($ClO_4^-$ and $NO_3^-$) | | | | BR Reactor (lactate, nutrient/minerals, and buffer) | | |
|---|---|---|---|---|---|---|---|
| Time (hour) | $ClO_4^-$ (mg/L) | $NO_3^-$ (mg/L) | % of $ClO_4^-$ biodegraded/hr | % of $NO_3^-$ biodegraded/hr | $ClO_4^-$ (mg/L) | $NO_3^-$ (mg/L) | Lactate (mg/L) |
| 0 | 0.96 | 11.09 | 0.00 | 0.00 | 0.00 | 0.00 | 6.73 |
| 2 | 0.94 | 9.50 | 1.04 | 7.17 | 0.00 | 0.00 | |
| 4 | 0.91 | 9.41 | 1.60 | 0.47 | 0.00 | 0.00 | |
| 6 | 0.92 | 8.93 | -0.55 | 2.55 | 0.00 | 0.00 | |
| 8 | 0.90 | 8.91 | 1.09 | 0.11 | 0.00 | 0.00 | |
| 9.5 | 0.89 | 8.86 | 0.74 | 0.37 | 0.00 | 0.00 | |
| 22.5 | 0.82 | 7.87 | 0.61 | 0.86 | 0.00 | 0.00 | |
| **25** | **0.81** | **7.74** | 0.49 | 0.66 | 0.00 | 0.00 | **0,52.17** |
| 27 | 0.73 | 7.75 | 4.94 | -0.06 | 0.00 | 0.00 | |
| 29 | 0.73 | 7.43 | 0.00 | 2.06 | 0.00 | 0.00 | |
| 31 | 0.71 | 6.83 | 1.37 | 4.04 | 0.00 | 0.00 | |
| 33 | 0.67 | 6.29 | 2.82 | 3.95 | 0.00 | 0.00 | |
| 46.5 | 0.44 | 4.20 | 2.54 | 2.46 | 0.00 | 0.00 | |
| **49** | **0.43** | **4.06** | 0.91 | 1.33 | 0.00 | 0.00 | **5.64** |
| 51 | 0.44 | 4.03 | -1.16 | 0.37 | 0.00 | 0.00 | |
| 53 | 0.44 | 3.98 | 0.00 | 0.62 | 0.00 | 0.00 | 1.92 |
| 55 | 0.43 | 3.83 | 1.14 | 1.88 | 0.00 | 0.00 | 1.69 |
| 56.5 | 0.42 | 3.75 | 1.55 | 1.39 | 0.015 | 0.00 | 0 |
| 70.5 | 0.37 | 3.30 | 0.85 | 0.86 | 0.064 | 0.00 | |
| 72 | 0.36 | 3.21 | 1.80 | 1.82 | | 0.00 | |
| 74 | 0.35 | 3.10 | 1.39 | 1.71 | 0.076 | 1.16 | |
| 77 | 0.35 | 2.94 | 0.00 | 1.72 | 0.081 | 1.07 | |
| 80 | 0.35 | 2.93 | 0.00 | 0.11 | 0.090 | 1.07 | |
| 94.5 | 0.31 | 2.51 | 0.79 | 0.99 | 0.120 | 1.17 | |
| 99 | 0.3 | 2.48 | 0.72 | 0.27 | 0.130 | 1.32 | |
| 104 | 0.29 | 2.43 | 0.67 | 0.40 | 0.140 | 1.62 | |
| 119 | 0.27 | 2.07 | 0.46 | 0.99 | 0.160 | 1.35 | 0 |

**Table 6.52**
**Summary of nitrate interference on perchlorate biodegradation**

| Initial perchlorate concentration (mg/L) | Initial nitrate concentration (mg/L) | Perchlorate biodegradation rate (mg/hr) | Nitrate biodegradation rate (mg/hr) | Ratio of nitrate to perchlorate biodegradation |
|---|---|---|---|---|
| 50.65 | 58.50 | 0.51 | 1.04 | 2.04 |
| 51.58 | 10.38 | 0.48 | 0.99 | 2.06 |
| 0.96 | 11.09 | 0.016 | 0.15 | 9.4 |
| 1.03 | 54.0 | 0.005 | 0.38 | 76 |

**Table 6.53**
**Interference of sulfate on perchlorate biodegradation (batch testing)**

| | $SO_4^{2-}$-40 test | | $SO_4^{2-}$-60 test | | $SO_4^{2-}$-140 test | | $SO_4^{2-}$-600 test | |
|---|---|---|---|---|---|---|---|---|
| Time (hour) | $ClO_4^-$ (mg/L) | $SO_4^{2-}$ (mg/L | $ClO_4^-$ (mg/L) | $SO_4^{2-}$ (mg/L) | $ClO_4^-$ (mg/L) | $SO_4^{2-}$ (mg/L) | $ClO_4^-$ (mg/L) | $SO_4^{2-}$ (mg/L) |
| 0 | 10.79 | 41.68 | 10.9 | 59.72 | 11.03 | 146.06 | 11.11 | 617.5 |
| 16 | 2.97 | 44.12 | 1.33 | 60.07 | 0 | 144.4 | 0 | 605.79 |
| 25 | 0 | 52.57 | 0 | 59.62 | 0 | 139.41 | 0 | 606.36 |
| 40 | | N/A | | 59.95 | | 140.36 | | 601.91 |
| 69 | | 41.77 | | 66.89 | | 139.66 | | 614.54 |
| 117 | | 40.8 | | 60.39 | | 138.4 | | 593.03 |
| Reduced % | 100% | 2.1% | 100% | ~ 0 % | 100% | 5.2% | 100% | 4.0% |

**Table 6.54**
**The interference of sulfate on perchlorate biodegradation (test 1)**

| | DR reactor ($ClO_4^-$ and $SO_4^{2-}$) | | | | | BR reactor (lactate, nutrient/minerals, and buffer) | | |
|---|---|---|---|---|---|---|---|---|
| Time (hour) | $ClO_4^-$ (mg/L) | $SO_4^{2-}$ (mg/L) | Lactate (mg/L) | % of $ClO_4^-$ biodegraded/hr | % of $SO_4^{2-}$ biodegraded/hr | $ClO_4^-$ (mg/L) | $SO_4^{2-}$ (mg/L) | Lactate (mg/L) |
| 0.00 | **10.78** | **13.60** | 0.00 | 0.00 | 0.00 | **0.02** | **46.49** | 224.40 |
| 8.00 | 10.22 | 11.80 | 14.71 | 0.65 | 1.65 | | 63.42 | 225.08 |
| 21.25 | 8.79 | 16.18 | 34.88 | 1.06 | -2.80 | 0.06 | 51.37 | 201.51 |
| 27.00 | **7.42** | **18.71** | 32.47 | 2.71 | -2.72 | 0.00 | **40.00** | 136.19 |
| 45.50 | 0.00 | **13.66** | 16.65 | 5.41 | 1.46 | 0.00 | **27.01** | 82.20 |
| 54.00 | 0.00 | 9.33 | 11.46 | | 3.73 | 0.00 | 13.68 | 67.25 |
| 70.00 | 0.00 | **10.15** | 0.00 | | -0.55 | 0.00 | **17.58** | 19.10 |
| 79.00 | 0.00 | 18.80 | 0.00 | | -9.47 | 0.00 | 14.09 | 0.00 |
| 94.00 | 0.00 | 18.09 | 0.00 | | 0.25 | 0.00 | 15.85 | 0.00 |
| 118.00 | 0.00 | 12.63 | 0.00 | | 1.26 | 0.00 | 13.19 | 0.00 |
| 142.00 | 0.00 | 12.71 | 0.00 | | -0.03 | 0.00 | 13.62 | 0.00 |
| 165.75 | 0.00 | 13.10 | 0.00 | | -0.13 | 0.00 | 14.06 | 0.00 |

**Table 6.55**
**The interference of sulfate on perchlorate biodegradation (test 2)**

| | DR reactor ($ClO_4^-$ and $SO_4^{2-}$) | | | | | BR reactor (lactate, nutrient/minerals, and buffer) | | |
|---|---|---|---|---|---|---|---|---|
| Time (hour) | $ClO_4^-$ (mg/L) | $SO_4^{2-}$ (mg/L) | Lactate (mg/L) | % of $ClO_4^-$ biodegraded/hr | % of $SO_4^{2-}$ biodegraded/hr | $ClO_4^-$ (mg/L) | $SO_4^{2-}$ (mg/L) | Lactate (mg/L) |
| 0.00 | **9.69** | **213.30** | 0.00 | 0.00 | 0.00 | 0.00 | **43.37** | 1119.84 |
| 8.00 | 8.90 | 206.05 | 110.34 | 1.02 | 0.42 | N/A | 52.57 | 1061.31 |
| 21.25 | 8.01 | 209.27 | 228.80 | 0.75 | -0.12 | 0.80 | 68.07 | 1020.54 |
| 27.00 | **7.09** | **208.34** | 222.61 | 2.00 | 0.08 | 0.00 | **62.71** | 937.34 |
| 45.50 | 0.00 | 173.70 | 235.43 | 5.41 | 0.90 | 0.00 | 64.89 | 649.77 |
| 54.00 | 0.00 | **158.43** | 303.53 | | 1.03 | 0.00 | **53.68** | 529.96 |
| 70.00 | 0.00 | 129.83 | 216.50 | | 1.13 | 0.00 | 51.46 | 220.70 |
| 79.00 | 0.00 | 113.11 | 146.22 | | 1.43 | 0.00 | 54.01 | 165.55 |
| 94.00 | 0.00 | 118.75 | 129.61 | | -0.33 | 0.00 | 57.16 | 145.55 |
| 118.00 | 0.00 | 108.34 | 78.14 | | 0.37 | 0.00 | 61.72 | 100.77 |
| 142.00 | 0.00 | 118.23 | 88.24 | | -0.38 | 0.00 | 72.85 | 90.58 |
| 165.75 | 0.00 | **112.21** | 54.01 | | 0.21 | 0.00 | **74.54** | 12.56 |

**Table 6.56**
**The interference of sulfate on perchlorate biodegradation (test 3)**

| | DR reactor ($ClO_4^-$ and $SO_4^{2-}$) | | | | | BR reactor (lactate, nutrient, and buffer) | | |
|---|---|---|---|---|---|---|---|---|
| Time (hour) | $ClO_4^-$ (μg/L) | $SO_4^{2-}$ (mg/L) | Lactate (mg/L) | % of $ClO_4^-$ biodegraded/hr | % of $SO_4^{2-}$ biodegraded/hr | $ClO_4^-$ (μg/L) | $SO_4^{2-}$ (mg/L) | Lactate (mg/L) |
| 0 | **116.63** | **49.77** | 0.00 | 0.00 | 0.00 | 0.00 | **42.15** | 423.92 |
| 2.5 | 109.16 | 48.60 | 7.96 | 2.56 | 0.94 | 0.00 | 42.93 | 432.78 |
| 6 | 93.01 | 49.00 | 20.38 | 4.23 | -0.24 | N/A | N/A | N/A |
| 16.5 | **40.04** | **52.38** | 37.92 | 5.42 | -0.66 | 0.00 | **41.13** | 331.94 |
| 23.5 | 0.00 | **45.60** | 40.83 | 14.29 | 1.85 | 0.00 | **37.72** | 276.24 |
| 30 | 13.12 | 45.13 | 38.34 | N/A | 0.16 | N/A | N/A | N/A |
| 40.5 | 0.00 | 40.65 | 39.55 | 9.52 | 0.95 | 0.00 | 34.92 | 124.74 |
| 64.5 | 0.00 | 42.95 | 33.07 | N/A | -0.24 | 0.00 | 34.90 | 60.01 |
| 71 | 0.00 | **49.15** | 24.76 | N/A | -2.22 | 0.00 | **32.89** | 14.98 |
| 88.5 | 0.00 | 48.33 | 0.97 | | 0.10 | N/A | N/A | N/A |
| 95 | 0.00 | 48.26 | 0 | | 0.02 | 0.00 | 36.27 | 0.00 |
| 113 | 0 | 46.46 | 0 | | 0.21 | 0.00 | 39.52 | 1.43 |
| 144 | 0 | 47.31 | 0 | | -0.06 | 0.00 | 41.52 | 0.00 |
| 144.5 | N/A | N/A | N/A | | N/A | 0 | 36.79 | 257.5 (add) |
| 168.5 | 0 | 42.85 | 0 | | 0.38 | 0 | 36.52 | 27.48 |
| 186 | 0 | 41.91 | 0 | | 0.13 | N/A | 37.57 | 0 |

**Table 6.57**
**The interference of sulfate on perchlorate biodegradation (test 4)**

| | DR reactor ($ClO_4^-$ and $SO_4^{2-}$) | | | | | BR reactor (lactate, nutrient, and buffer) | | |
|---|---|---|---|---|---|---|---|---|
| Time (hour) | $ClO_4^-$ (mg/L) | $SO_4^{2-}$ (mg/L) | Lactate (mg/L) | % of $ClO_4^-$ biodegraded/hr | % of $SO_4^{2-}$ biodegraded/hr | $ClO_4^-$ (mg/L) | $SO_4^{2-}$ (mg/L) | Lactate (mg/L) |
| 0 | **50.00** | **47.55** | 0.00 | 0.00 | 0.00 | **0.54** | **40.41** | 630.86 |
| 6 | 46.96 | 50.92 | 23.32 | 1.01 | -1.18 | 2.10 | 37.08 | 612.90 |
| 16.5 | 38.30 | 46.17 | 37.63 | 1.76 | 0.89 | 1.44 | 32.42 | 531.78 |
| 23.5 | **12.62** | **40.43** | 21.93 | 9.58 | 1.78 | 0.00 | **29.27** | 489.46 |
| 30 | 0.00 | **38.46** | 17.48 | 15.38 | 0.75 | 0.00 | **22.29** | 431.52 |
| 40.5 | 0.00 | 42.90 | 22.97 | | -1.10 | | | |
| 47.5 | 0.00 | 39.77 | 33.97 | | 1.04 | 0.00 | 14.90 | 185.61 |
| 71 | 0.00 | **45.18** | 32.18 | | -0.58 | 0.00 | **12.52** | 52.90 |
| 95 | 0.00 | 41.08 | 34.46 | | 0.38 | 0.00 | 16.95 | 0.00 |
| 113 | 0.00 | 42.7 | 17.78 | | -0.22 | 0.00 | 19.67 | 1.83 |
| 144 | 0 | 40.31 | 0 | | 0.18 | | | |
| 144.5 | | | | | | 0 | 24.16 | 238.68 |
| 168.5 | 0 | 38.22 | 0 | | 0.21 | | 2.05 | 18.71 |
| 186 | 0 | 39.18 | 0 | | -0.14 | | 10.23 | 0.83 |

**Table 6.58**
**Biodegradation rate for perchlorate in the presence of sulfate**

| | Test 1 | | Test 2 | | Test 3 | | Test 4 | |
|---|---|---|---|---|---|---|---|---|
| | $ClO_4^-$ | $SO_4^{2-}$ | $ClO_4^-$ | $SO_4^{2-}$ | $ClO_4^-$ | $SO_4^{2-}$ | $ClO_4^-$ | $SO_4^{2-}$ |
| Initial concentration (mg/L) | 10 | 10 | 10 | 200 | 0.1 | 50 | 50 | 50 |
| Biodegradation rate (mg/hr) (before all $ClO_4^-$ was biodegraded) | 0.13 | 0.05 | 0.1 | 0.00 | 0.005 | 0.00 | 1.6 | 0.78 |
| Biodegradation rate (mg/hr) (after all $ClO_4^-$ was biodegraded) | | 0.51 | | 0.23 | | 0.03 | | 0.074 |

**Table 6.59**
**Specific growth rate under different salinity conditions**

| Salinity concentration (%) | μ ($day^{-1}$) | $R^2$ | Fraction of μ as compared to 0% salinity |
|---|---|---|---|
| 0% | 0.26 | 0.93 | 100% |
| 0.5% | 0.19 | 0.78 | 73% |
| 1.0% | 0.096 | 0.61 | 37% |
| 1.5% | 0.072 | 0.544 | 28% |
| 2.0% | 0.096 | 0.69 | 37% |
| 2.5% | 0.096 | 0.70 | 37% |
| 3.0% | 0.072 | 0.72 | 28% |
| 3.5% | 0.048 | 0.67 | 18% |
| 4.0% | No growth | | |
| > 5% | No growth | | |

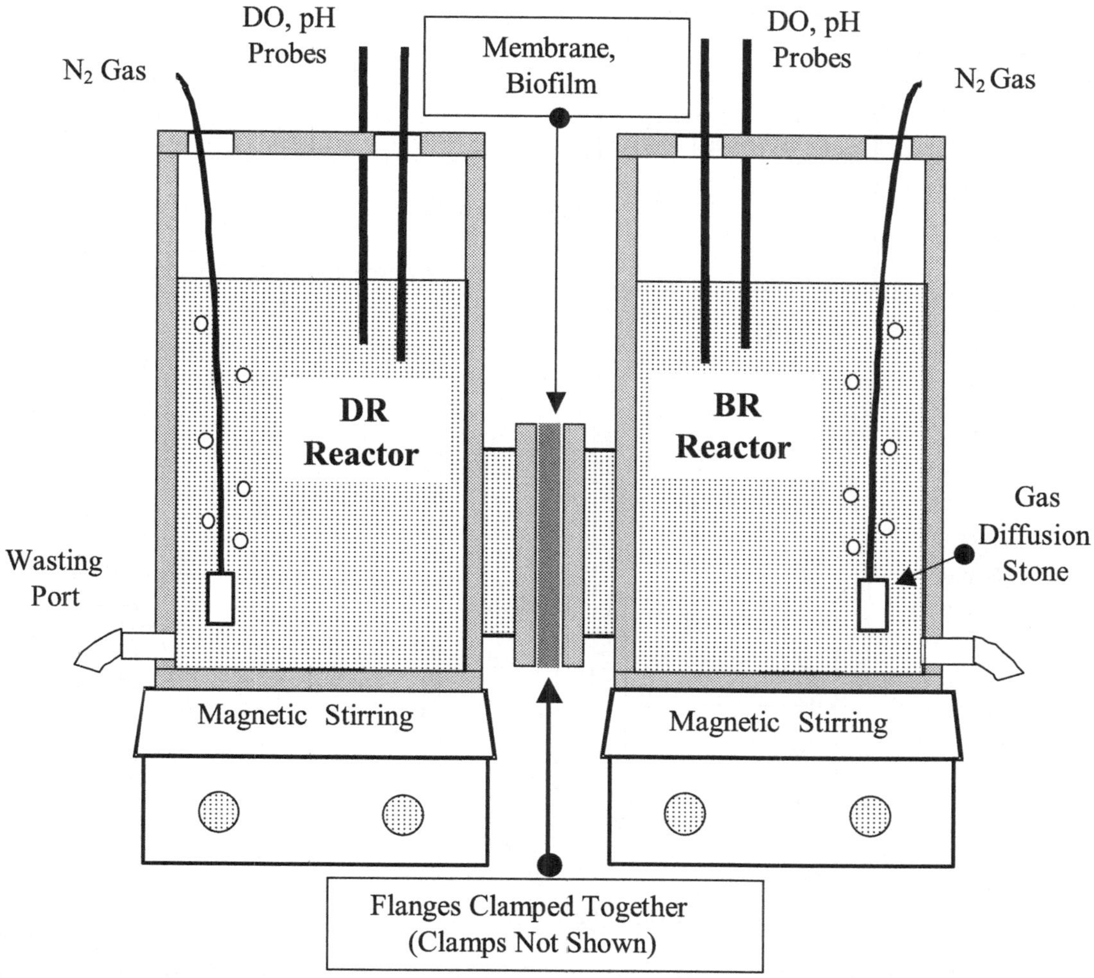


**Figure 6.1 Schematic representation of the membrane-immobilized biofilm reactor**

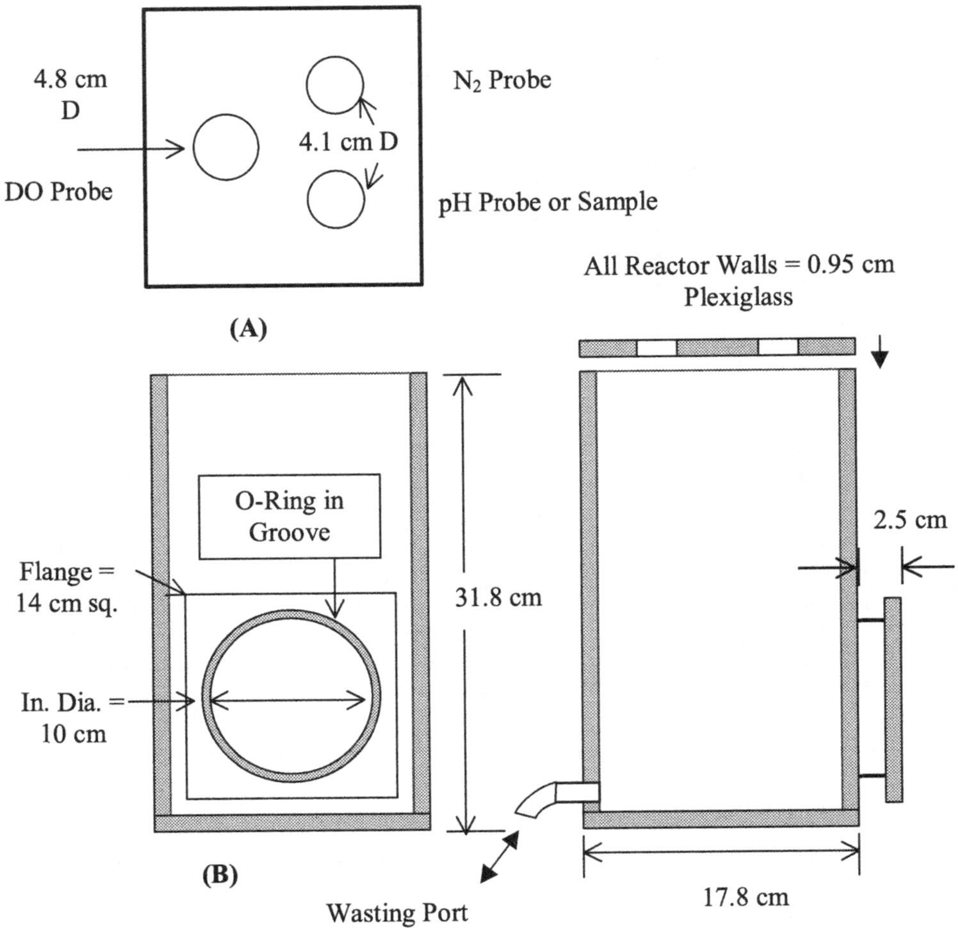


**Figure 6.2 Dimension of the membrane-immobilized biofilm reactor showing: (A) Top cover with three openings for sampling and monitoring; (B) Side view showing connecting plate and membrane opening**

**Figure 6.3 Biofilm established on the microporous membrane at UNLV**

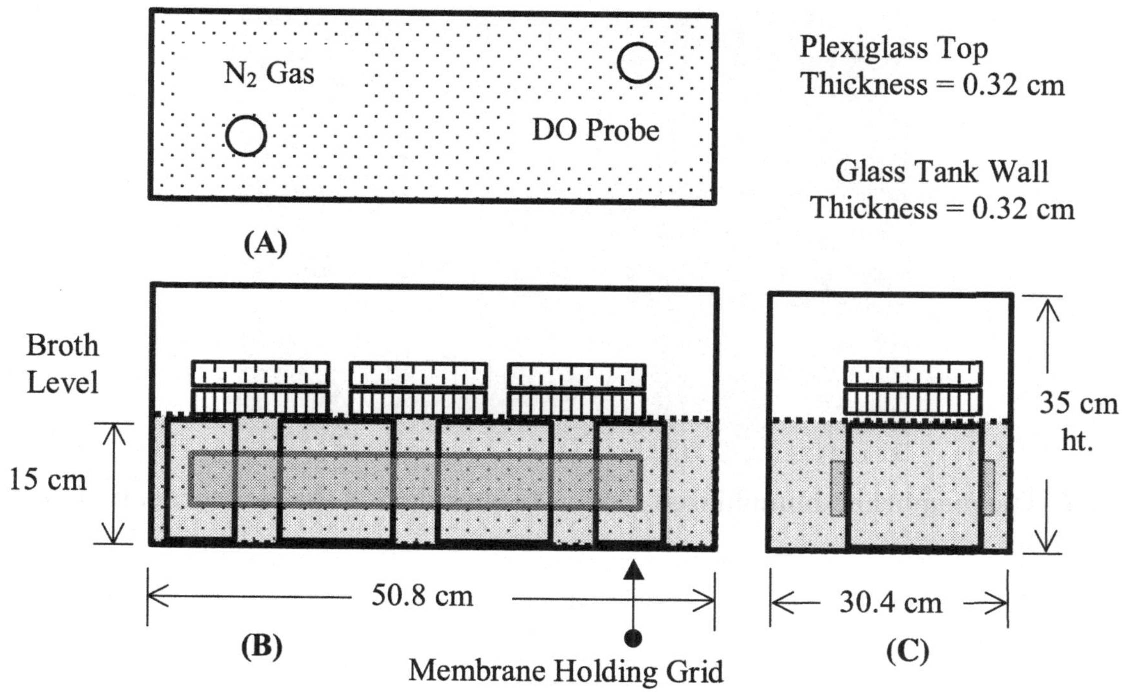


**Figure 6.4 Membrane biofilm growth chamber showing: (A) Top view with DO and nitrogen ports; (B) Front view showing holding grid and membrane plate, and (C) Side view showing membrane plate and holding grid**

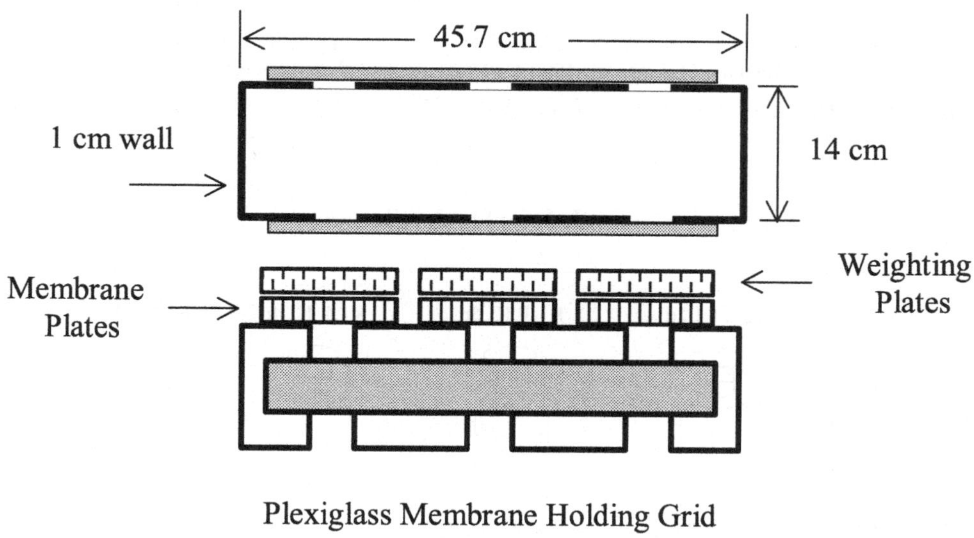


**Figure 6.5 Construction of modified biofilm growth reactor**

**Figure 6.6 The membrane-immobilized biofilm reactor during the biodegradation mode**

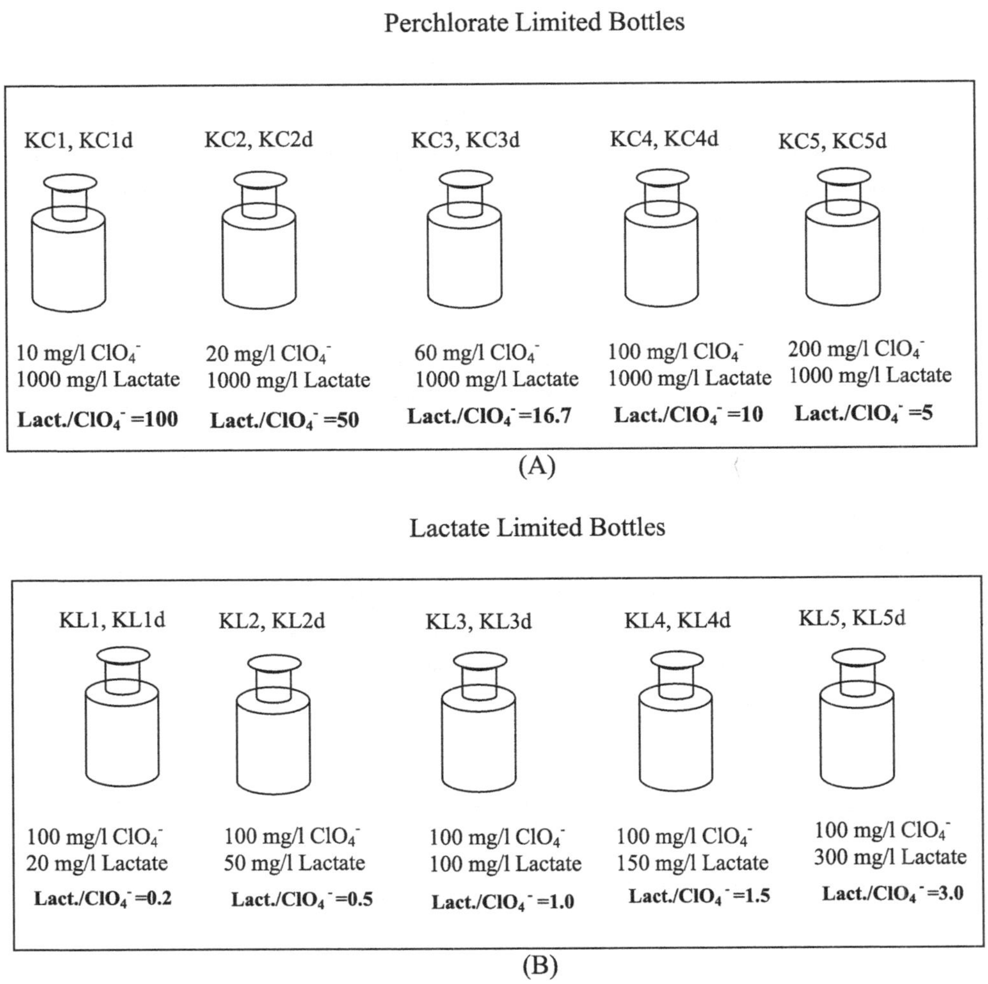


**Figure 6.7 Preliminary kinetics study with the "BALI" culture performed under perchlorate and lactate limited conditions**

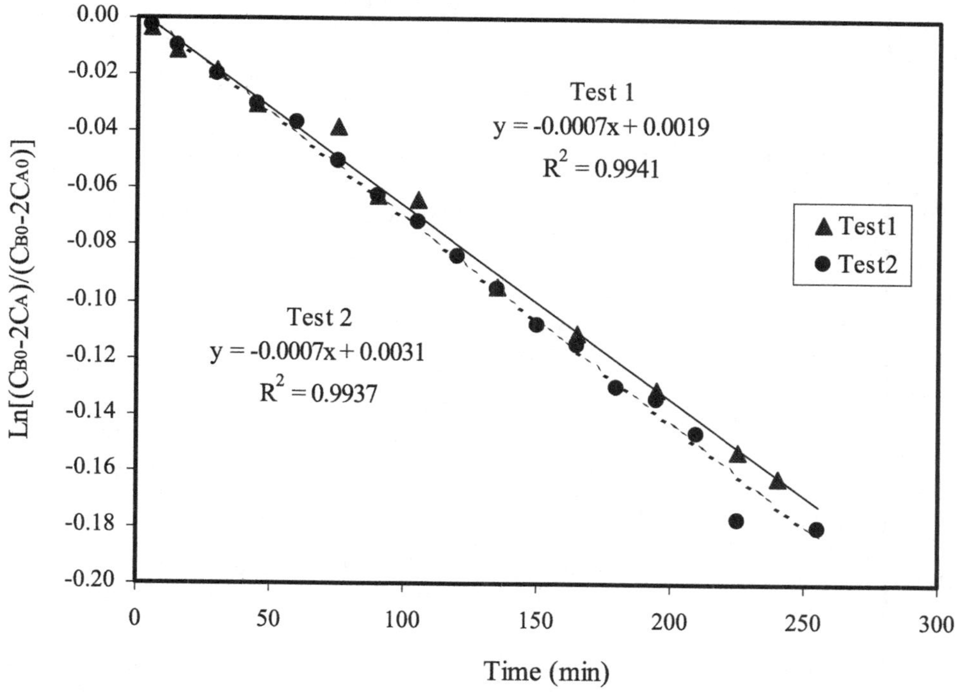


**Figure 6.8 Perchlorate diffusion coefficient by a Memcor BTS-55 membrane**

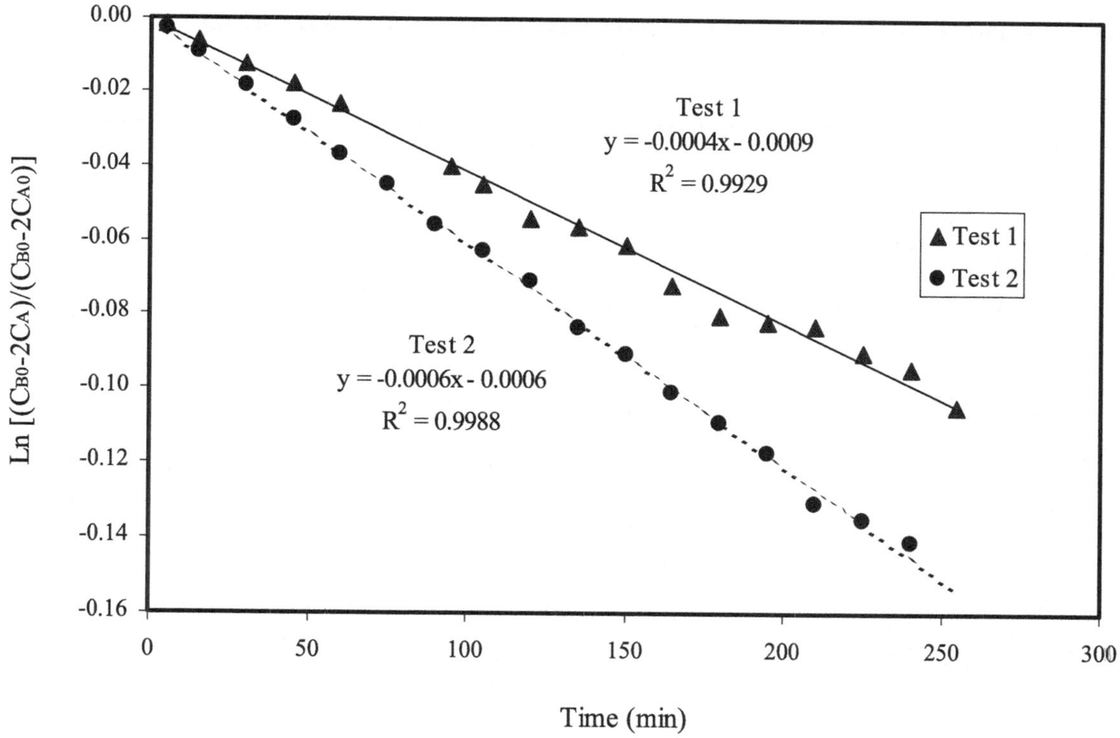


**Figure 6.9 Perchlorate diffusion through a Memcor PVDF membrane**

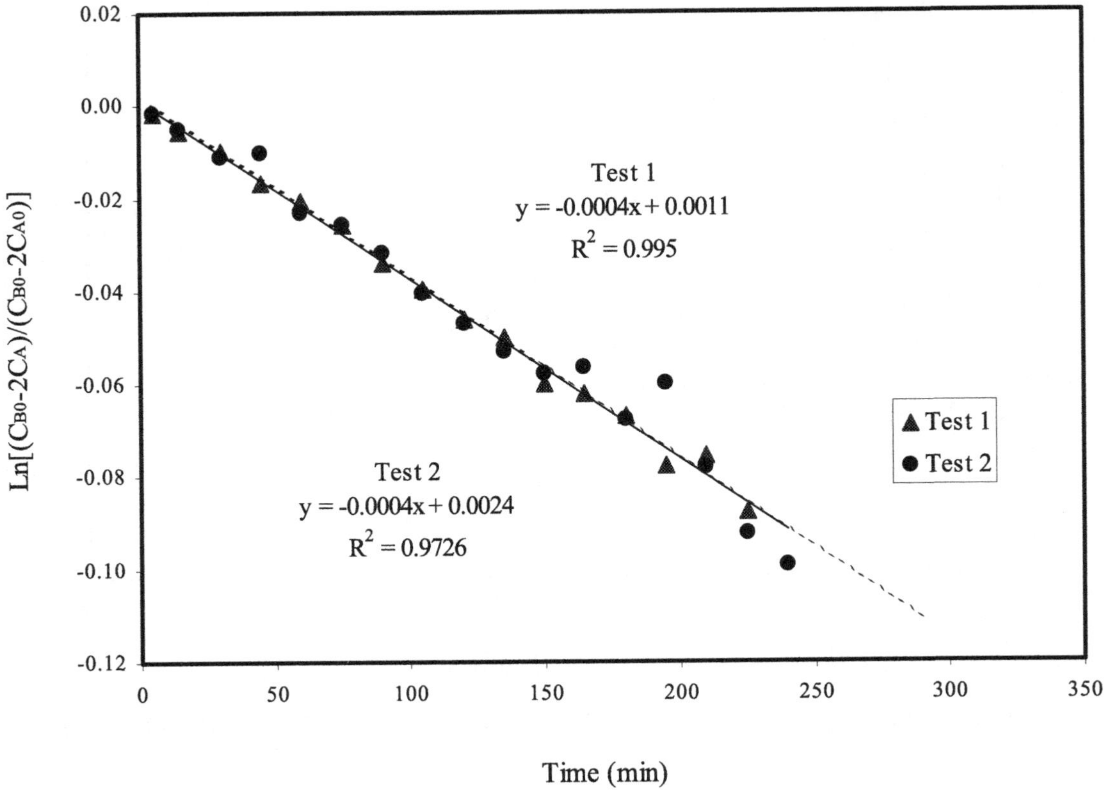


**Figure 6.10 Perchlorate diffusion through a Millipore FGLP membrane**

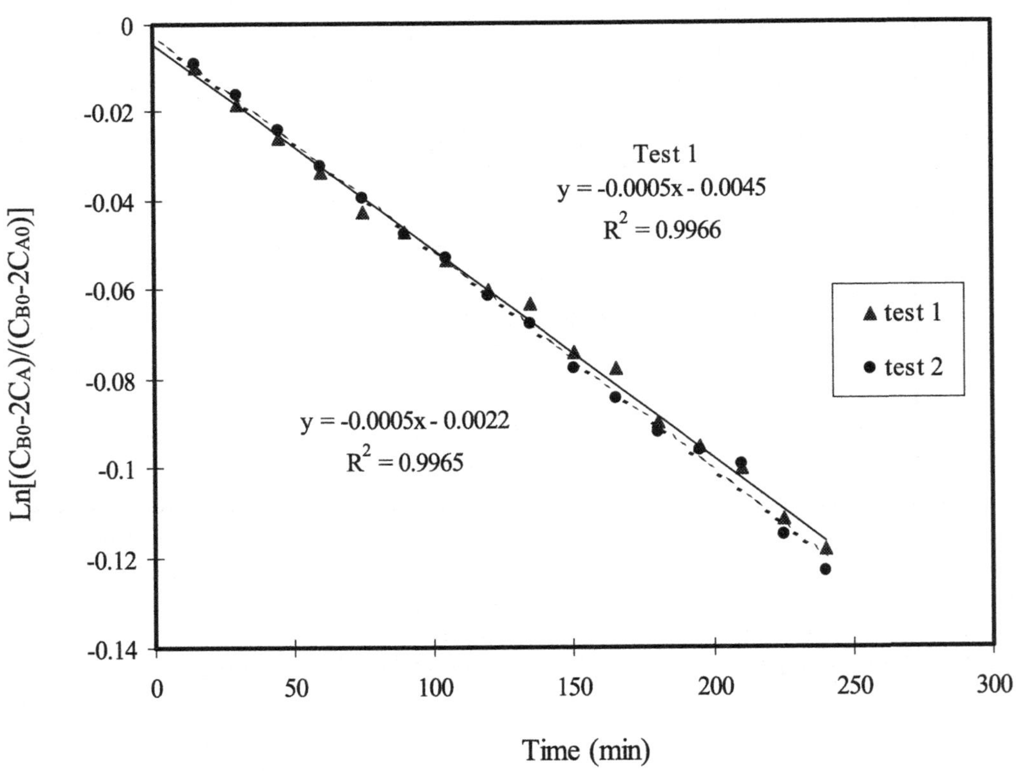


**Figure 6.11 Nitrate diffusivity testing through Memcor BTS-55 membrane**

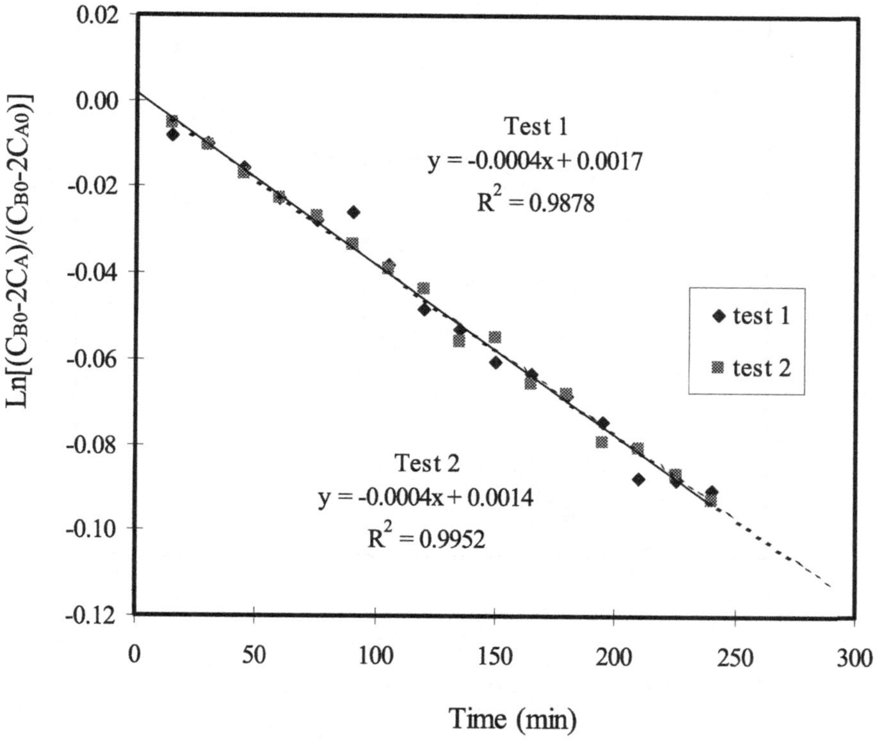


**Figure 6.12 Sulfate diffusivity testing by a Memcor BTS-55 membrane**

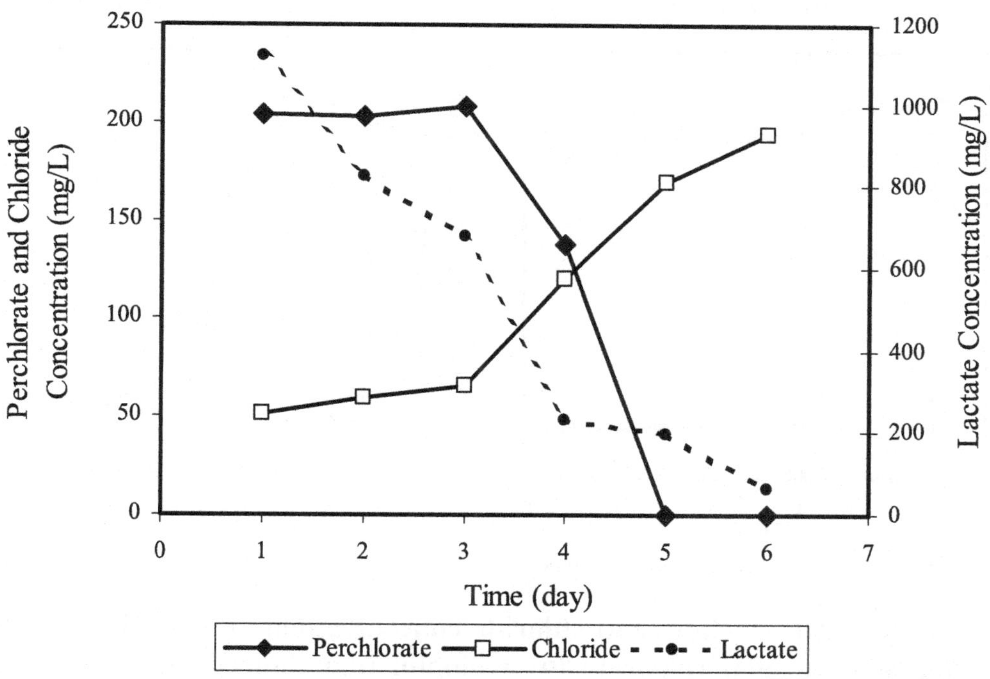


**Figure 6.13 Anaerobic consumption of perchlorate and lactate in the BR**

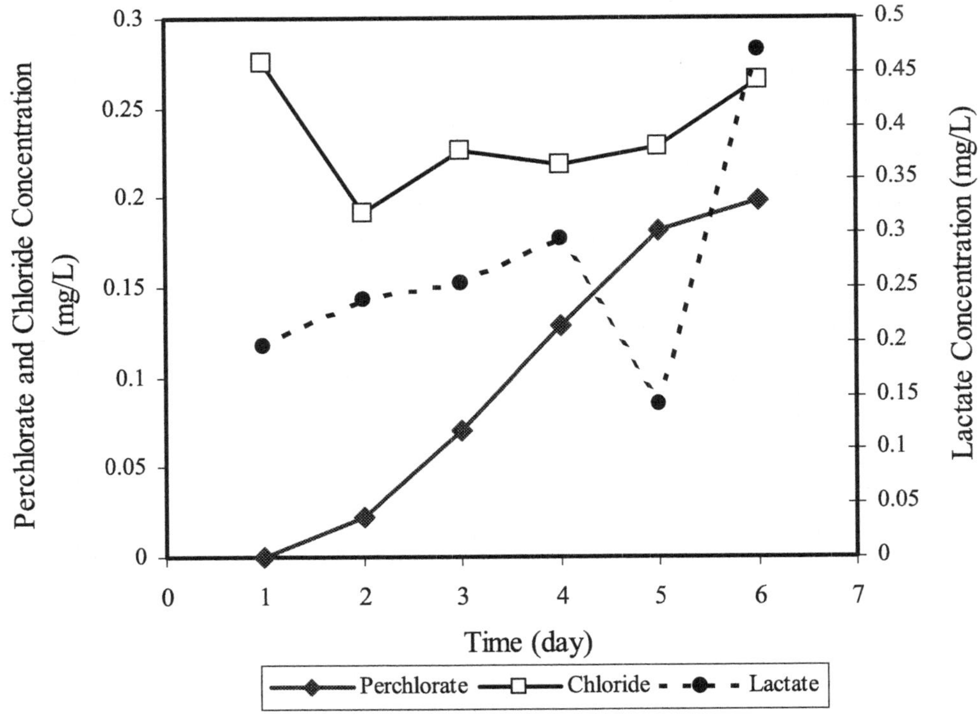


**Figure 6.14 Anaerobic consumption of perchlorate and lactate in the DR**

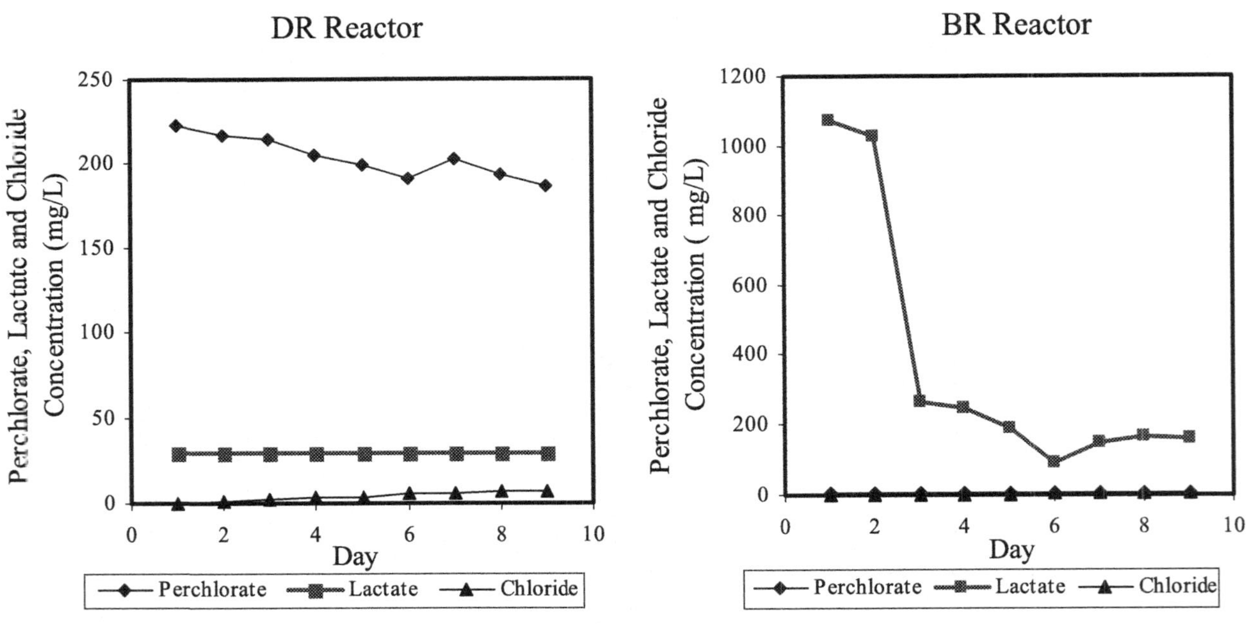


**Figure 6.15 Perchlorate, lactate, and chloride concentrations in the DR and BR reactors during biodegradation of perchlorate by a biofilm immobilized on a FGLP membrane (1st cycle)**

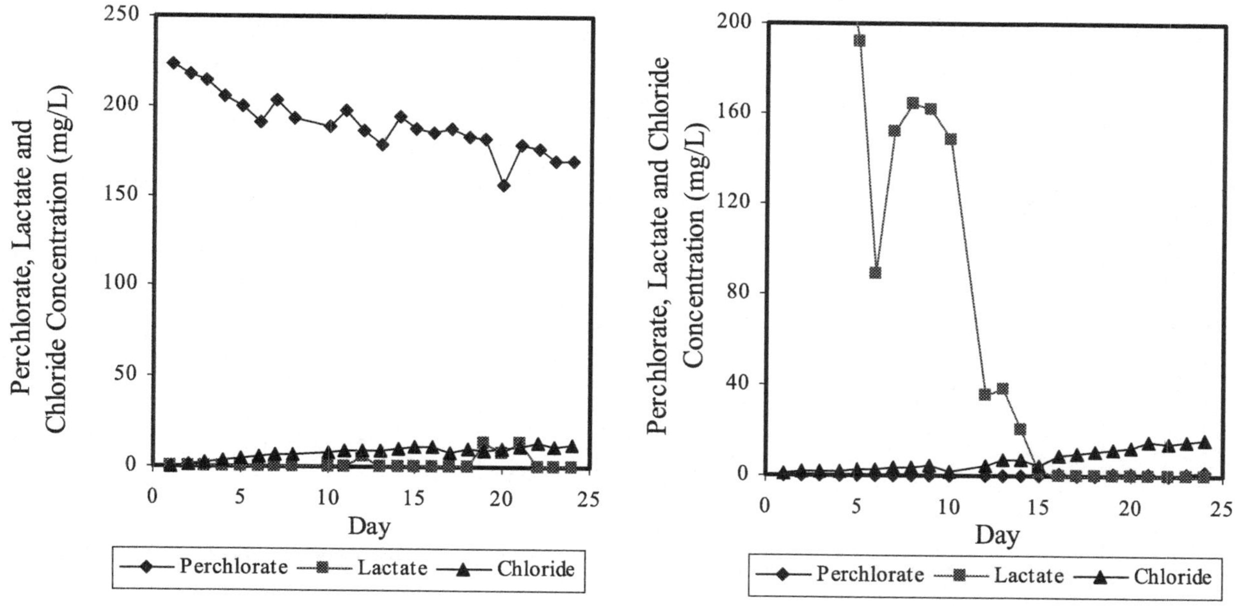


**Figure 6.16 Perchlorate, lactate, and chloride concentrations in the DR and BR reactors during biodegradation of perchlorate by a biofilm immobilized on a FGLP membrane (2nd cycle)**

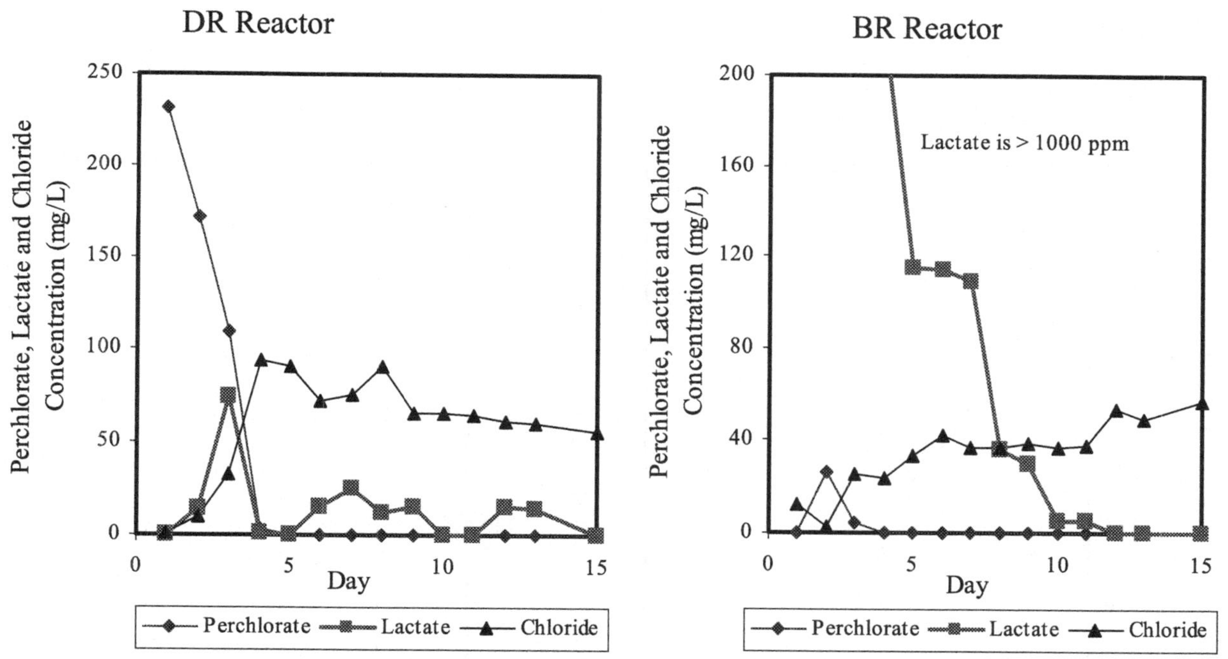


**Figure 6.17 Perchlorate, lactate, and chloride concentrations in the DR and BR reactors during biodegradation of perchlorate by a biofilm immobilized on a BTS-55 membrane (1st cycle)**

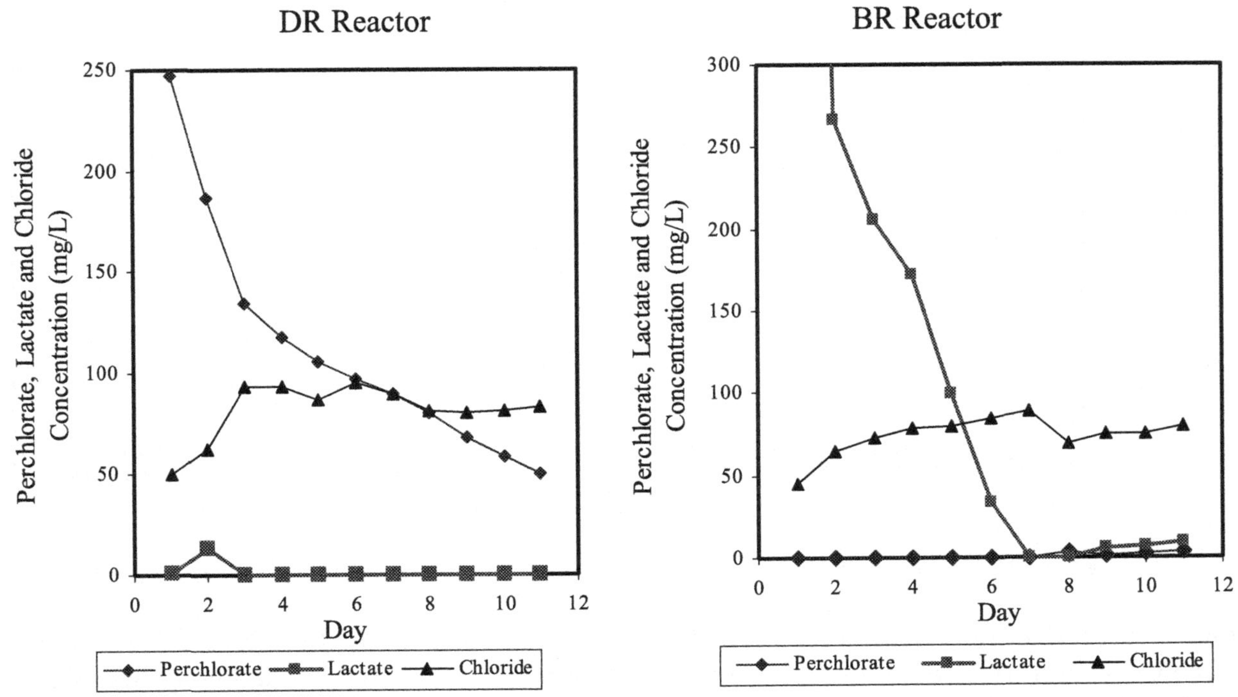


**Figure 6.18 Perchlorate, lactate, and chloride concentrations in the DR and BR reactors during biodegradation of perchlorate by a biofilm immobilized on a BTS-55 membrane (2nd cycle)**

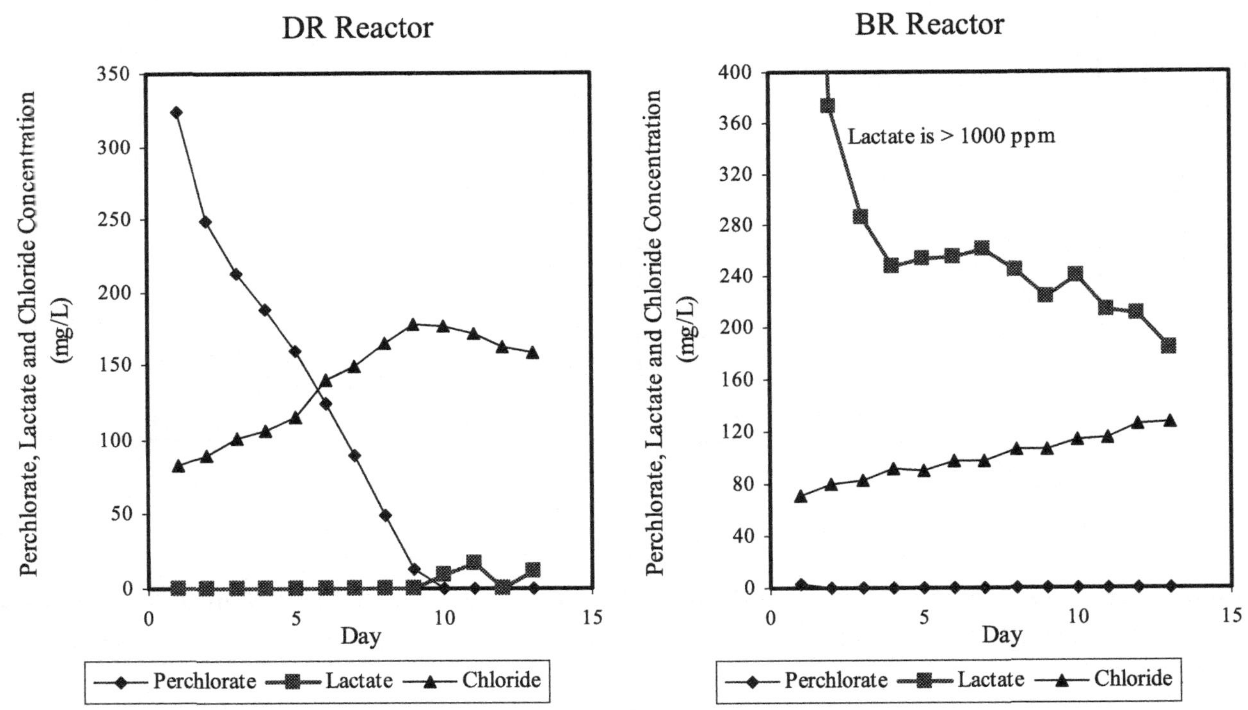


**Figure 6.19 Perchlorate, lactate, and chloride concentrations in the DR and BR reactors during biodegradation of perchlorate by a biofilm immobilized on a BTS-55 membrane (3rd cycle)**

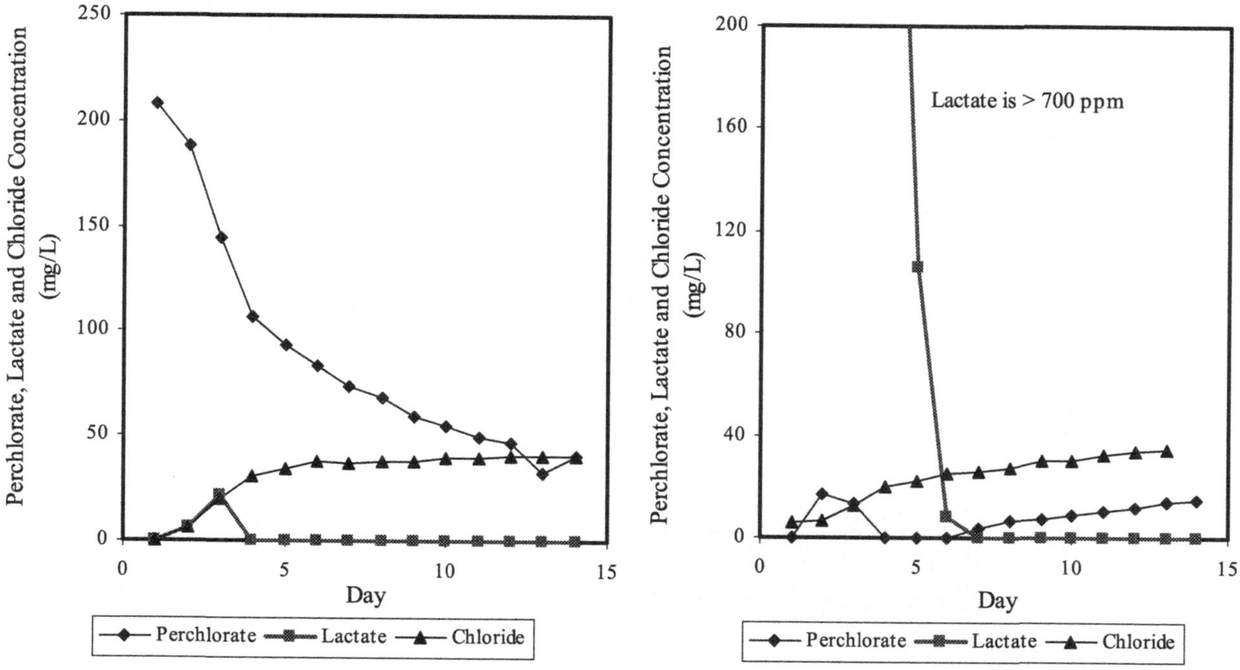


**Figure 6.20 Perchlorate, lactate, and chloride concentrations in the DR and BR reactors during biodegradation of perchlorate by a biofilm immobilized on a PVDF membrane**

**Figure 6.21 Master culture reactors containing the BALI red culture (on the right side) and another perchlorate degrading culture which does not have the red coloration (on the left side)**

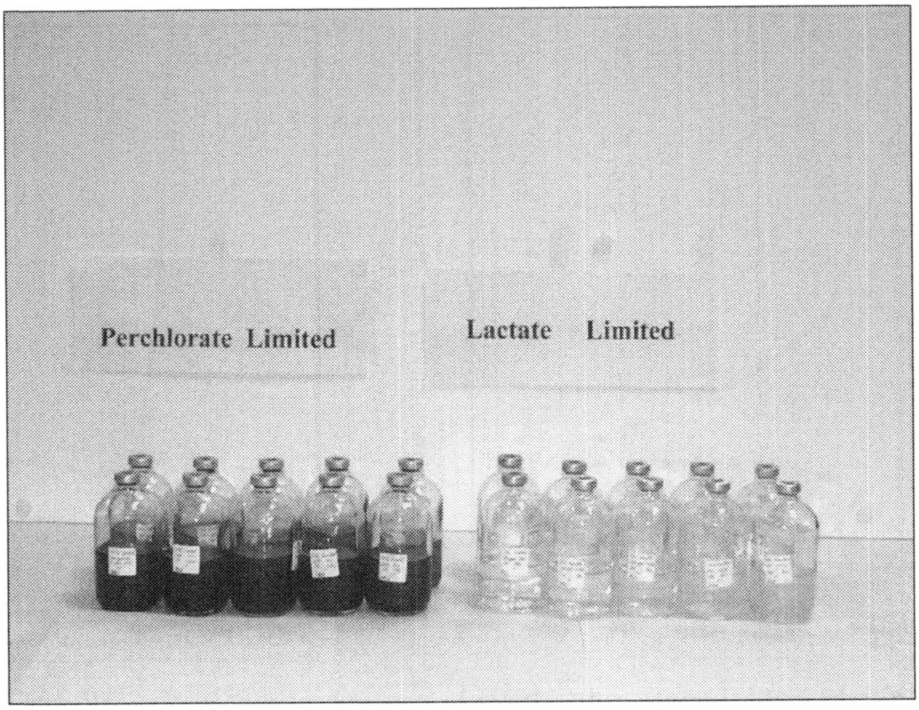


**Figure 6.22 Kinetic studies with the "BALI" culture showing microbial growth in perchlorate limited bottles (left) and no growth in lactate limited bottles (right)**

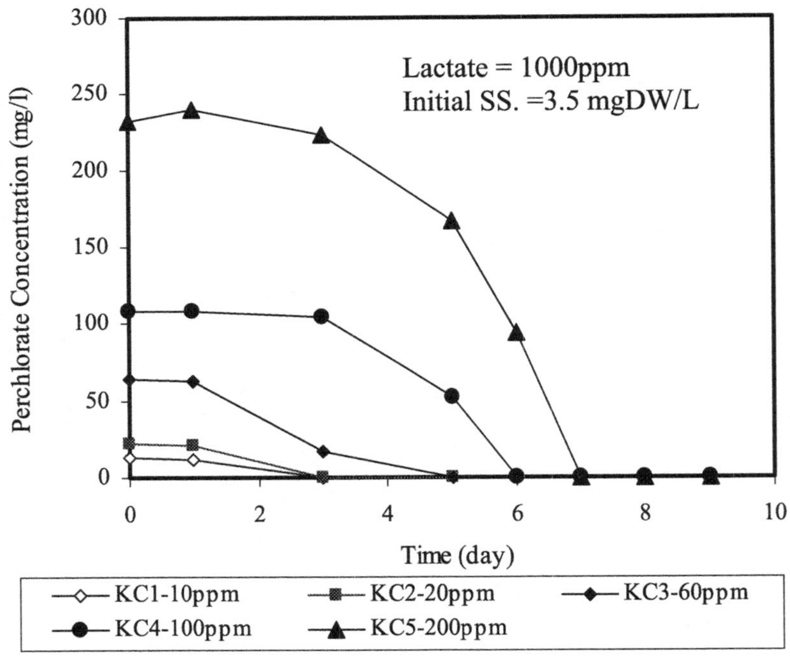


**Figure 6.23 Perchlorate biodegradation in perchlorate limited bottles**

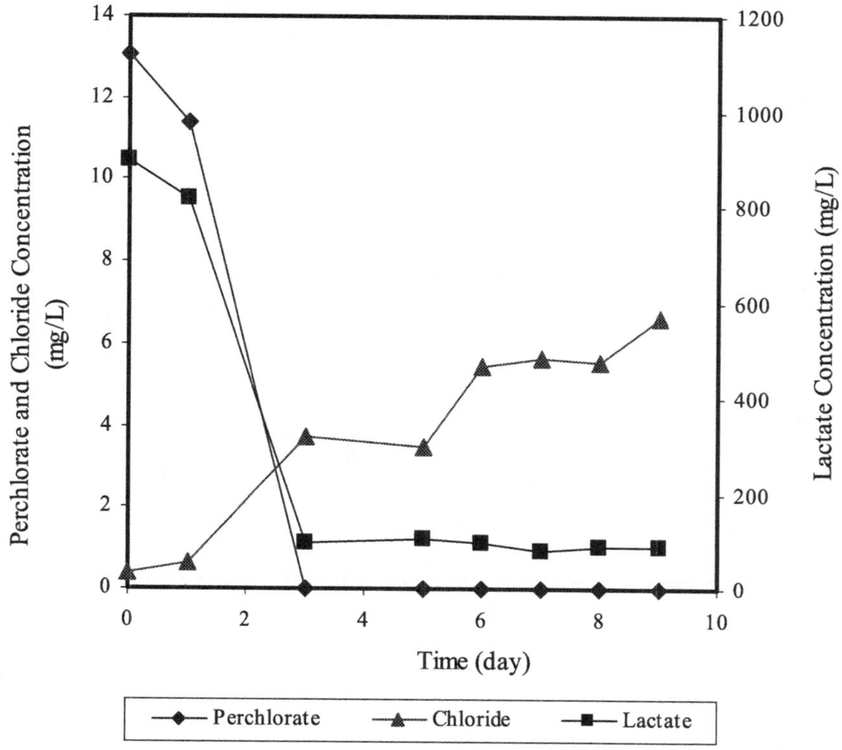


**Figure 6.24 Microbial reduction of perchlorate to chloride when lactate/$ClO_4^-$ = 100**

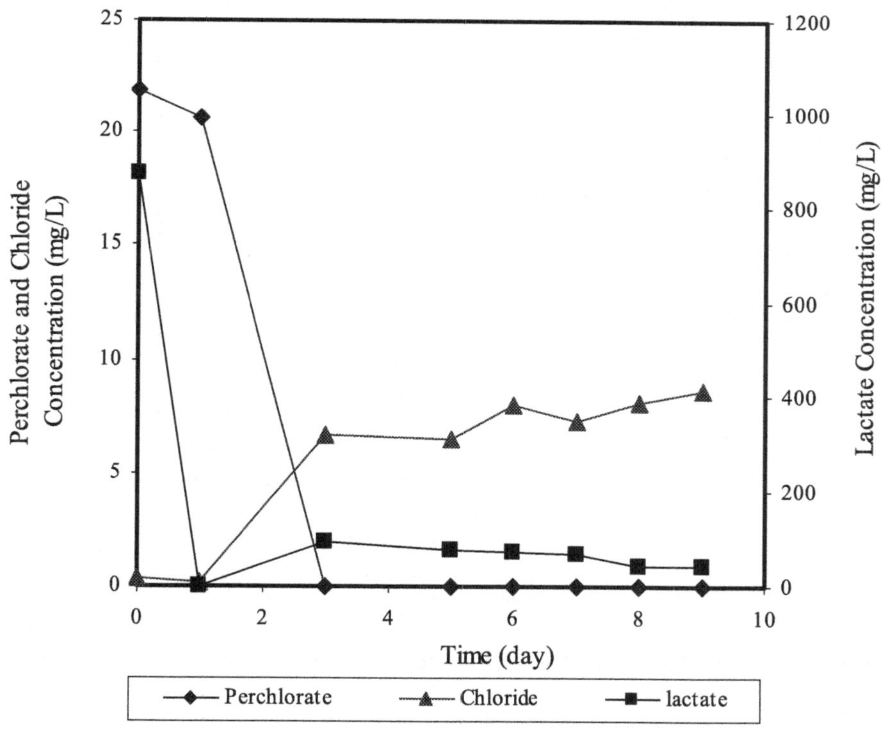


**Figure 6.25 Microbial reduction of perchlorate to chloride when lactate/$ClO_4^-$ = 50**

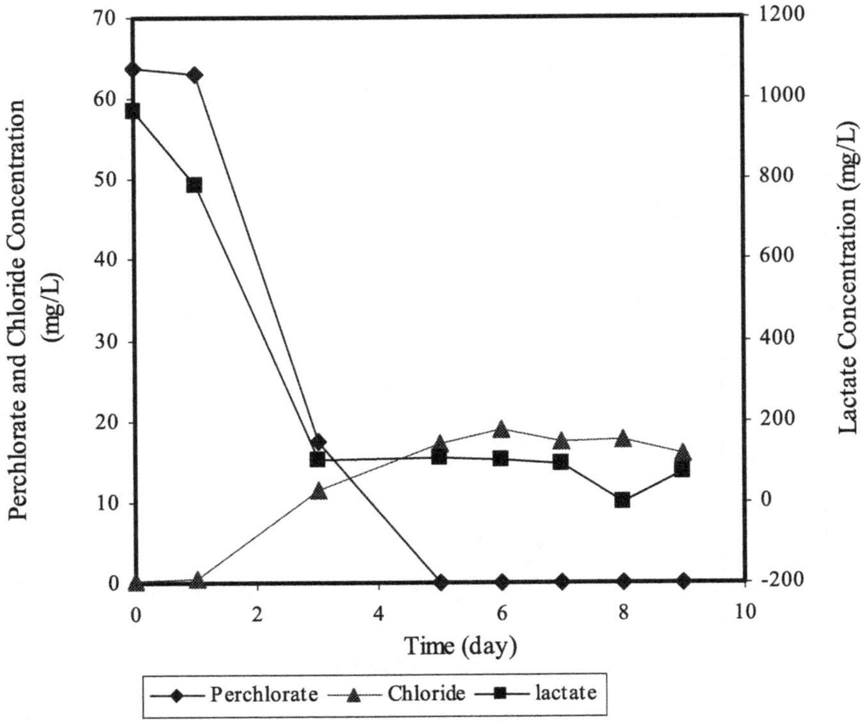


**Figure 6.26 Microbial reduction of perchlorate to chloride when lactate/$ClO_4^-$ = 16.7**

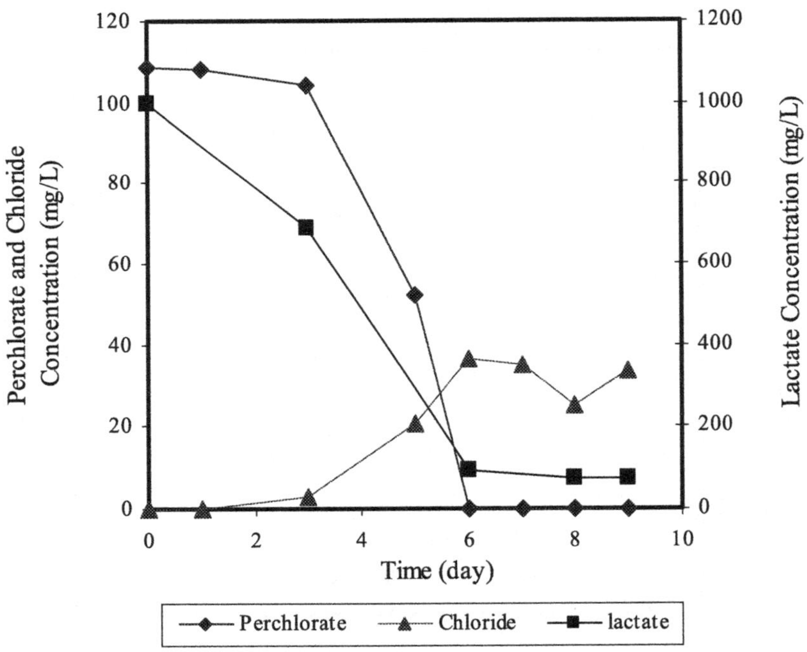


**Figure 6.27 Microbial reduction of perchlorate to chloride when lactate/$ClO_4^-$ = 10**

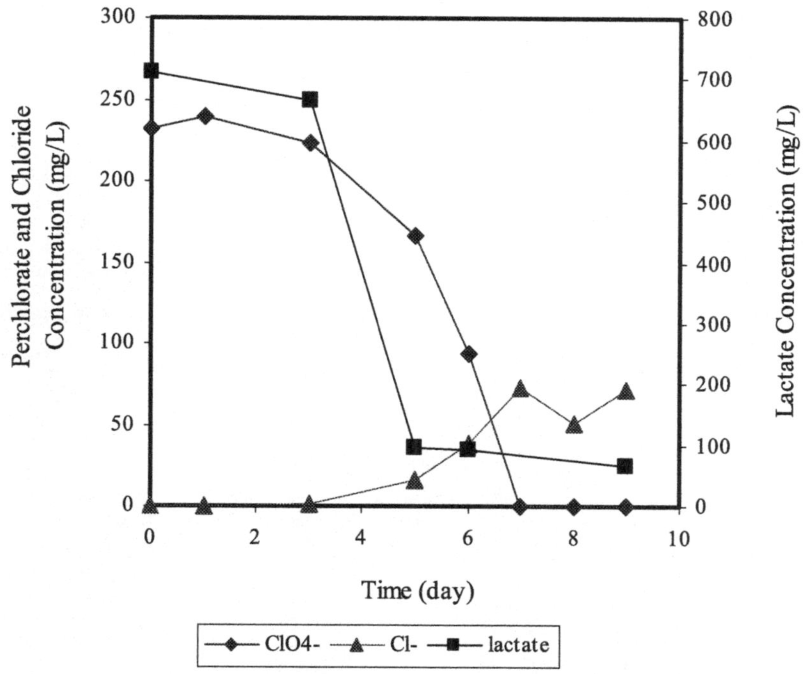


**Figure 6.28 Microbial reduction of perchlorate to chloride when lactate/$ClO_4^-$ = 5**

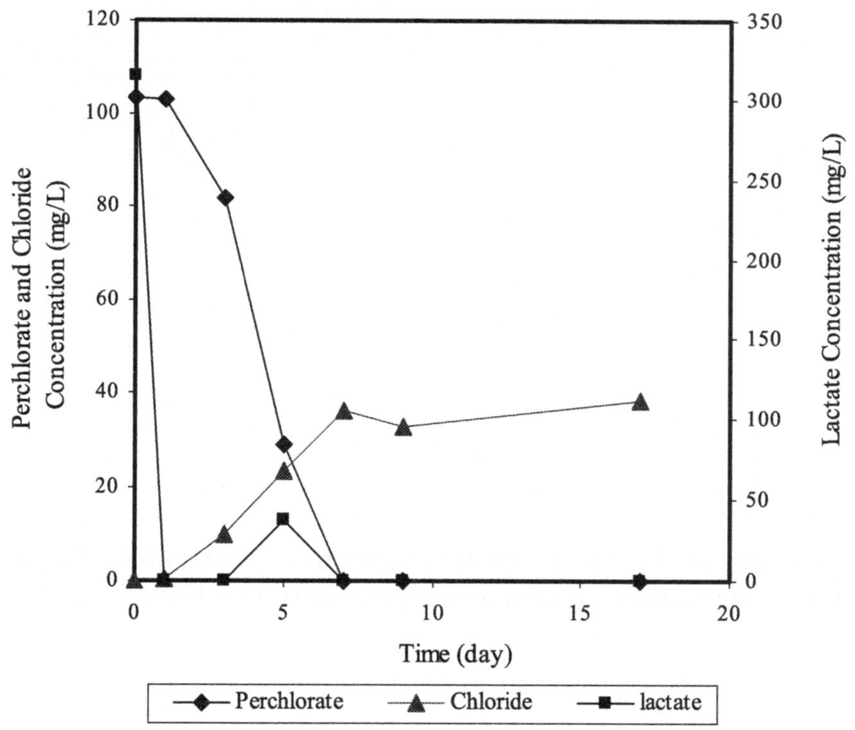


**Figure 6.29 Microbial reduction of perchlorate to chloride when lactate/$ClO_4^-$ = 3**

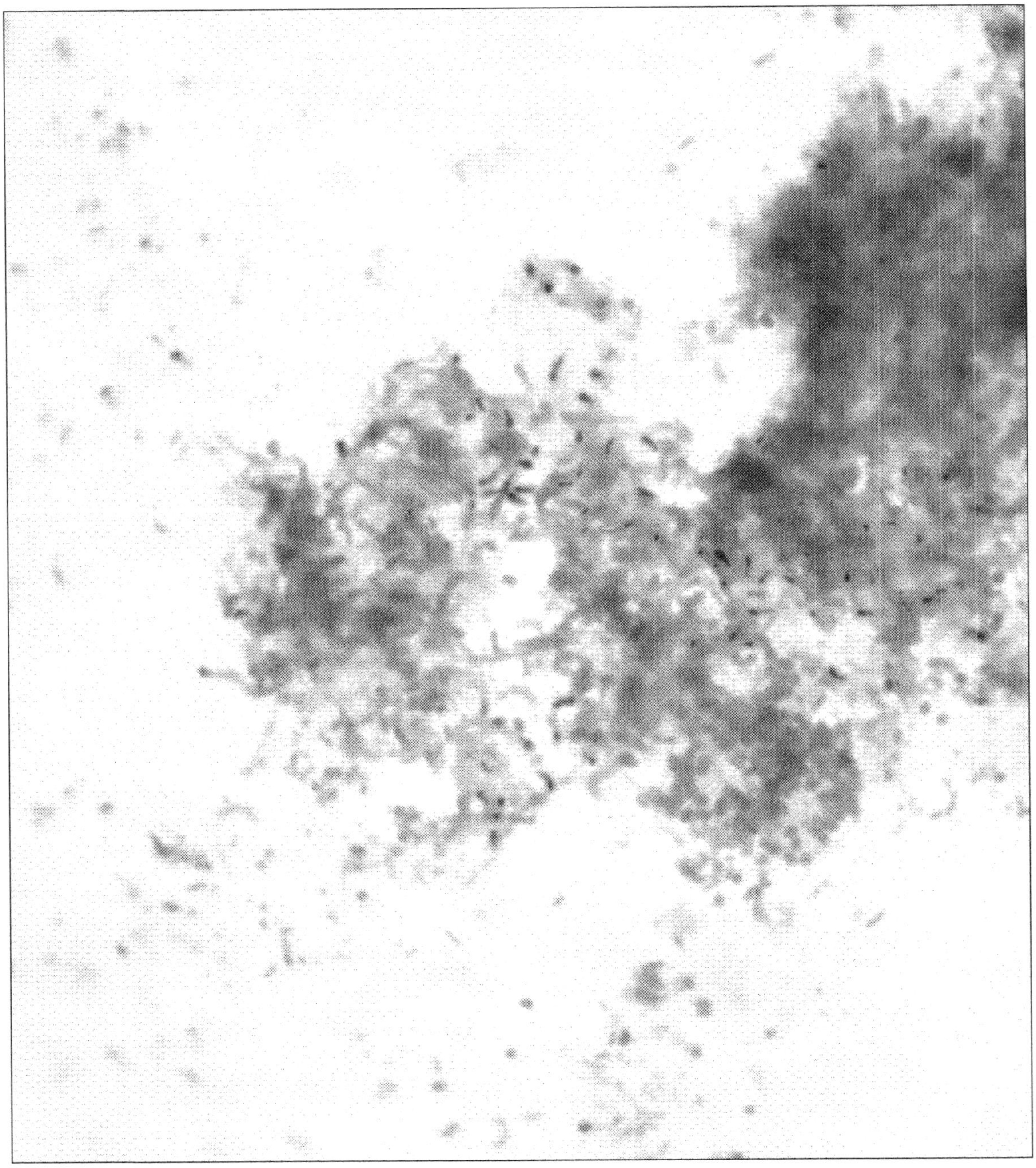

**Figure 6.30 Gram-stain of "BALI" taken from growth reactor (TK1) showing the predominance of rod and helical gram-negative microorganisms (magnification: 1000×)**

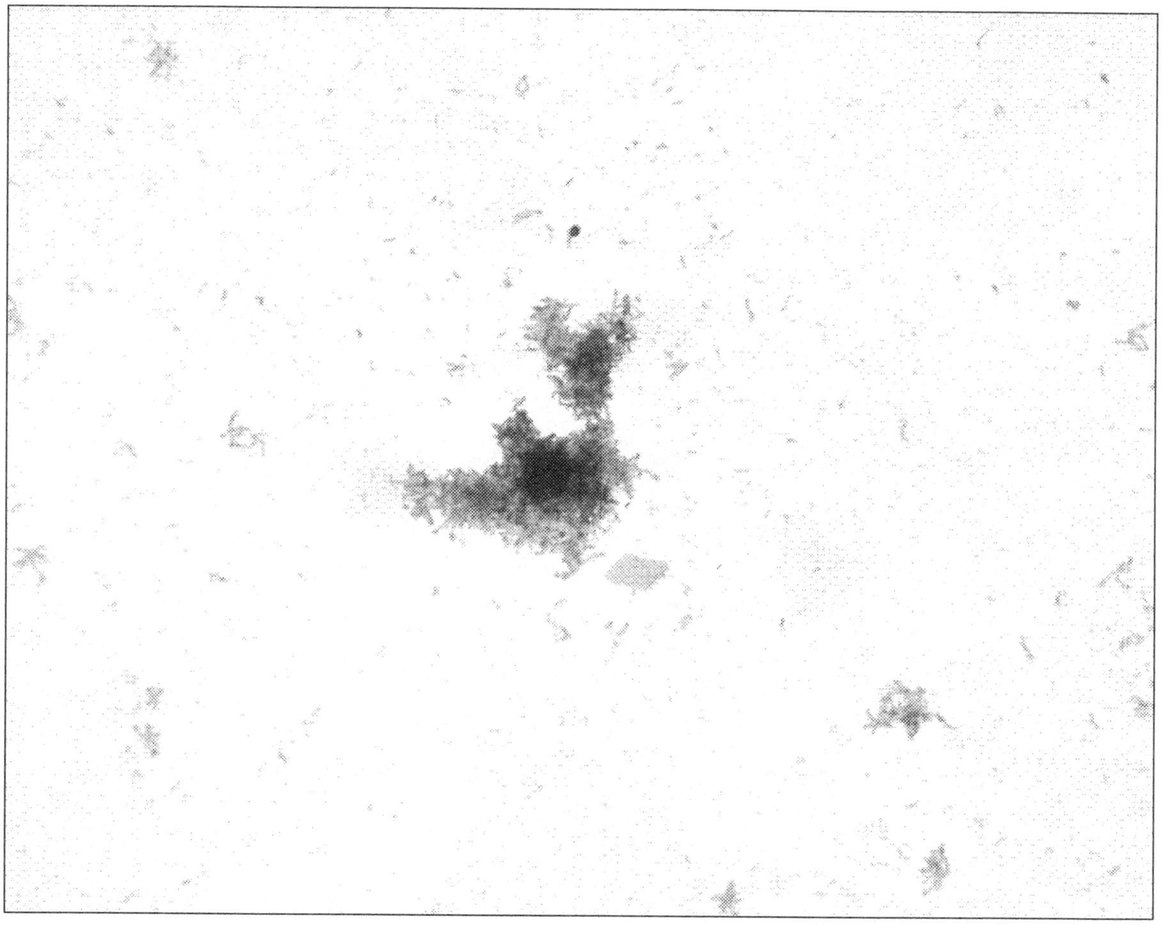

**Figure 6.31 Gram-stain of "BALI" taken from kinetics study (KC5d) showing helical morphology of growing microorganisms (magnification: 1000×)**

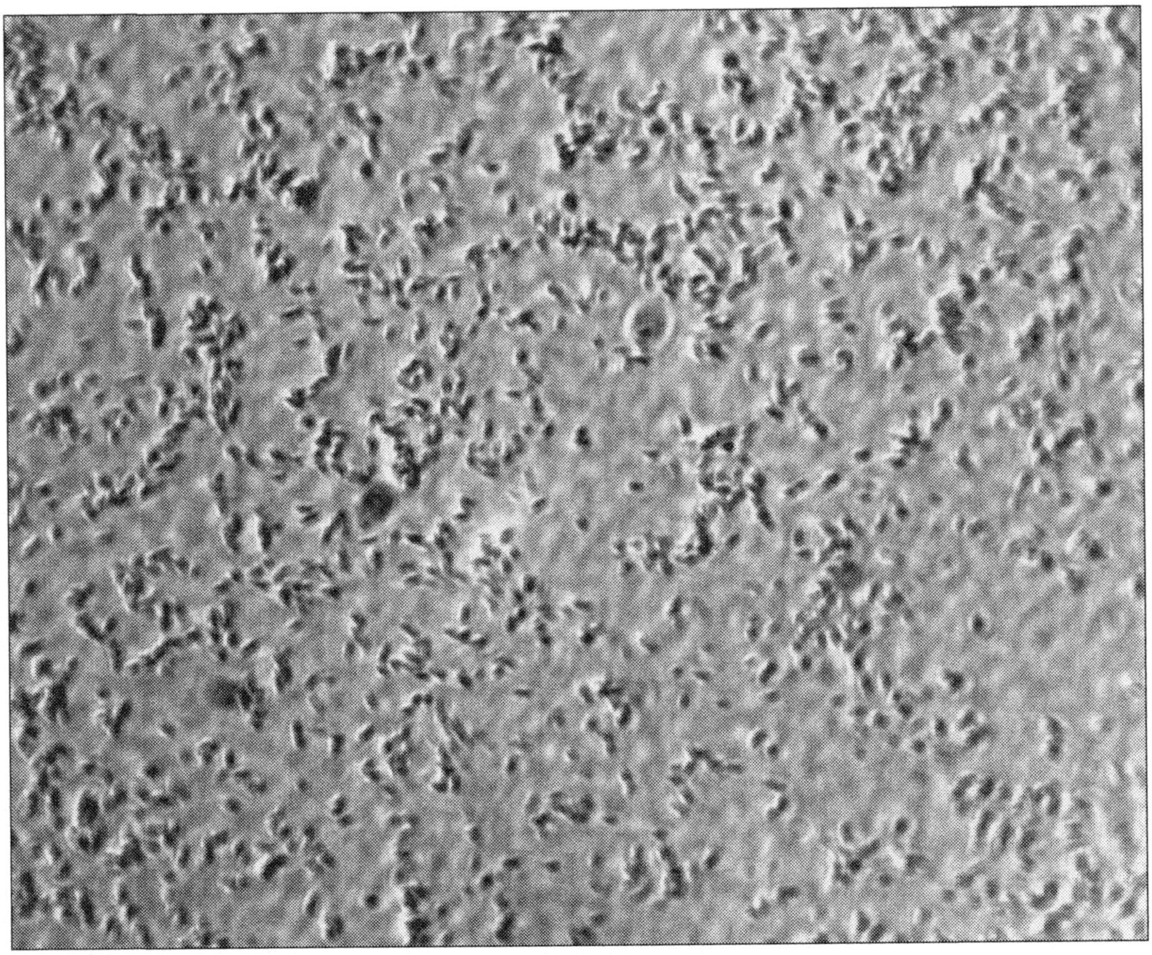

**Figure 6.32 Micrograph of isolate KA2-1**

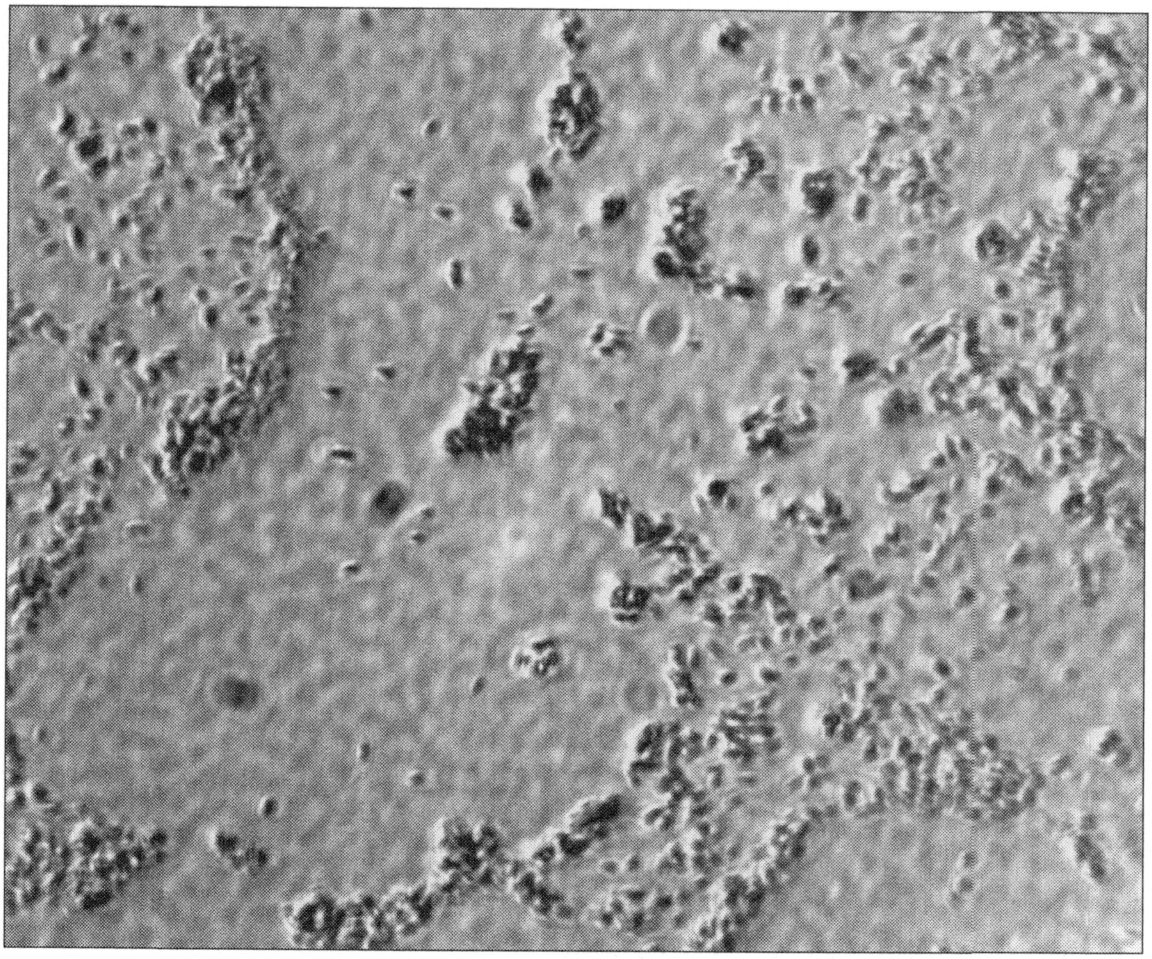

**Figure 6.33 Micrograph of isolate KL1-A**

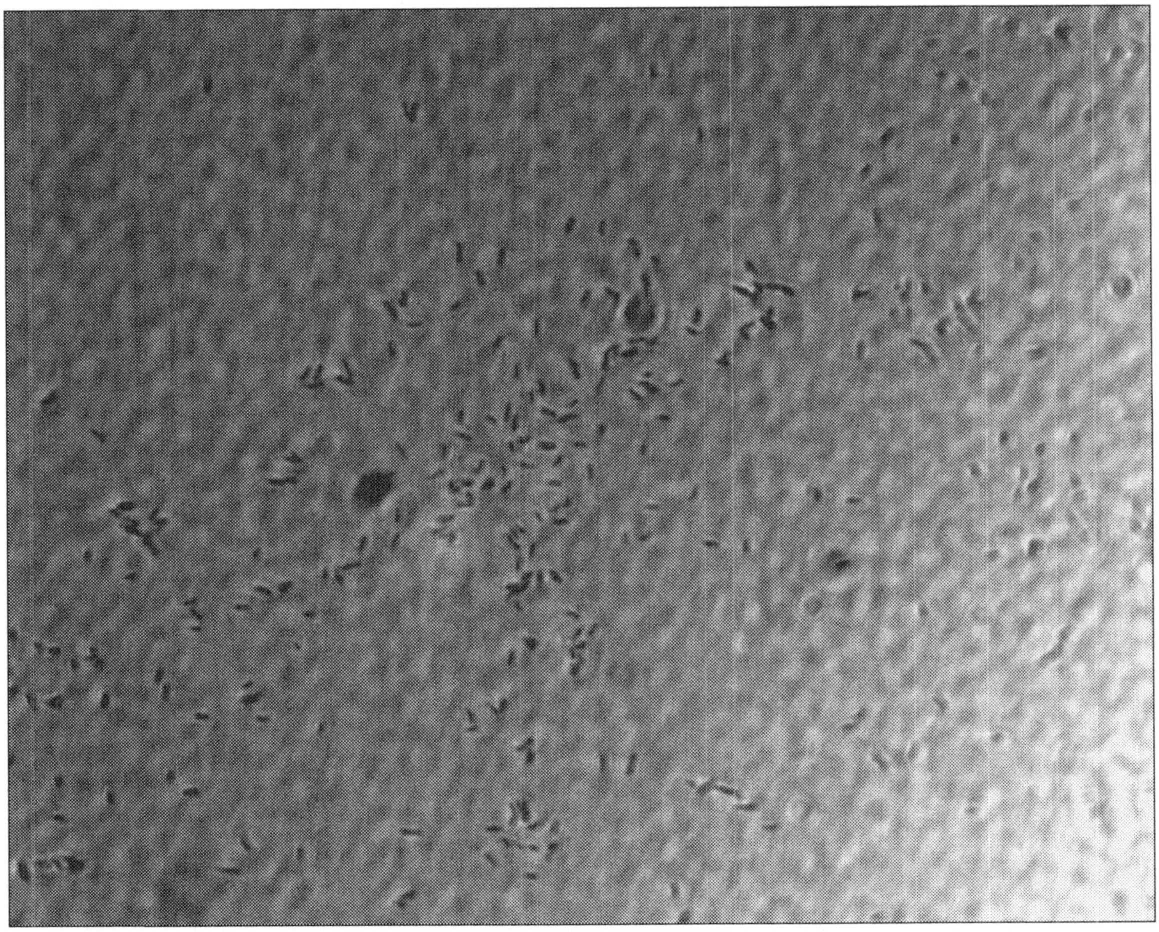

**Figure 6.34 Micrograph of isolate RAS-CCSD**

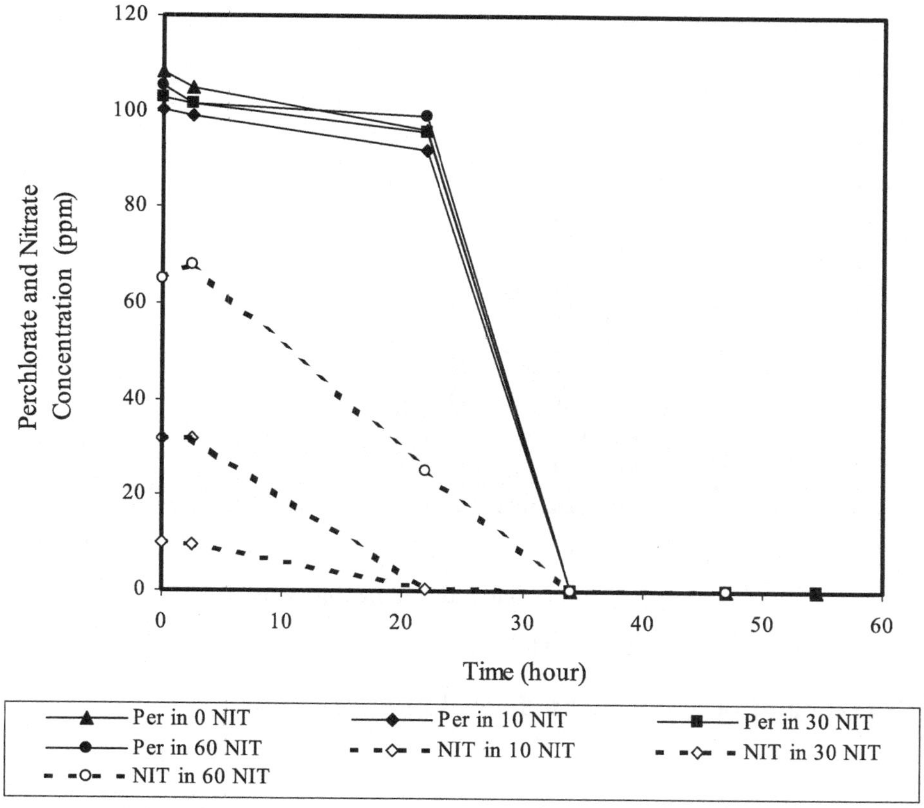


**Figure 6.35 Batch testing of nitrate interference on perchlorate biodegradation using "BALI" culture**

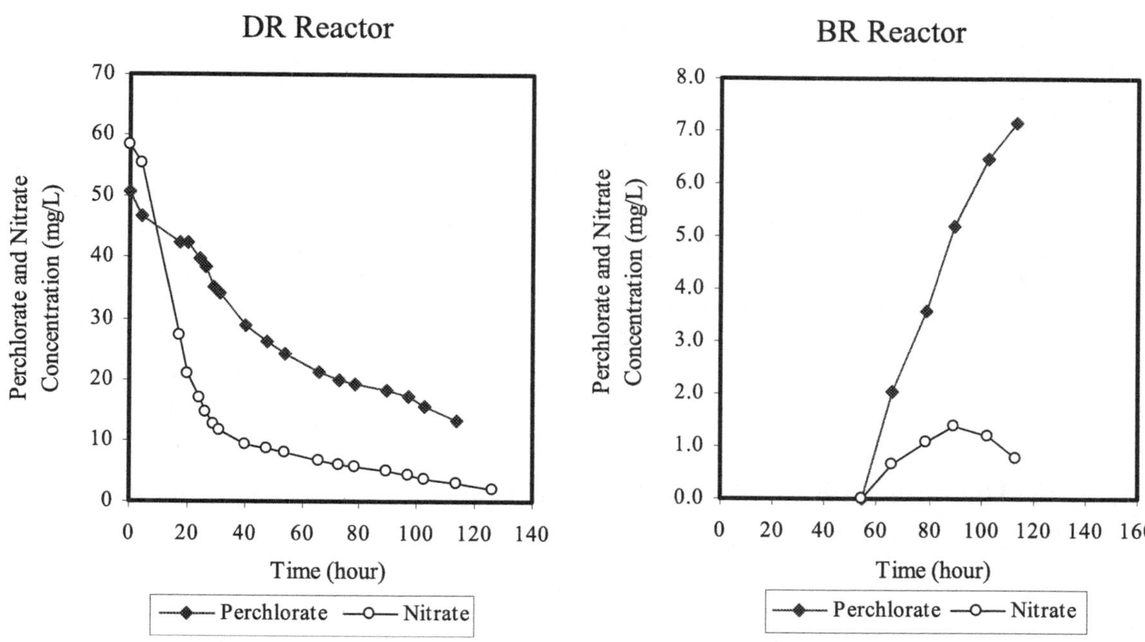


**Figure 6.36 The interference of nitrate on perchlorate biodegradation by BTS-55 membrane-immobilized biofilm (test 1)**

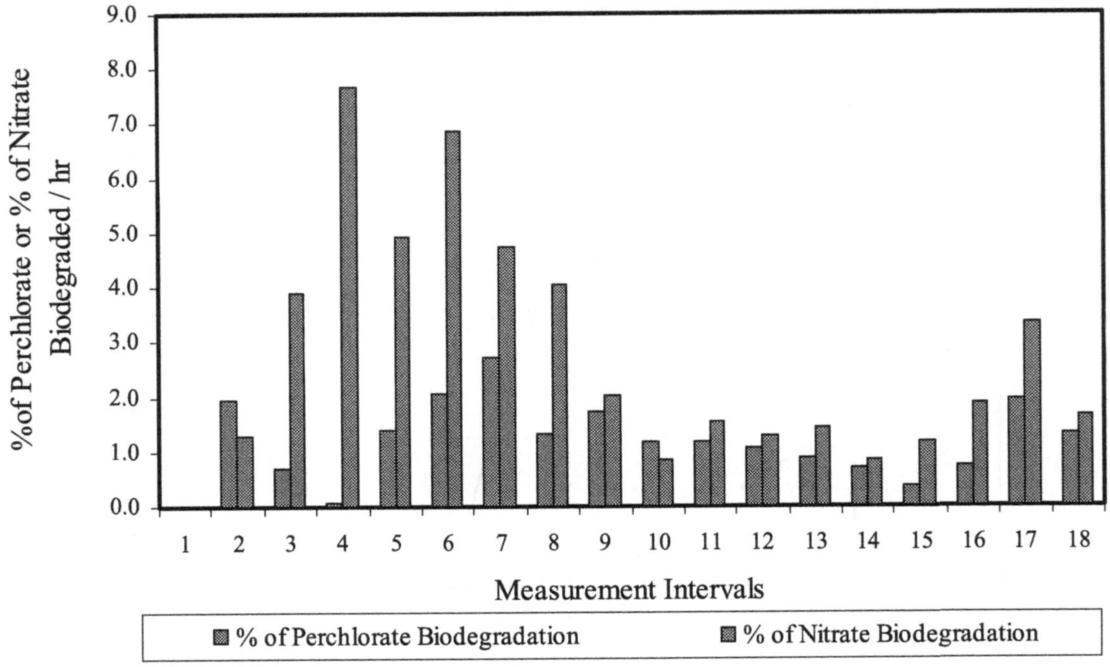


**Figure 6.37 Nitrate and perchlorate biodegradation rate (test 1)**

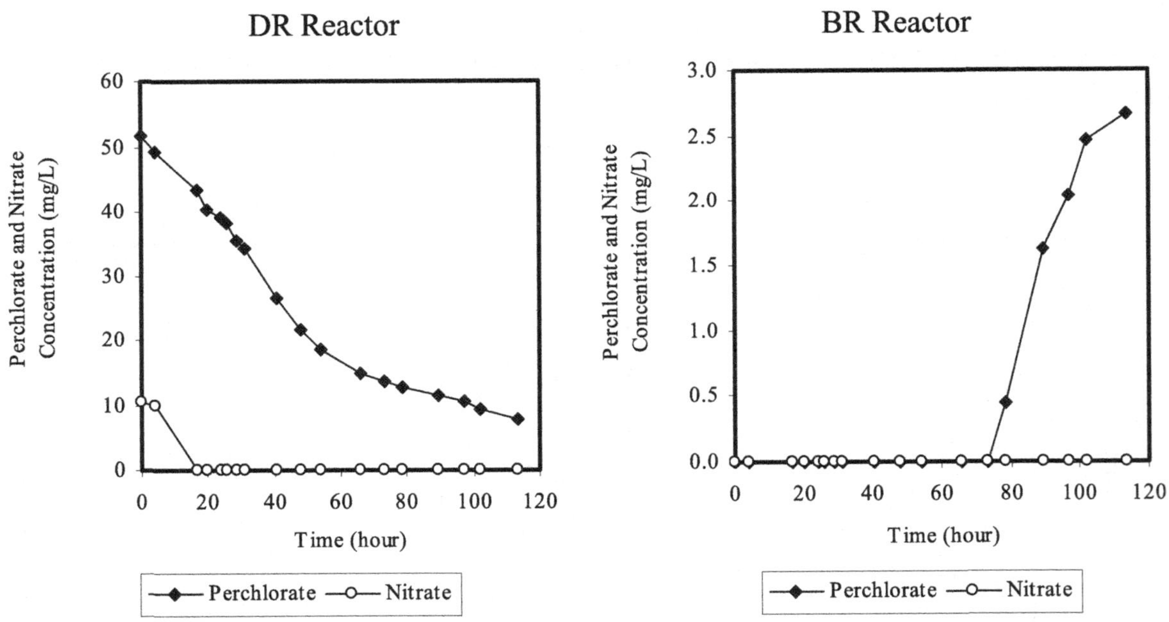


**Figure 6.38 The interference of nitrate on perchlorate biodegradation by a BTS-55 membrane-immobilized biofilm (test 2)**

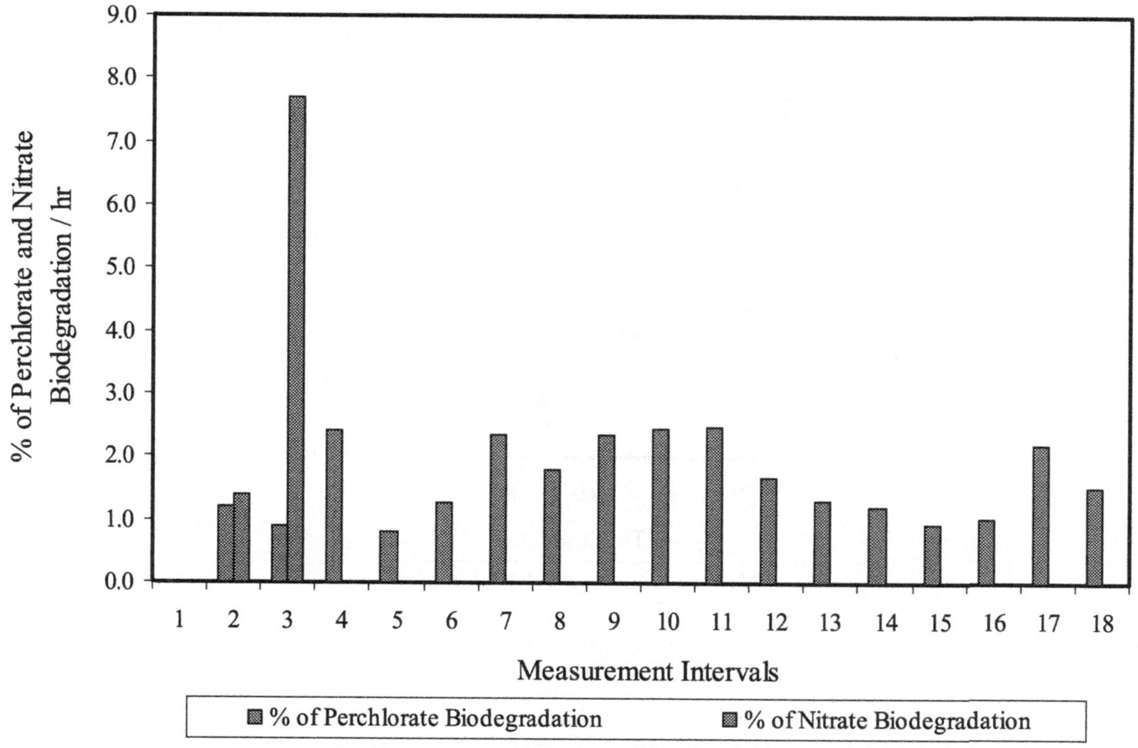


**Figure 6.39 Nitrate and perchlorate biodegradation rate (test 2)**

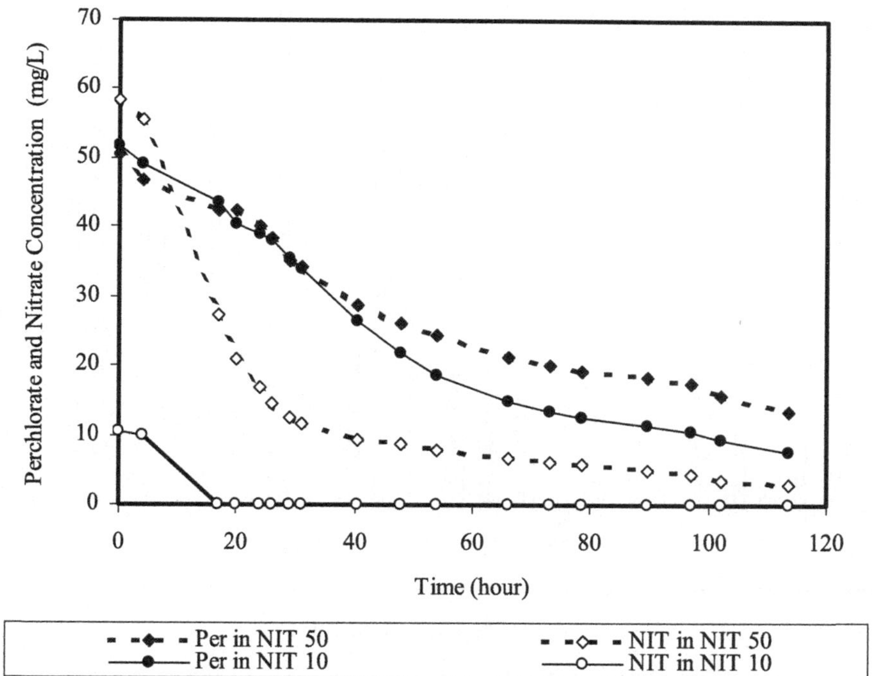


**Figure 6.40 Interference of various concentration of nitrate on perchlorate biodegradation in BTS-55 membrane-immobilized biofilm reactors (test 1 & test 2)**

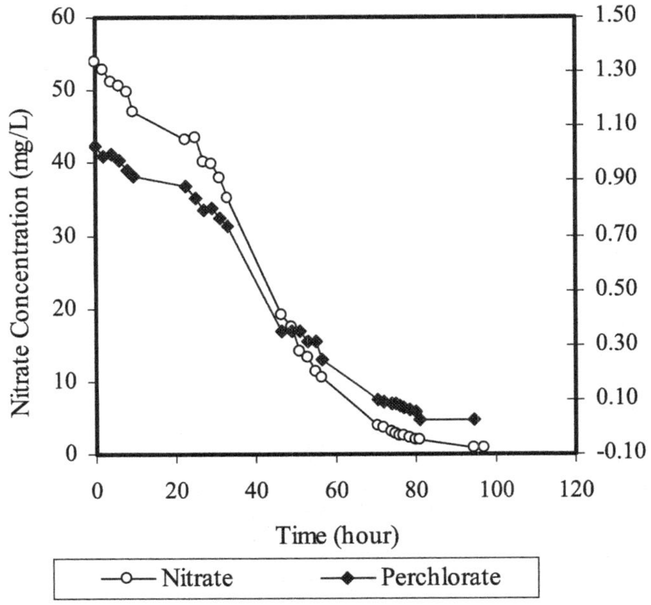


**Figure 6.41 The interference of nitrate on perchlorate biodegradation (test 3)**

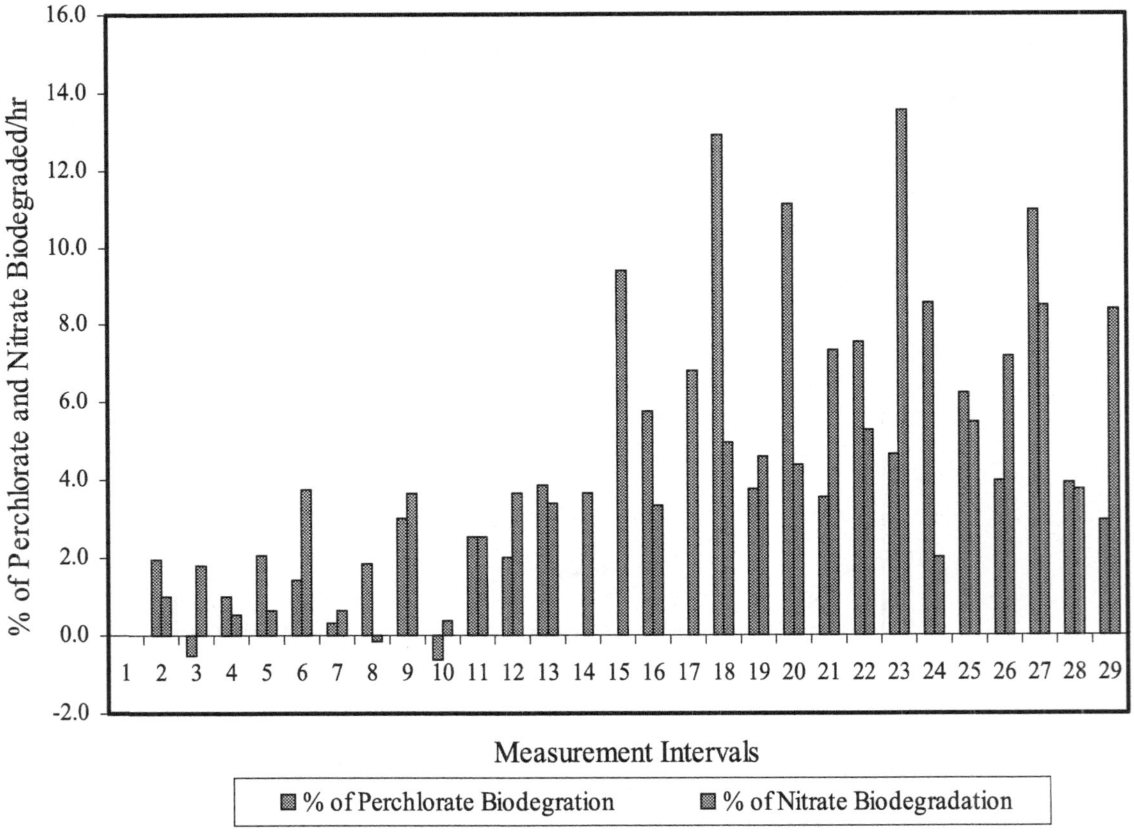


**Figure 6.42 Biodegradation rates of perchlorate and nitrate (test 3)**

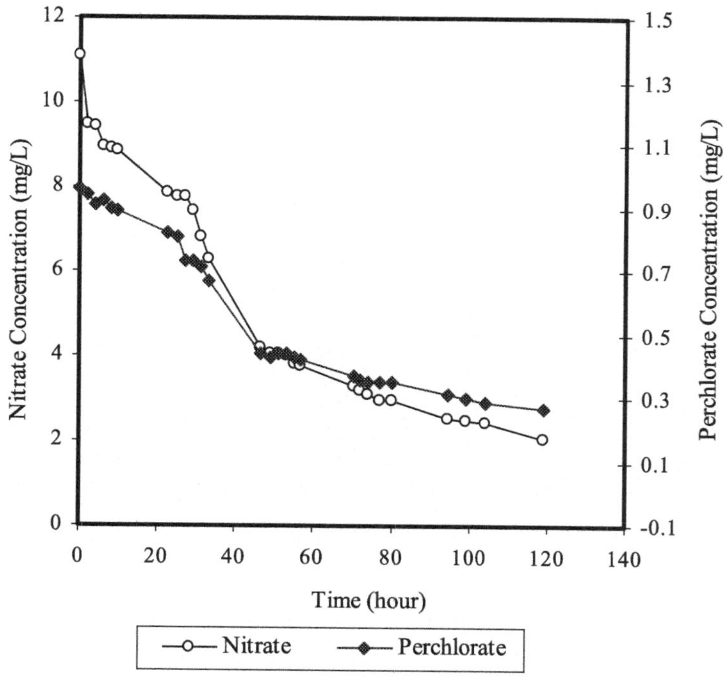


BR Reactor

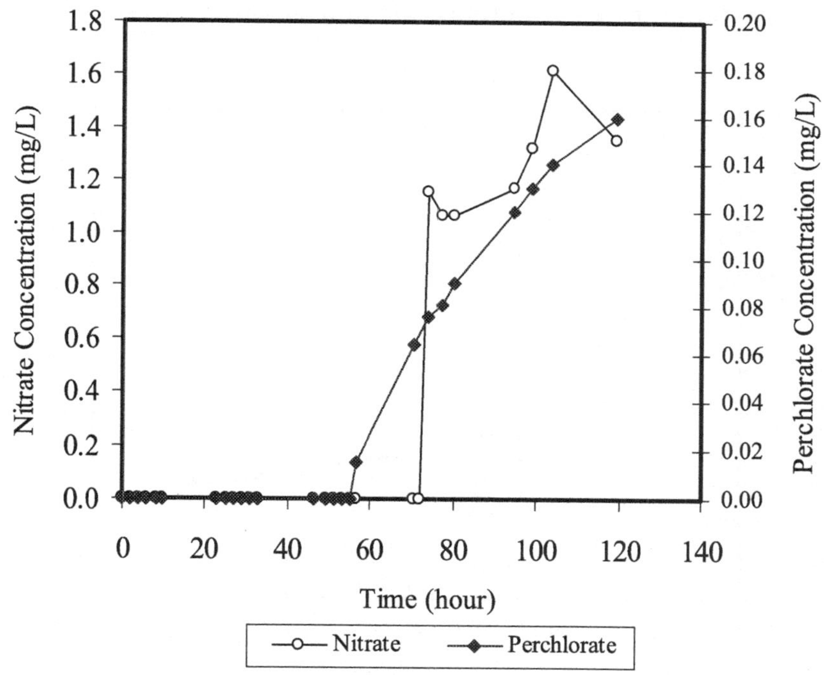


**Figure 6.43 The interference of nitrate on perchlorate biodegradation (test 4)**

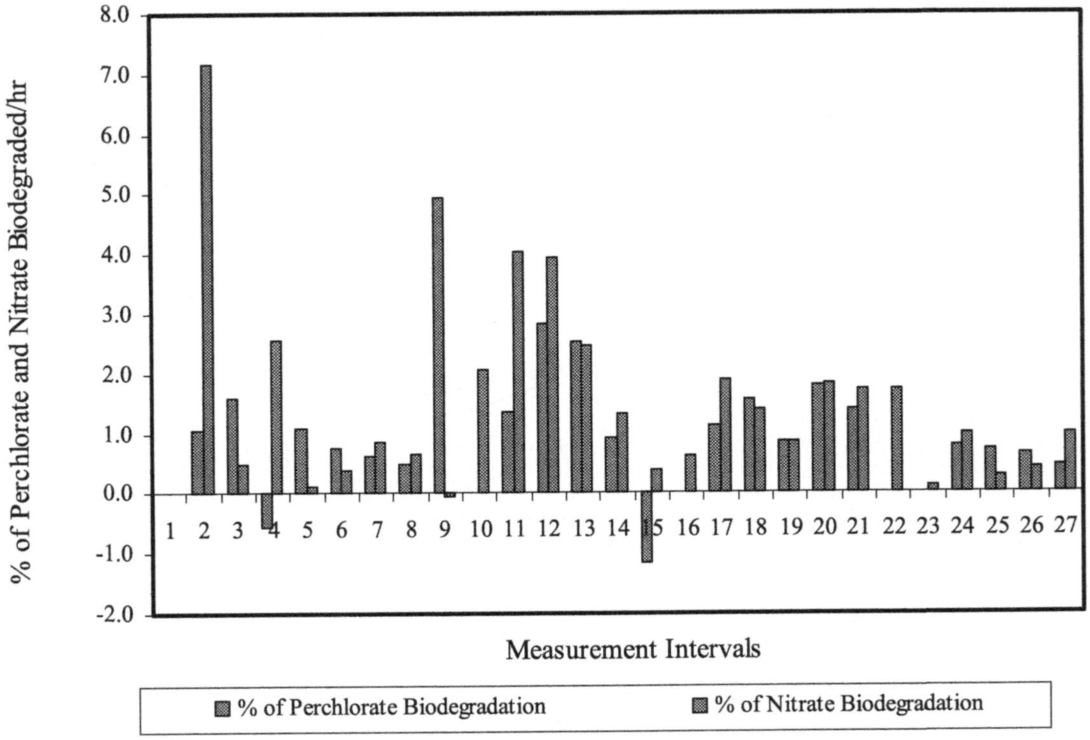


**Figure 6.44 Biodegradation rates of perchlorate and nitrate (test 4)**

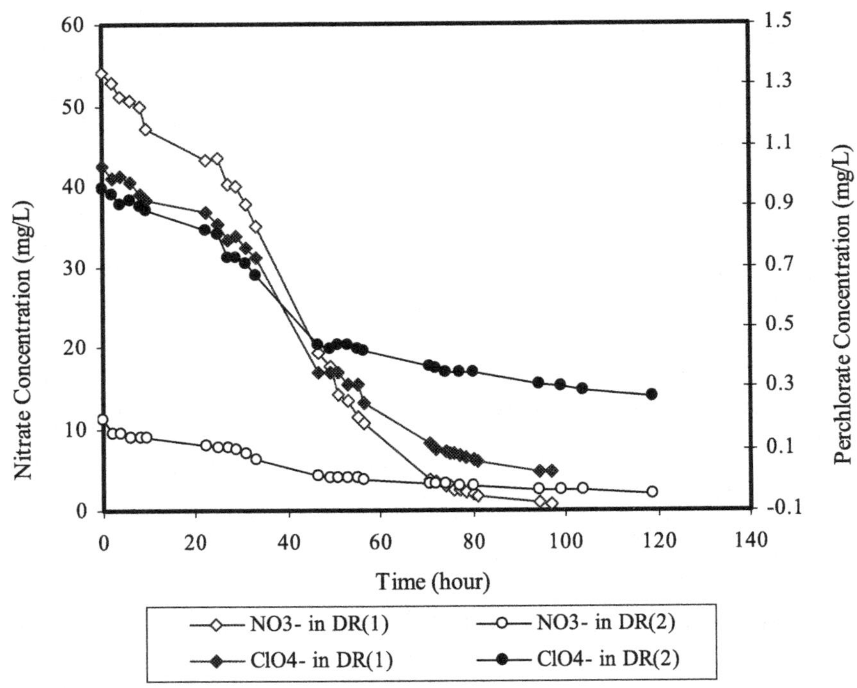


**Figure 6.45 Nitrate and perchlorate biodegradation in BTS-55 membrane-immobilized biofilm reactors (test 3 & test 4)**

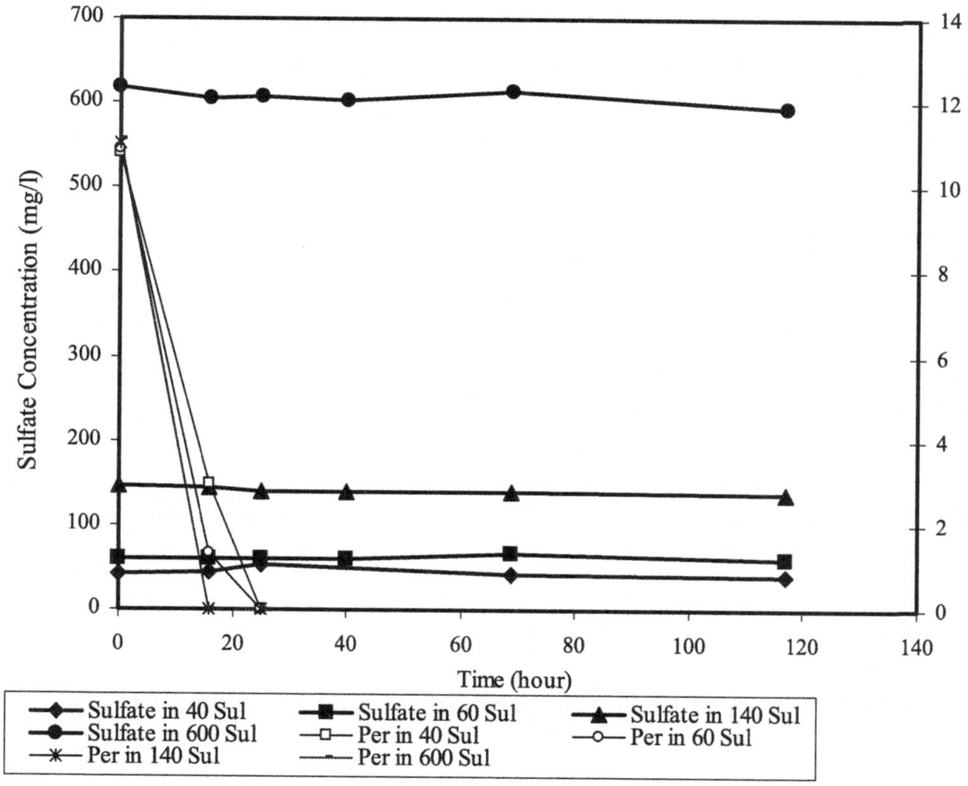


**Figure 6.46 Interference of sulfate on perchlorate biodegradation (batch testing)**

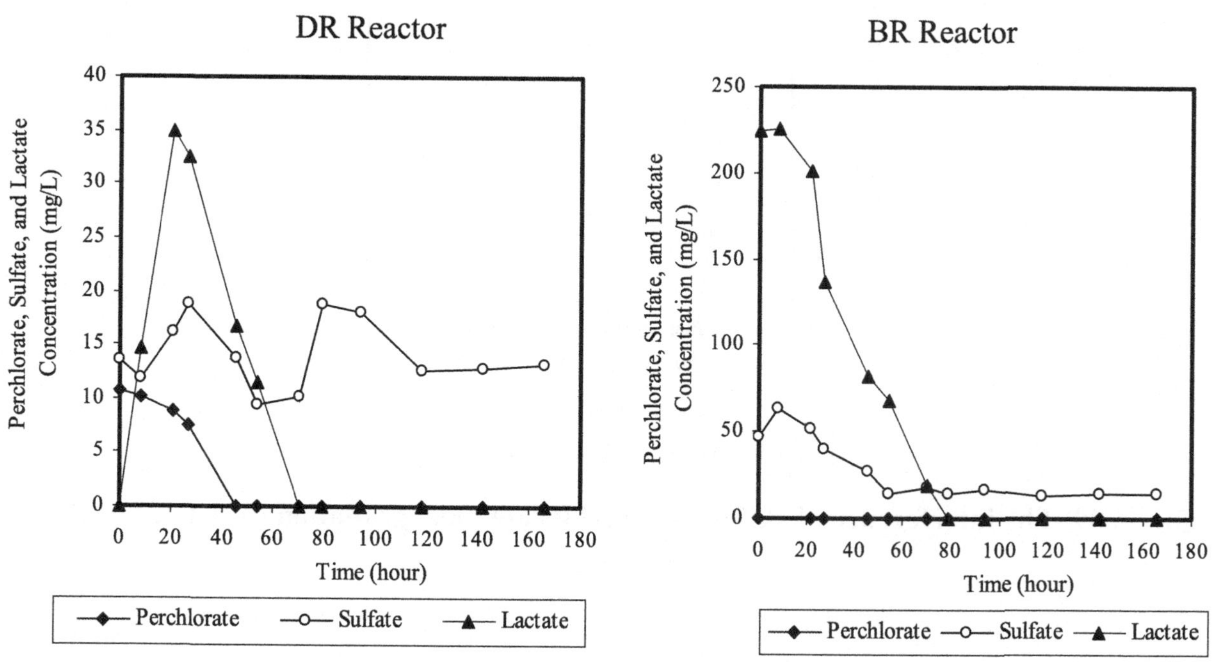


**Figure 6.47 The interference of sulfate on perchlorate biodegradation (test 1)**

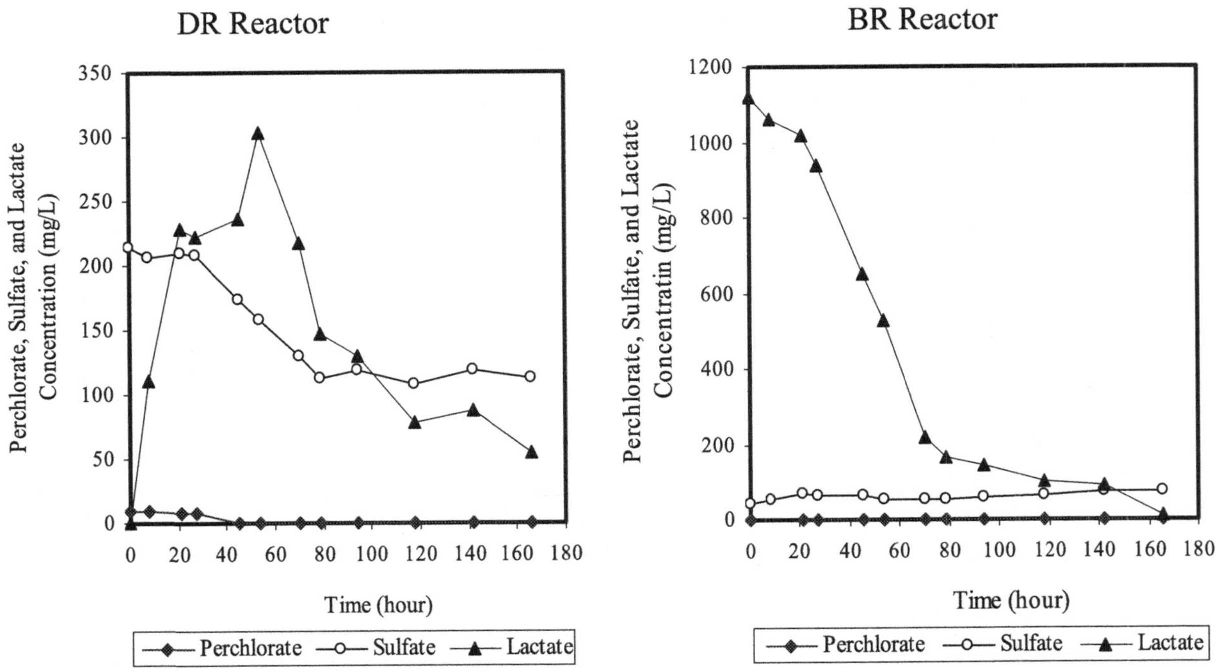


**Figure 6.48 The interference of sulfate on perchlorate biodegradation (test 2)**

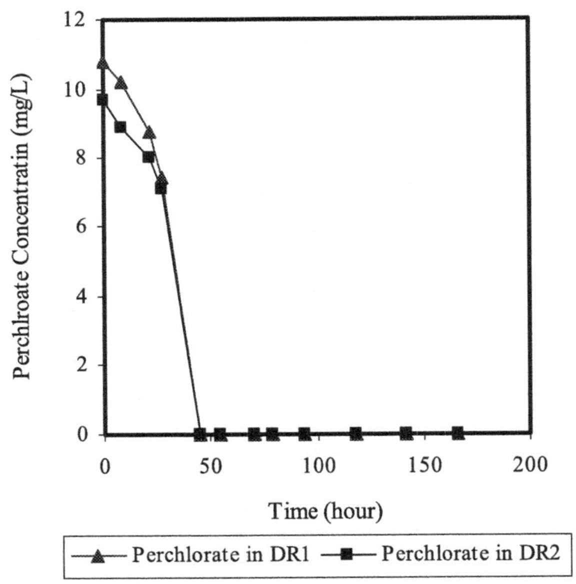


**Figure 6.49 The interference of sulfate on perchlorate biodegradation (test 1 & test 2)**

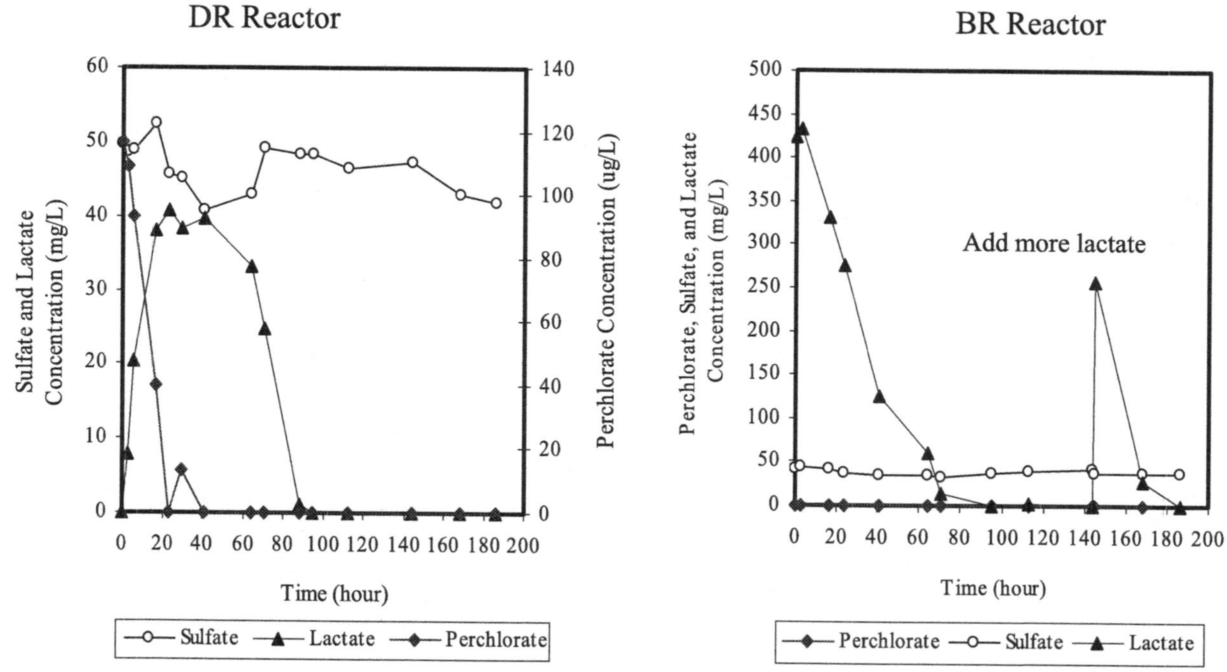


**Figure 6.50 The interference of sulfate on perchlorate biodegradation (test 3)**

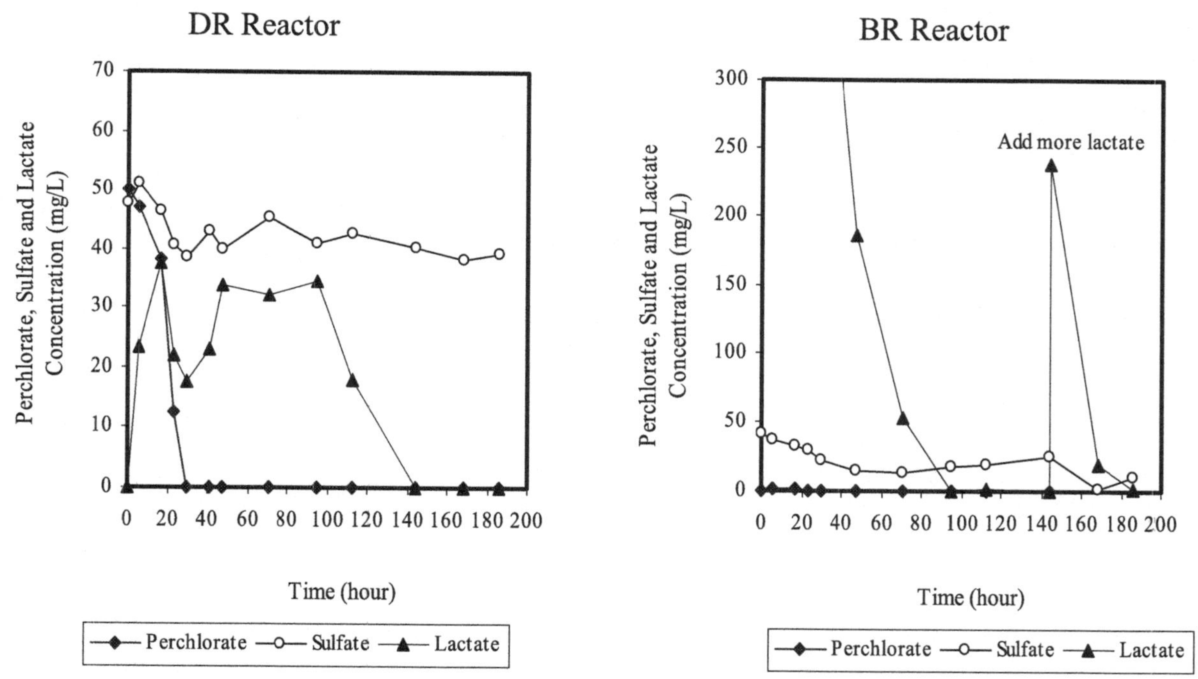


**Figure 6.51 The interference of sulfate on perchlorate biodegradation (test 4)**

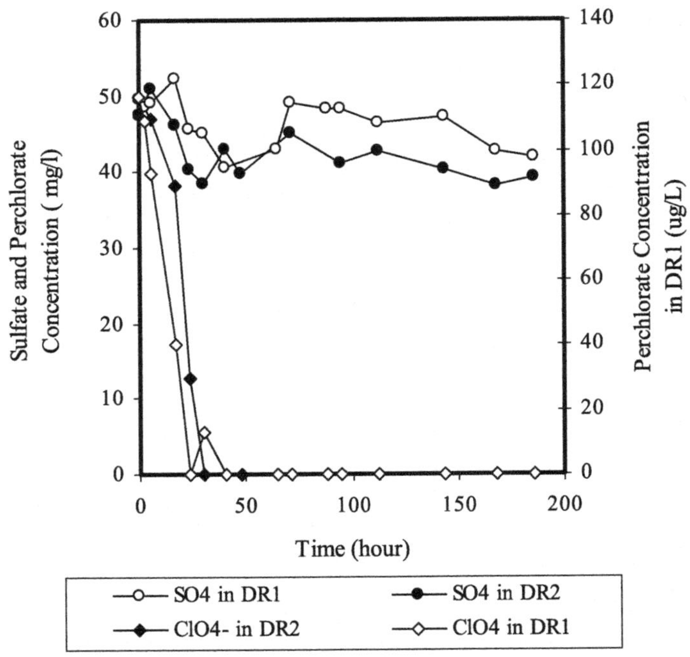


**Figure 6.52 The interference of sulfate on perchlorate biodegradation (test 3 & test 4)**

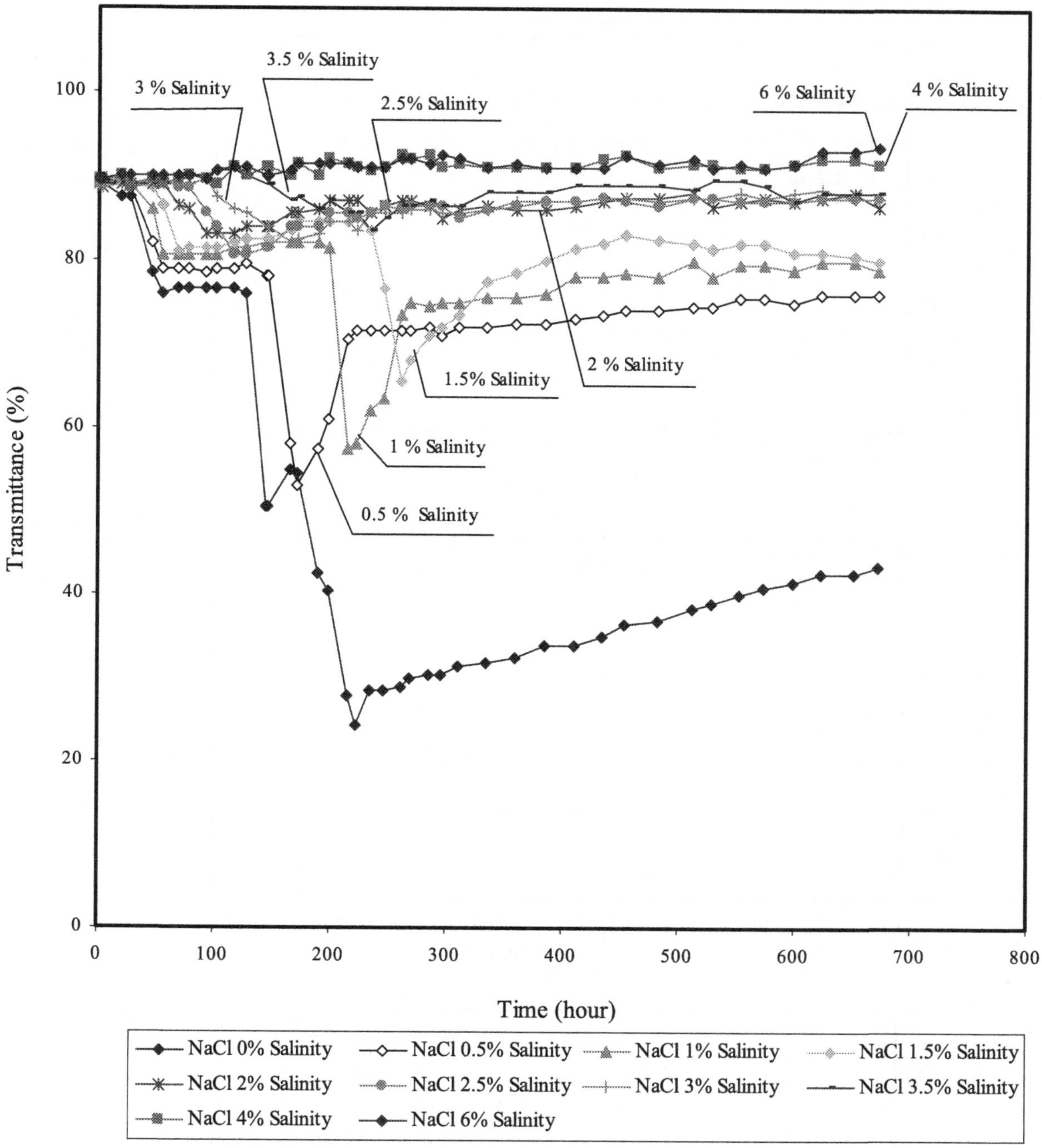


**Figure 6.53 Transmittance of different salinity concentrations**

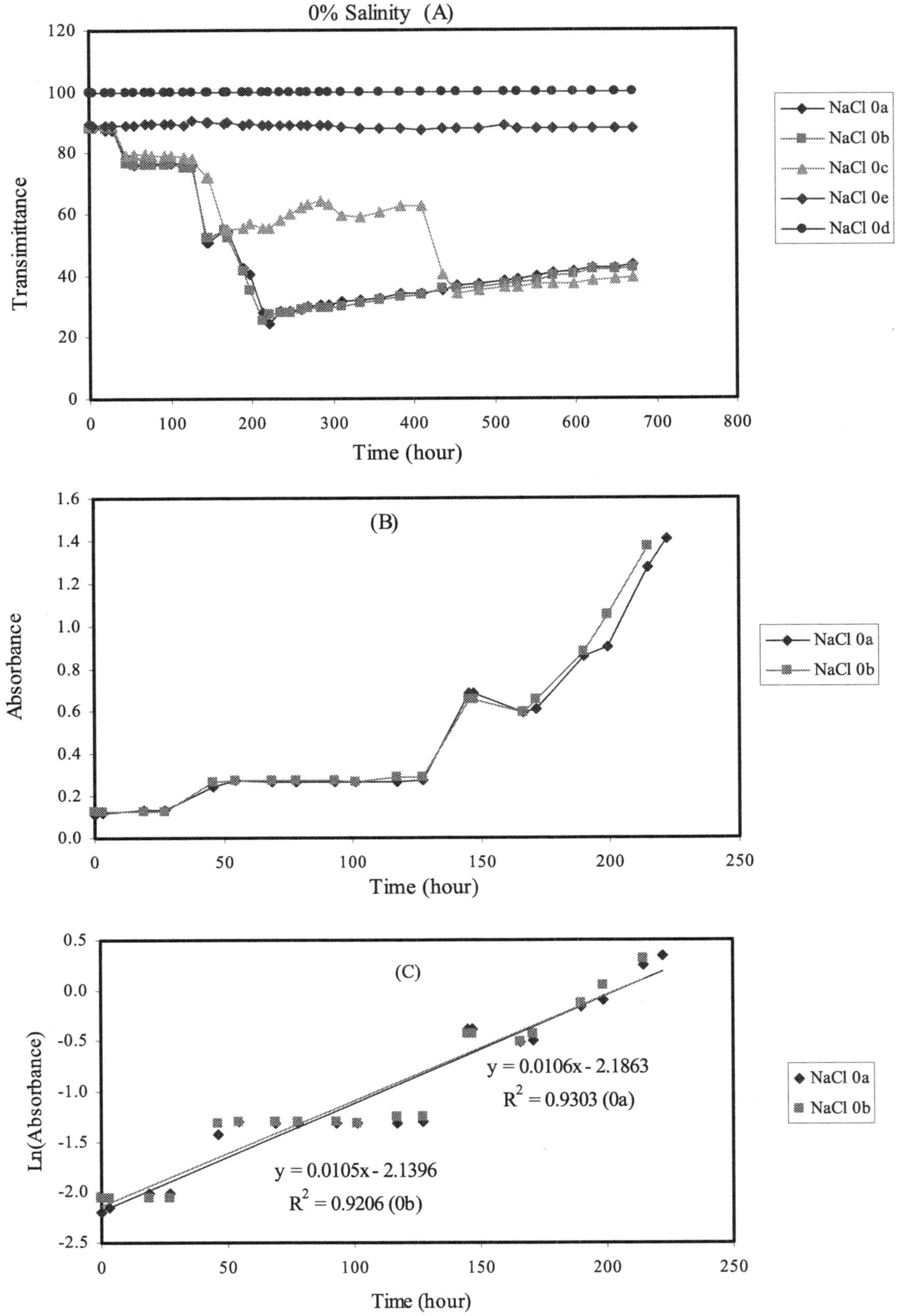


**Figure 6.54 Interference of salinity (0%) on perchlorate biodegradation**

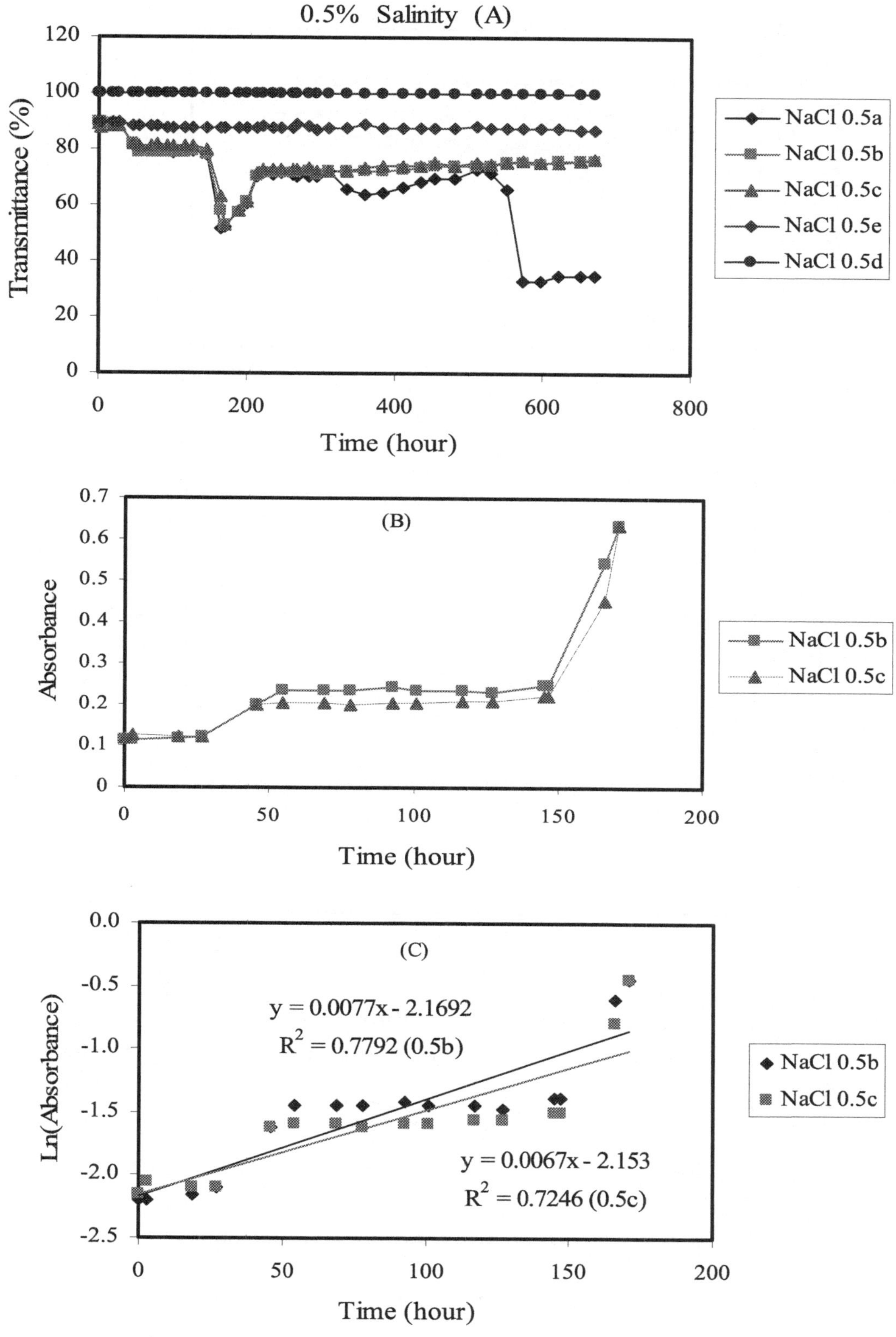


**Figure 6.55 Interference of salinity (0.5%) on perchlorate biodegradation**

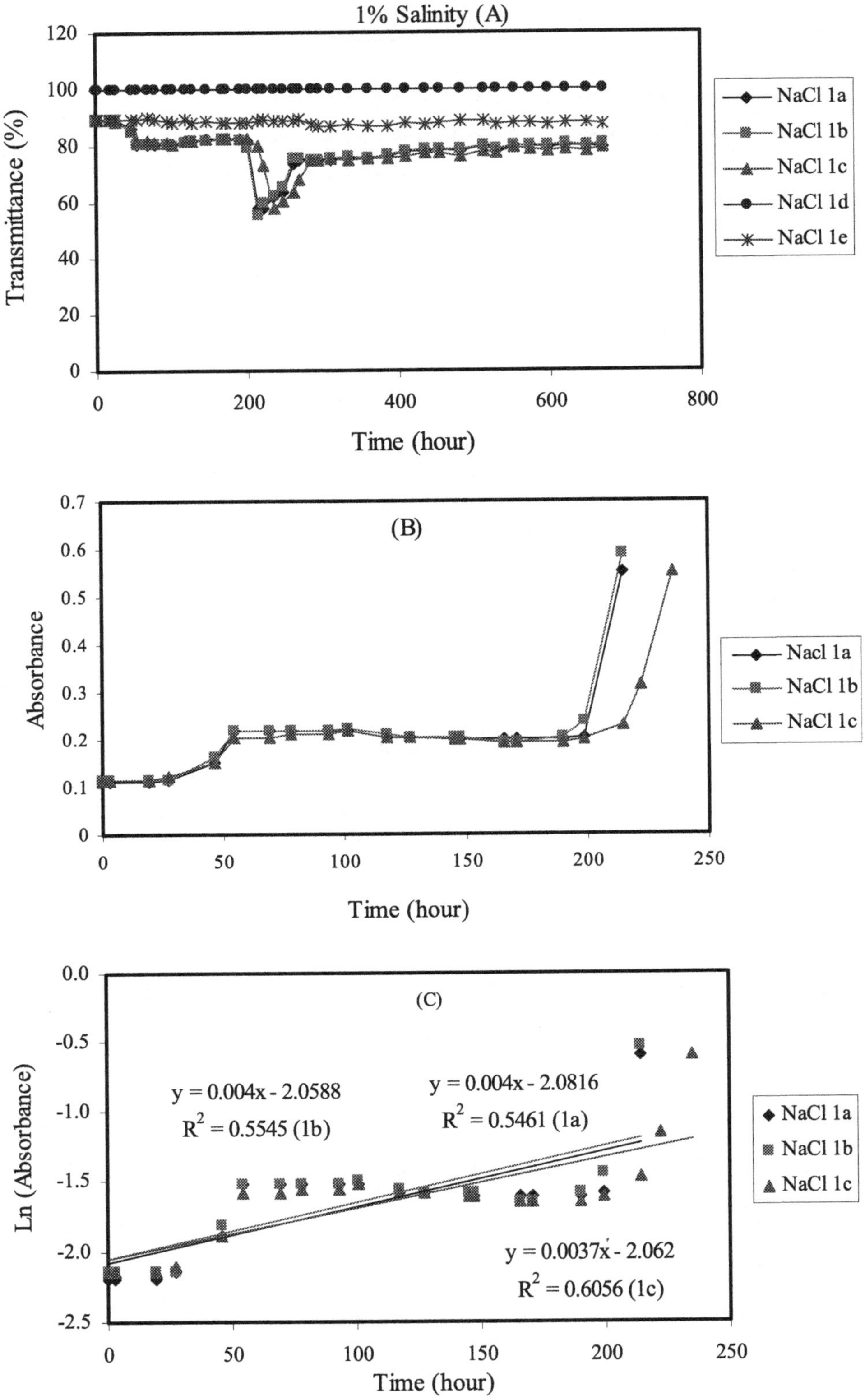


**Figure 6.56 Interference of salinity (1.0%) on perchlorate biodegradation**

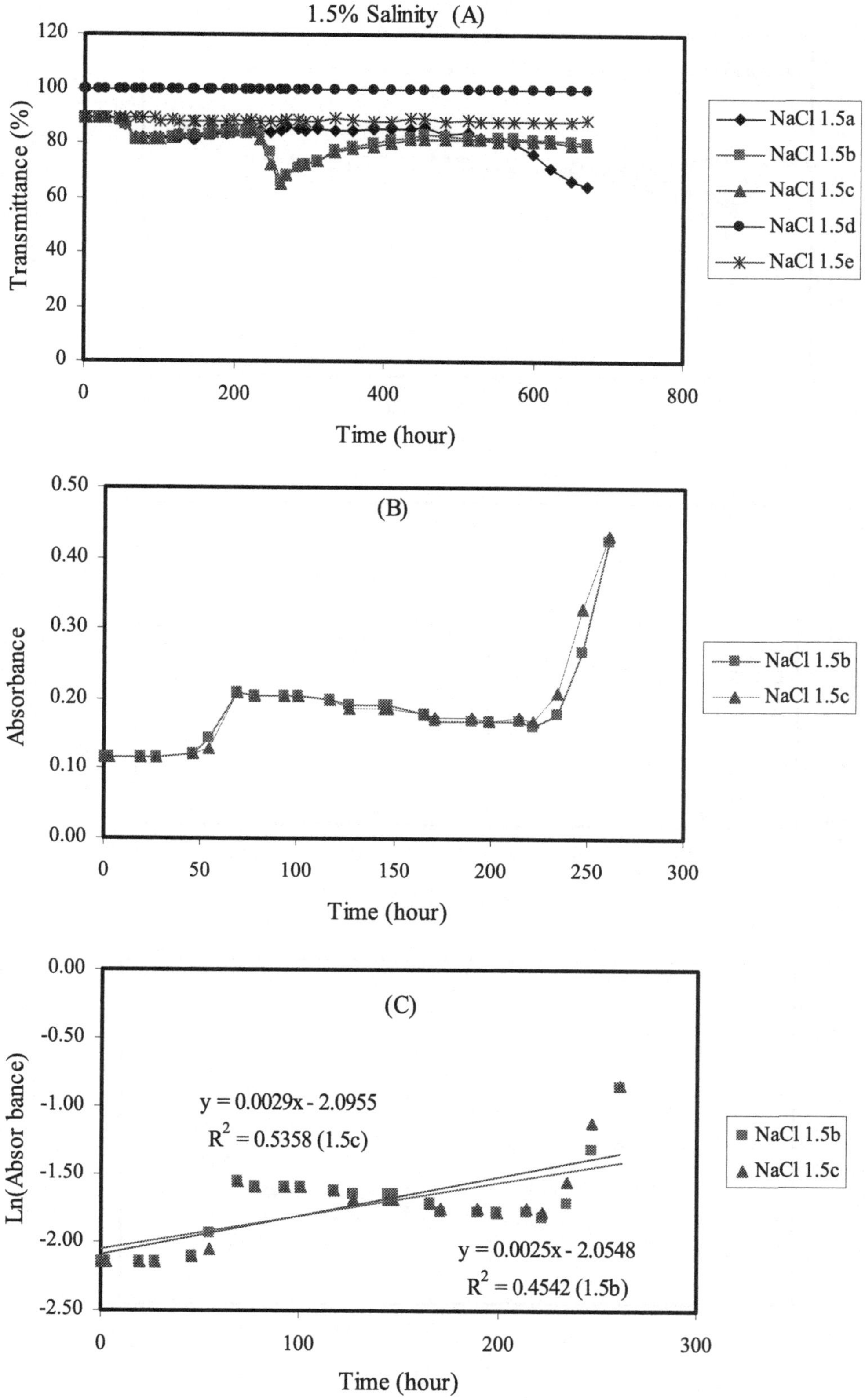


**Figure 6.57 Interference of salinity (1.5%) on perchlorate biodegradation**

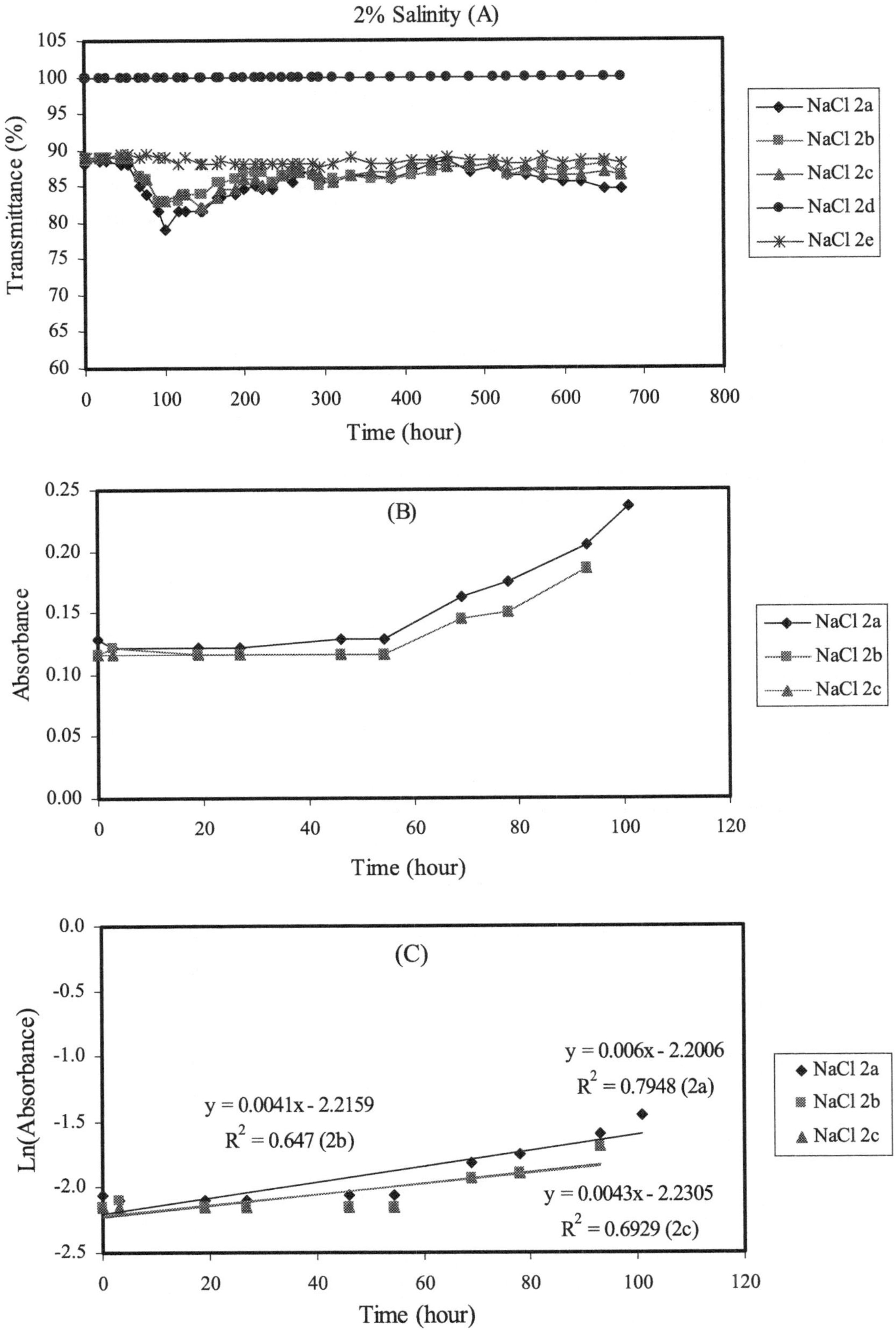


**Figure 6.58 Interference of salinity (2.0%) on perchlorate biodegradation**

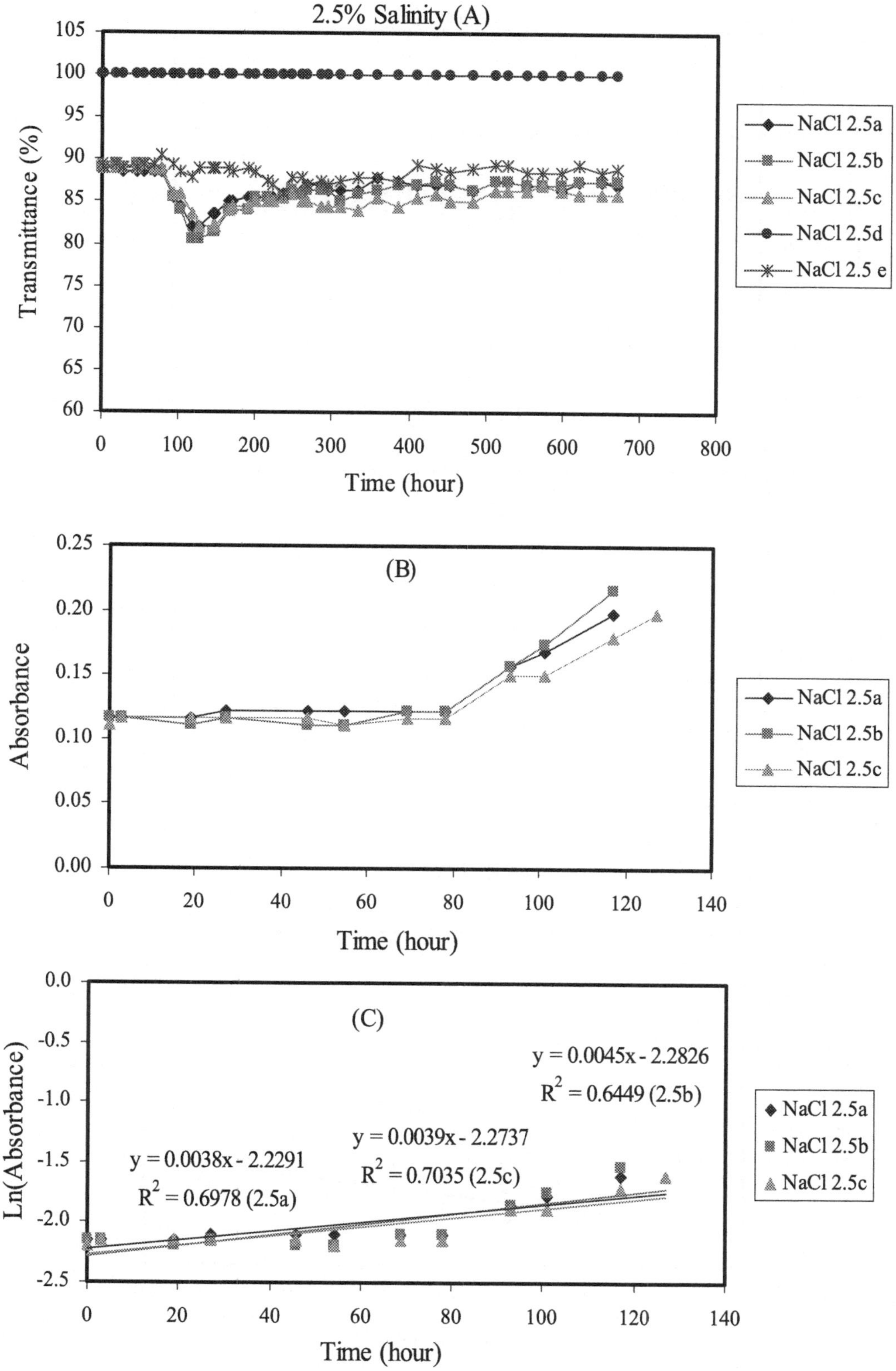


**Figure 6.59 Interference of salinity (2.5%) on perchlorate biodegradation**

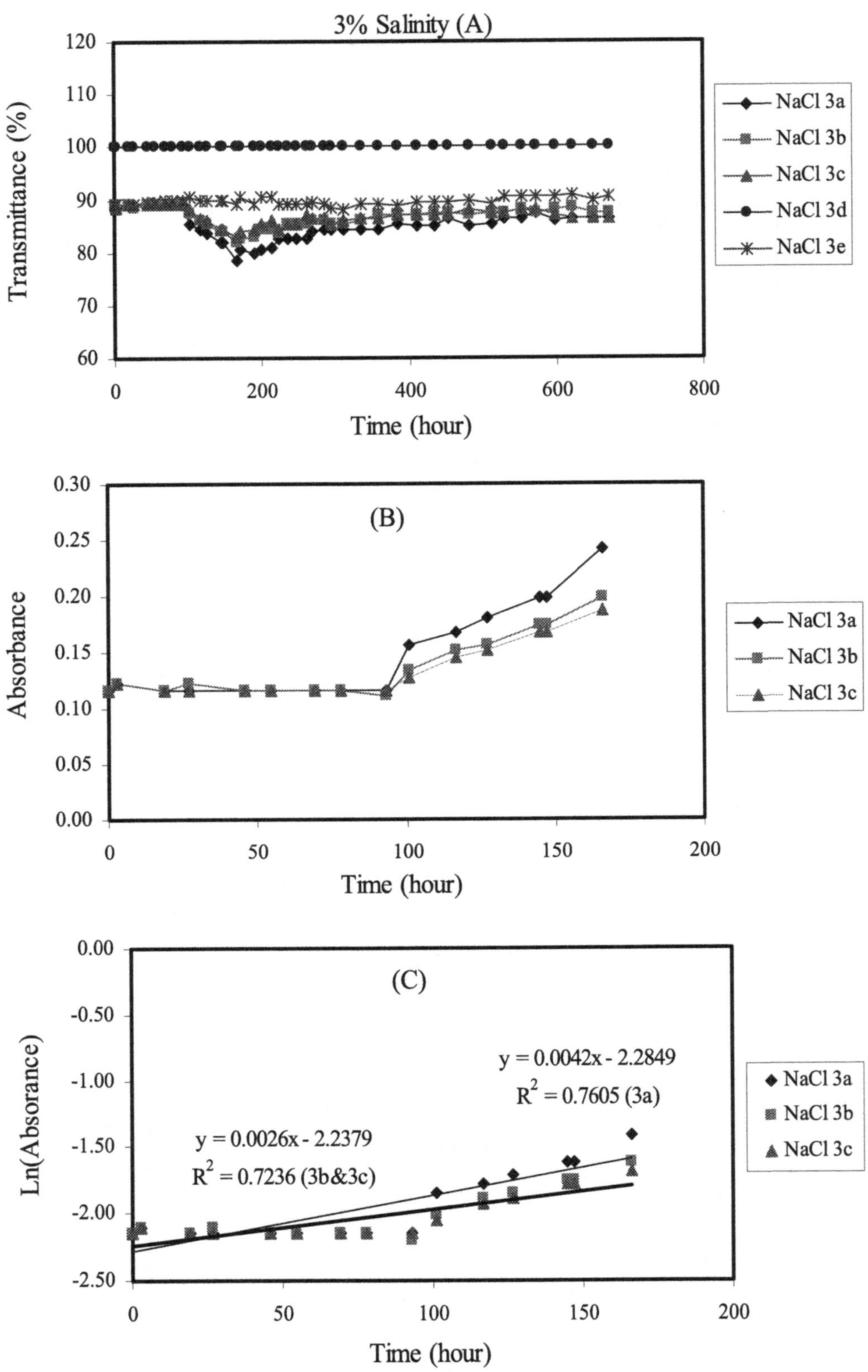


**Figure 6.60 Interference of salinity (3.0%) on perchlorate biodegradation**

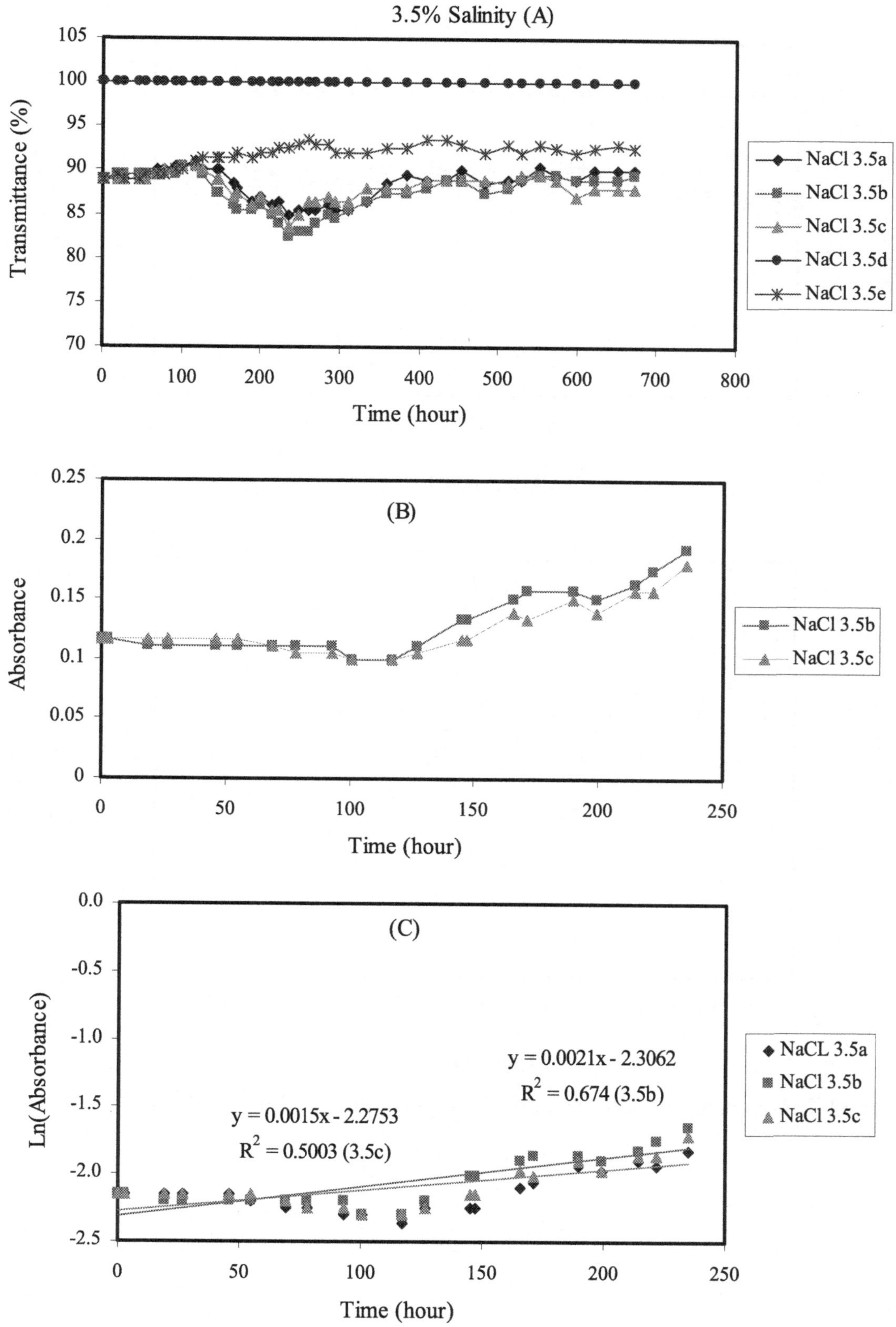


**Figure 6.61 Interference of salinity (3.5%) on perchlorate biodegradation**

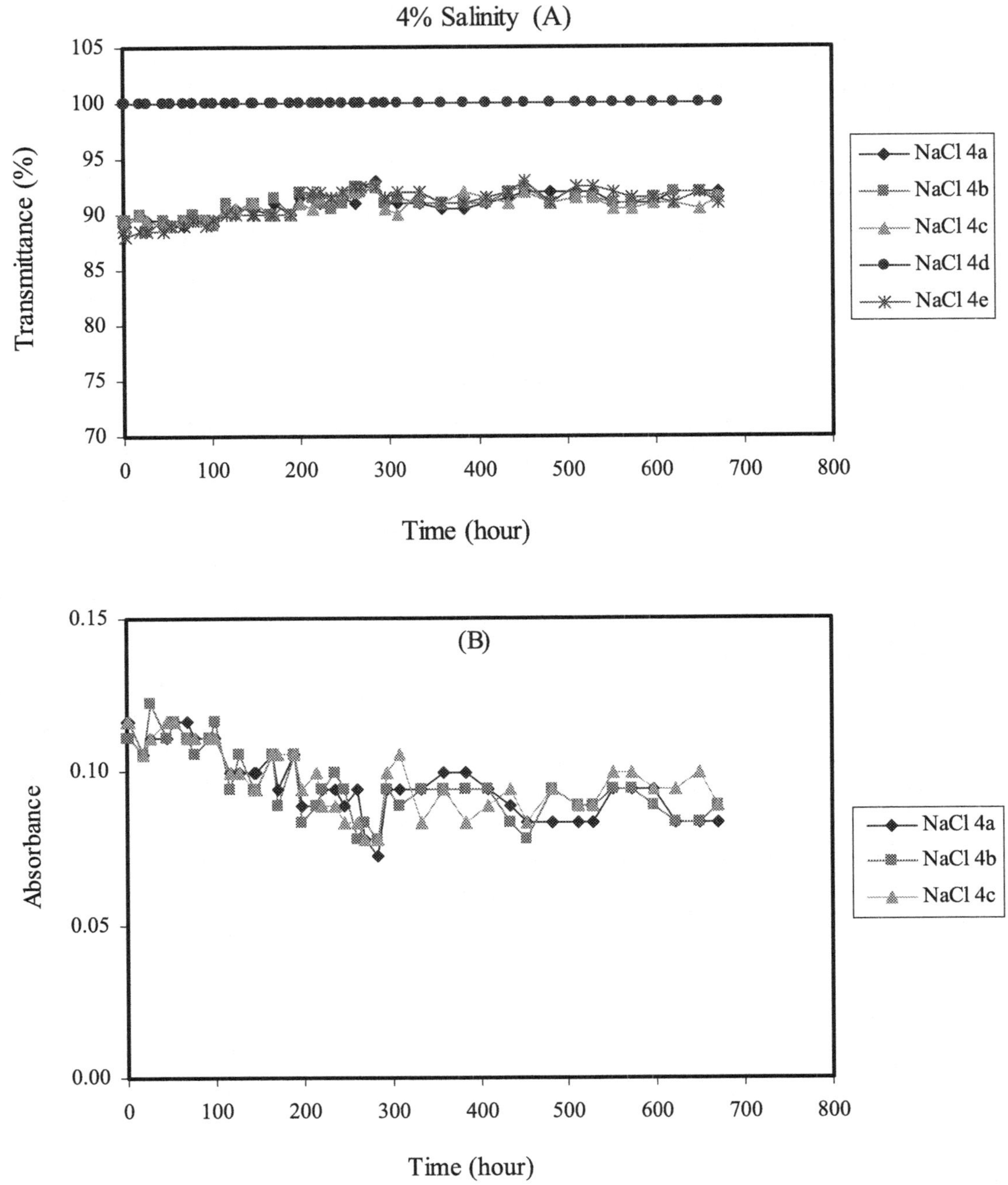


**Figure 6.62 Interference of salinity (4.0%) on perchlorate biodegradation**

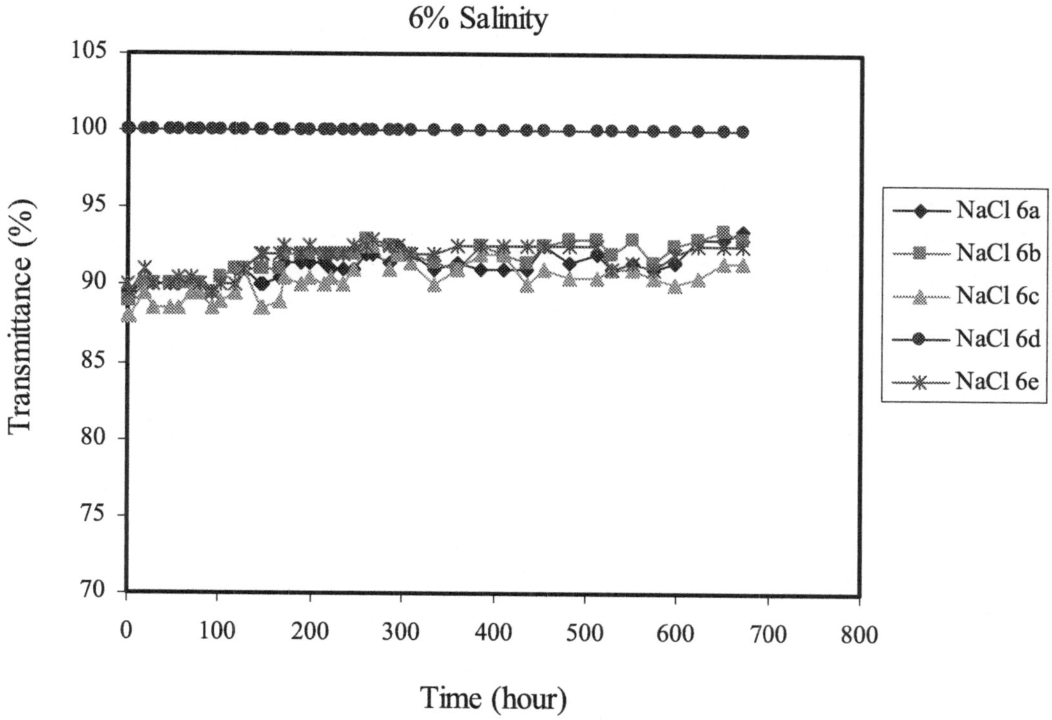


**Figure 6.63 Interference of salinity (6.0%) on perchlorate biodegradation**

# CHAPTER 7
# PILOT SCALE TESTS

## INTRODUCTION

In May of 2001, the pilot reactor system for perchlorate removal (the PSU O4 system) was delivered to the Redlands site (the Texas Street Well Facility), inoculated, and operated by personnel from Penn State University and the subcontractor (Camp, Dresser and McKee Inc.). The plant consisted of two parallel trains feeding Redlands water (containing approximately 70 to 80 ppb of perchlorate) to packed beds containing either sand or plastic media. Acetic acid (50%) was added along with nutrients (ammonium phosphate) to the reactor feed. The plant was operated until mid-December of 2001, and then was shut down. The data from the pilot study is now being fully analyzed and the results are being put into a format suitable for a final AwwaRF report.

In this chapter, we provide our findings of the pilot plant study using two different reactors, a sand medium reactor and a plastic medium reactor. In the next chapter, we provide an economic analysis of the reactor system with cost estimates for installing a system at the Redlands site.

## METHODS FOR THE PILOT STUDY

### Inoculum

A perchlorate-reducing bacterium, *Dechlorosoma* sp. KJ, isolated from a wastewater sample obtained from an activated sludge aeration basin at the Penn State University (PSU) Wastewater Treatment Plant, was used to inoculate the pilot plant reactors. The isolate KJ was enriched in an Erlenmeyer flask containing medium consisting of 300 g/L $NaClO_4$, 450 mg/L $NaCH_3COO$, buffer solution, and trace metal solution (see below). The pure culture (1 mL) was transferred weekly into freshly prepared medium (200 mL) in an anaerobic glove box (Coy Scientific Products) in the PSU Kappe laboratories. For inoculating the pilot-scale bioreactor, 1 L of the culture was prepared and poured into a 55-gallon drum containing the same composition of medium used at laboratory except that groundwater from field site was used instead of deionized water. The headspace of the 55-gallon drum was purged with nitrogen gas (100%) before and after inoculation to prevent oxygen from dissolving into media, and it was mixed well after being tightly sealed. After one-week incubation, perchlorate degradation was observed and the culture was pumped into the reactor.

### Media

In the laboratory, the medium used for growing KJ was prepared with 300 mg/L $NaClO_4$, 450 mg/L $NaCH_3COO$, buffer solution, and metal solution using deionized water (Milli-Q System, Millipore Corp., New Bedford, MA, USA). The buffer solution consisted of 1.92 g $K_2HPO_4$, 0.98 g $NaH_2PO_4H_2O$, and 0.5 g $NH_4H_2PO_4$ per liter of medium. The metal solution contained the following constituents per liter of a medium: 55 mg $MgSO_47H_2O$, 0.6 mg $H_3BO_3$, 1 mg $CaCl_26H_2O$, 0.4 mg $CoCl_26H_2O$, 0.2 $CuSO_46H_2O$, 3 mg EDTA, 4 mg $FeSO_47H_2O$, 1 mg $MnCl_24H_2O$, 0.1 mg $NiCl_26H_2O$, 0.4 $NaMoO_42H_2O$, 0.15 mg $NaSeO_35H_2O$, and 2 mg $ZnSO_47H_2O$. Medium used in laboratory was autoclaved using steam sterilizer (STERIS

Corporation, Heisley Road Mentor, Ohio) before it was inoculated with bacteria. Medium prepared at the field site and put into the 55-gallon drum was not sterilized.

### Bioreactors

Two side-by-side packed-bed bioreactors were constructed by CDM and transported to the Texas Street Well Facility to investigate the performance of perchlorate removal. One reactor was filled with sand (1 mm diameter) and the other was filled with tri-pack plastic media (Jaeger Product Inc., Texas; 3.175 cm diameter). These two media were packed to a height of 4 ft in the reactor that was 7 ft in height, 2 ft in width, and 1 ft in length. Because the density of the plastic media was less than that of water, a perforated plate was placed on the top of media to hold it down. However, there was still a small space below the media at the bottom of the plastic media reactor. Nine taps were positioned on the side of wall of each reactor in order to allow sampling of the reactor water along the length of the reactor. Samples from each tap were collected from the top to the bottom tap. The two reactors were operated identically in up-flow mode. Groundwater was pumped into the media from the bottom it overflowed a weir located at the top of reactor and dropped by gravity into a drainage system.

For startup of the sand media reactor, the KJ culture was pumped from the drum (55-gal.) into the empty reactor after complete perchlorate degradation was observed in the drum. Perchlorate (300 mg/L of $NaClO_4$) and acetate (450 mg/L of $NaCH_3CO_2$) were added into the tank immediately before culture was pumped into the reactor. After the reactor was filled with KJ culture, sand was poured into the reactor from the top of the reactor allowing sand particles to mix well with culture. The plastic media was packed into the reactor and then the KJ culture was pumped into the bottom of the reactor. Both reactors inoculated with bacteria were left undisturbed for one day to allow bacteria to attach to the surfaces of media. In order to enhance the development biofilm in the reactors, the system was operated in full recirculation mode using a 55-gallon drum containing groundwater and high concentrations of acetate and perchlorate (150 mg/L of $NaCH_3CO_2$; 200 mg/L of $NaClO_4$). After 11 days of operation in recirculation-mode, the reactor was switched to continuous-feed mode by pumping groundwater at 1 gallon per min into the media and wasting the effluent. The operation in continuous-feed mode started on May 20, 2001 and continued to November 2, 2001 for the sand media reactor and to November 20 for plastic media reactor. The pumping rate ranged from 1 gpm to 2 gpm during the operation for both reactors.

Feed to the reactors consisted of acetic acid (50%) that was pumped (Masterflex pump) into the reactors into the feed line to serve as a carbon source and electron donor for bacteria. The final acetic acid concentration was about 55 mg/L after it was mixed with groundwater. For N and P sources for the bacteria, ammonium phosphate ($NH_4H_2PO_4$) was also pumped, at C: N ratio of 5:1, into the influent line and mixed with groundwater before entering into the reactor.

### Analytical Procedures

Samples were collected from influent and effluent, and taps of the reactor to obtain daily and weekly measurements, respectively, of the properties listed in Table 7.1. Before storing a sample in a plastic bottle (Nalgene® Labware), each sample was filtered using 0.2 μm syringe filter (Fisher Scientific). The sample was then stored at 4°C temperature at the Redlands Wastewater Treatment Plant laboratory. For most samples, the anions (perchlorate, acetate and nitrate) were measured at the Redlands laboratory using ion chromatography (IC) and the following Dionex IC (Dionex Corp., Sunnyvale, CA, USA) and column systems: perchlorate, DX-600 with AS 16 column; acetate and nitrate, DX 450 with AS4A-SC column. After day 85, perchlorate and acetate were measured at Penn State University using the following IC systems and columns: acetate, DX 500 with column AS 11; perchlorate, DX-20 with AS 16 column.

The hydraulic retention time of each reactor was determined by spiking a high concentration of a NaCl tracer solution (about 150 mg/L) into the reactor. The salt concentration in the reactor effluent was measured using a conductivity meter (YSI 600XL; YSI Incorporated, Yellow Springs, OH, USA).

## RESULTS AND DISCUSSION OF THE PILOT STUDY

### Sand Media Reactor

The pilot scale reactor is shown in Figure 7.1. Perchlorate concentrations were measured several times each week in the influent and effluent at the reactor over the 166 days of operation (Figure 7.2). The average influent perchlorate concentration was 76 μg/L. The effluent perchlorate concentrations varied depending on the conditions of the reactor. The start up period was defined as the first 26 days of reactor operation based on the time for perchlorate concentration to reach non-detectable levels (<4 μg/L) in the reactor effluent. The initial groundwater pumping flowrate was 1 gpm, which corresponded to an estimated detention time of 28 min based on the assumed porosity of the sand media of 0.4. The flow rate was increased on day 33 to 2 gpm to determine the perchlorate removals at a higher flow rate. During several days of operation at 2 gpm, high concentrations of perchlorate were often detected from the effluent. Many bubbles, which were determined later to be 99% nitrogen gas (using gas chromatography), were observed to be flowing out of the top of the media in the reactor. It appeared (by looking through the glass window on the side of the reactor) that there was short channeling in the sand media, especially at the bottom of reactor. On day 64 (on a Sunday), there was a power interruption due to a failure in the well pump, and no water was pumped into the water storage tank. As a result, water was depleted from the storage tank and then air was pumped into the reactor instead of water. Full scale systems should be designed to contain a check-valve to prohibit draining of the reactor in the event of a power outage. This would help reduce the chance for oxygen contamination of the system. The effect of oxygen was obvious when the reactor was restarted, because when flow resumed to the reactor, the perchlorate concentration in the effluent was initially nearly the same as that in the reactor influent. The reactor performance then recovered over several days.

The gray circles used for the effluent data in Figure 7.2 indicate days when the media was backwashed. Unfortunately, the pump used for backwashing the media was not powerful enough to lift the sand media agglomerated by biofilm for complete mixing, and therefore complete mixing of the sand reactor was not possible until a new backwash pump was installed on day 79. The inability to completely mix the sand media led to channeling in the sand and incomplete perchlorate removal. After the new backwash pump was installed on day 79, the perchlorate concentrations from the effluent of the sand column were consistently non-detectable level (4 μg/L)—except when there were other operation problems as noted in the figure. When the flow rate was increased to 2 gpm on day 148, perchlorate concentrations in the effluent remained non-detectable.

The differences in perchlorate removals before and after backwashing are clearly shown in Figure 7.3 as a function of the different flow rates used in this reactor. Only data from days where there were not other operational problems with the reactor were used in this comparison. Operational problems consisted of days when groundwater, acetate, and nutrients were added properly into the reactor or there were no mechanical problem such as leakage from the reactor or breaks in the piping (Figure 7.4). Based on the same days as in the Figure 7.4, the nitrate concentrations and its removal are shown in Figure 7.5.

The groundwater at the Redlands site contained several potential electron acceptors including perchlorate, oxygen, and nitrate. In order to investigate the removal of these compounds at the media, profiles of each of these chemicals were obtained at a frequency of about once a week. Perchlorate was completely removed by tap 2 when the flow was 1 gpm, and was completely removed by tap 5 at a flow of 2 gpm (Figure 7.6). At the 1 gpm flow rate, it was determined that the oxygen and nitrate were also completely removed at the same point in the column where the perchlorate was removed. However, this result was obtained based on samples collected at certain taps. Therefore, it is not exactly determined where these electron acceptors were removed at the media. In case of the 2 gpm flow rate, oxygen and nitrate compounds were completely removed before the complete removal of perchlorate.

In order to determine the actual retention time of the groundwater in the reactor, a high concentration of salt (NaCl) was injected quickly into the line normally used for pumping nutrients. An analysis of the retention data shows that there was a significant difference in retention times before and after backwashing. Two examples of the effect of backwashing at flowrates of 1 gpm and 2 gpm are shown in Figure 7.7. The retention time was about 30 min at 1 gpm after backwashing, while it was approximately 21 min before backwashing at the same flowrate. There represents a 9 min difference or about a 30% reduction in detention time prior to backwashing. In case of the 2 gpm flow rate, the retention time measured after backwash was about 18 min while it was 10 min prior to backwashing, or a retention time reduction of 44% before backwashing. This data highlights the importance of regular backwashing to obtain detention times necessary for complete perchlorate removal.

### Plastic Media Reactor Operation

The plastic medium bioreactor is shown in Figure 7.8. Plastic media reactor was operated for 185 days, at flow rates of 1, 1.5, or 2 gpm. The average influent concentration of perchlorate was 75 μg/L, and the effluent concentration was usually below the detection limit except on days when there was no nutrient input or there were mechanical problems (Figure 7.9). The plastic media reactor was placed into continuous-mode operation at a flow rate of 1 gpm after 11 days of recycling groundwater between a reactor and 55-gal drum. The start up period was the same as for the sand reactor (26 days) and was defined as a period when perchlorate was completely removed from the effluent. When the flow rate was changed from 1 to 2 gpm, the effluent perchlorate concentration at first increased into the detectable region, but decreased again to non-detectable concentrations after a few days of operation. Because the storage tank was used for both sand and plastic media reactors, the power outage that affected the sand media reactor also affected the plastic media reactor. Following the power outage on 64 day, the effluent perchlorate concentration increased up to 63 μg/L, but non-detectable perchlorate concentrations in the effluent were quickly achieved after power was restored to the system. Changing the backwash pump on day 79 did not affect perchlorate removal as it did with the sand media reactor (Figure 7.10).

The complete data set was reduced to only days when operation was not affected by external conditions such as power or pump failures. This smaller, selected data set is shown in Figures 7.11 and 7.12 for effluent perchlorate and nitrate concentrations. Analysis of this smaller data set shows that complete perchlorate removal could be achieved in the plastic media reactor at flow rates of 1 gpm and 1.5 gpm, but consistent perchlorate removal was not achieved at a flow rate of 2 gpm.

Chemical concentrations inside the plastic media reactor are shown in Figure 7.13 for perchlorate, acetate, nitrate and oxygen, at flow rates of 1 gpm or 2 gpm. The tap numbers are listed in the direction of flow up through the reactor. At a flow rate of 1 gpm, complete removal of perchlorate was achieved after tap 6. The other electron acceptors, oxygen and nitrate, were completely removed after tap 4. When the flow rate was increased to 2 gpm, the perchlorate concentration was completely removed at tap 7 on this sampling day. Oxygen and nitrate were completely removed at tap 3 and tap 7, respectively. Oxygen was always the first electron acceptor that was removed, followed by nitrate and perchlorate. It was not necessary to completely remove any electron acceptor to see removal of the other electron acceptors.

The actual retention time in the plastic media reactor was measured on the days when the media was backwashed both before and after backwashing (Figure 7.14). At a flow rate of 1 gpm, the retention time was approximately 60 min before backwashing, and increased to 70 min after backwashing. At a flow rate of 2 gpm, the retention times were 32 min and 37 min before and after backwashing, respectively. These are relatively smaller changes in detention time for than were observed for the plastic media reactor, but for both flowrates backwashing increased retention times.

**Table 7.1**
**List of measurements**

| Measurement | Sample location | Frequency | Instrument |
|---|---|---|---|
| $ClO_4^-$ | Influent, Effluent | Daily | DX 600 |
| $ClO_4^-$ | Taps | Weekly | DX 600 |
| Acetate | Influent, Effluent | Daily | |
| Acetate | Taps | Weekly | |
| $NO_3^-$ | Influent, Effluent | Daily | DX 450 |
| $NO_3^-$ | Taps | Weekly | DX 450 |
| pH | Influent, Effluent | Daily | YSI 600XL |
| DO | Influent, Effluent | Daily | YSI 600XL |
| DO | Taps | Weekly | YSI 600XL |
| ORP | Influent, Effluent | Weekly | YSI 600XL |
| Temperature | Influent, Effluent | Daily | YSI 600XL |
| Backpressure | Influent | Daily | Pressure Transmitter |
| GW Flow Rates | Influent | Daily | Rotameter |
| AP and AA Flow Rates | Influent | M, W, F | Graduated cylinder |
| Ammonia | Influent, Effluent | Weekly | Titration |
| RTD | Effluent conductivity | Weekly | YSI 600XL |
| Conductivity | Influent, Effluent | Daily | YSI 600XL |
| Turbidity | Effluent | Daily | HACH |
| $SO_4^-$ | Influent, Effluent | Weekly | DX 450 |
| DOC | Raw, Influent, Effluent | Weekly | |
| TCE/PCE | Raw, Effluent | Monthly | GC |

**Figure 7.1 Pilot-scale reactor with sand media**

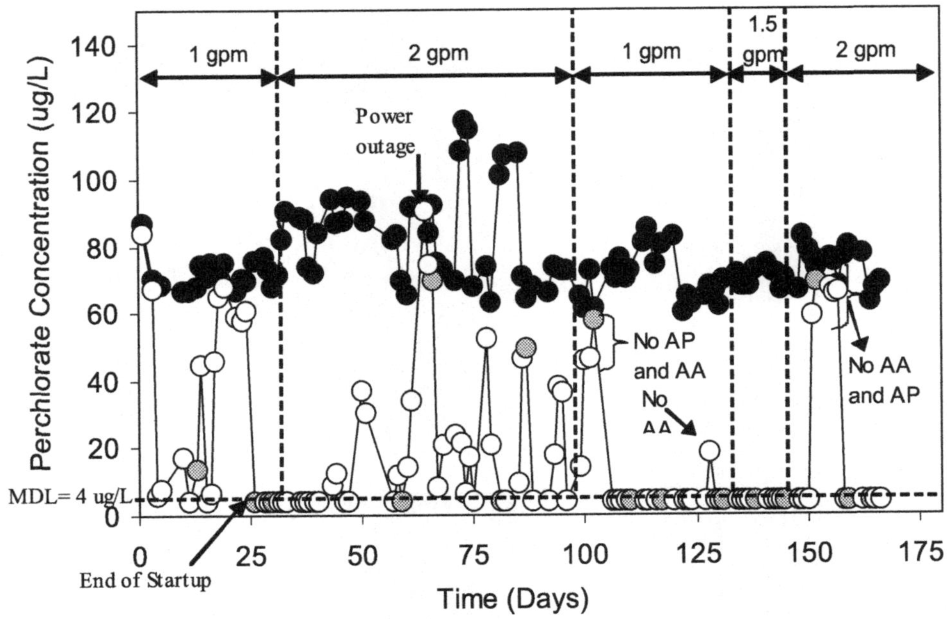


**Figure 7.2 Raw perchlorate data from influent (closed circle) and effluent (open circle; gray circle for indicating backwash time) of the sand media reactor**

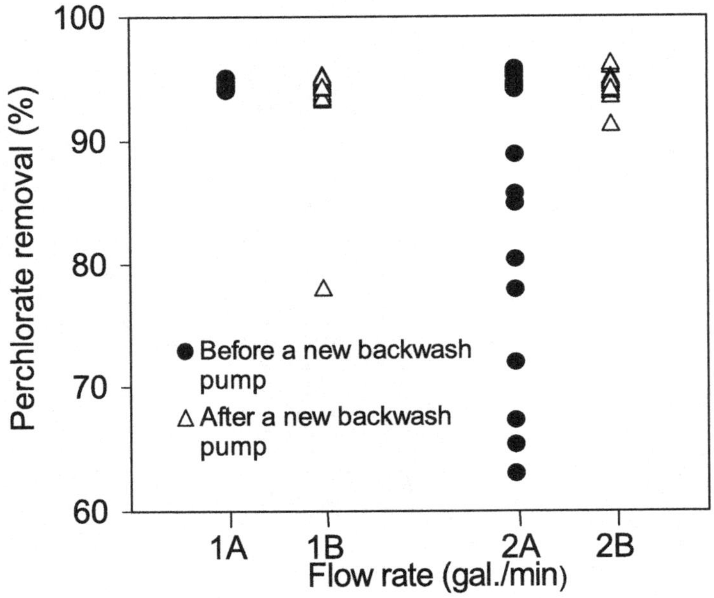


**Figure 7.3 Comparison of perchlorate removal before and after using a new backwash pump**

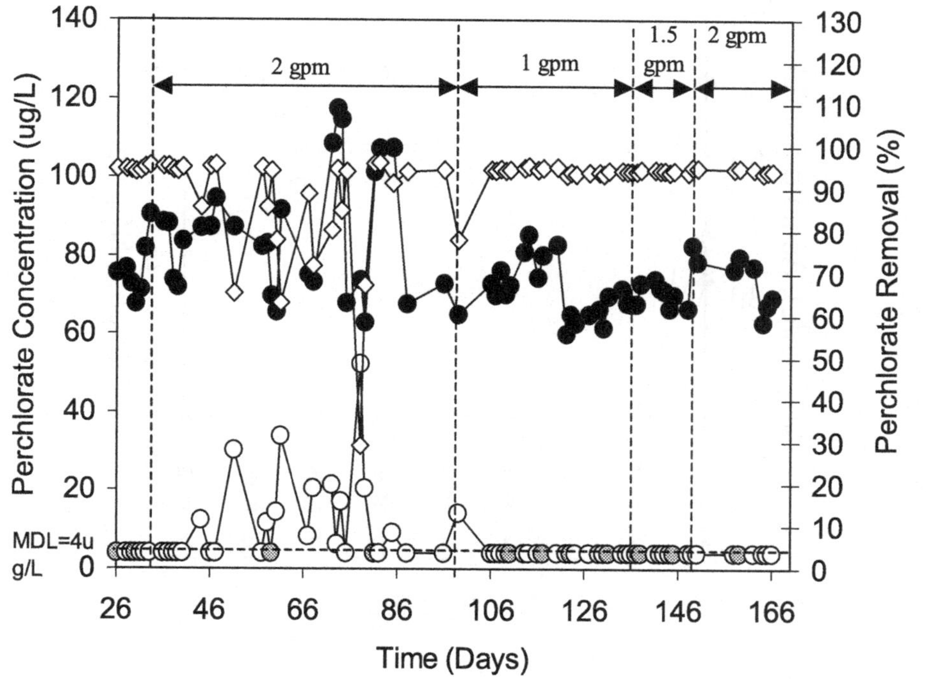


**Figure 7.4 Data for days of acceptable operation (no mechanical failures) of perchlorate from influent (closed circle) and effluent (open circle; gray circle for indicating backwash time) at the sand media reactor (open diamond for perchlorate removal)**

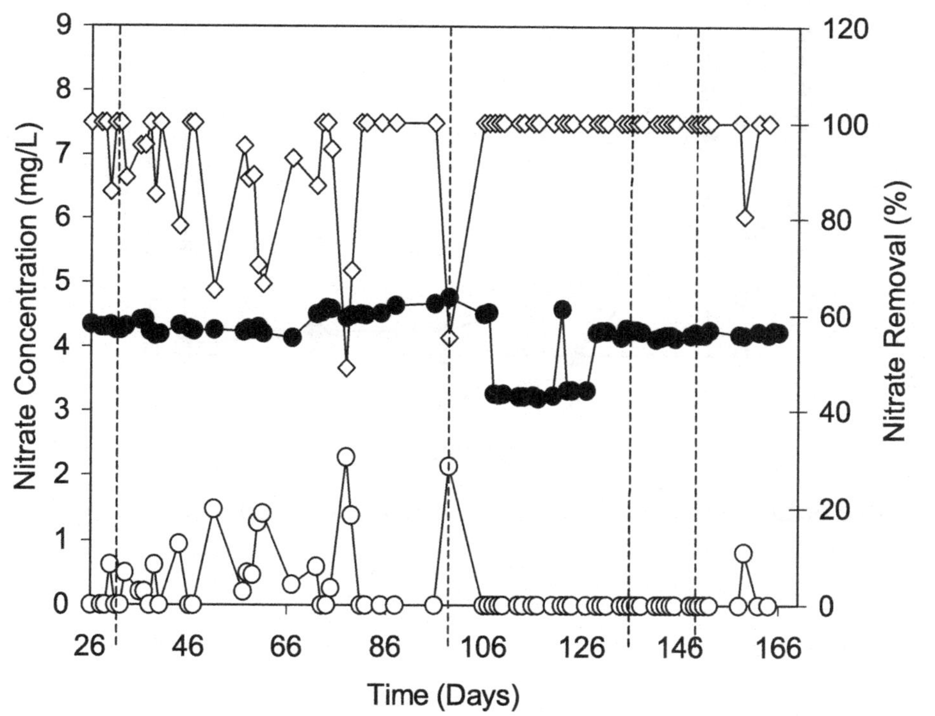


**Figure 7.5 Nitrate concentrations from influent (closed circle) and effluent (open circle) at the sand media reactor (open diamond for nitrate removal)**

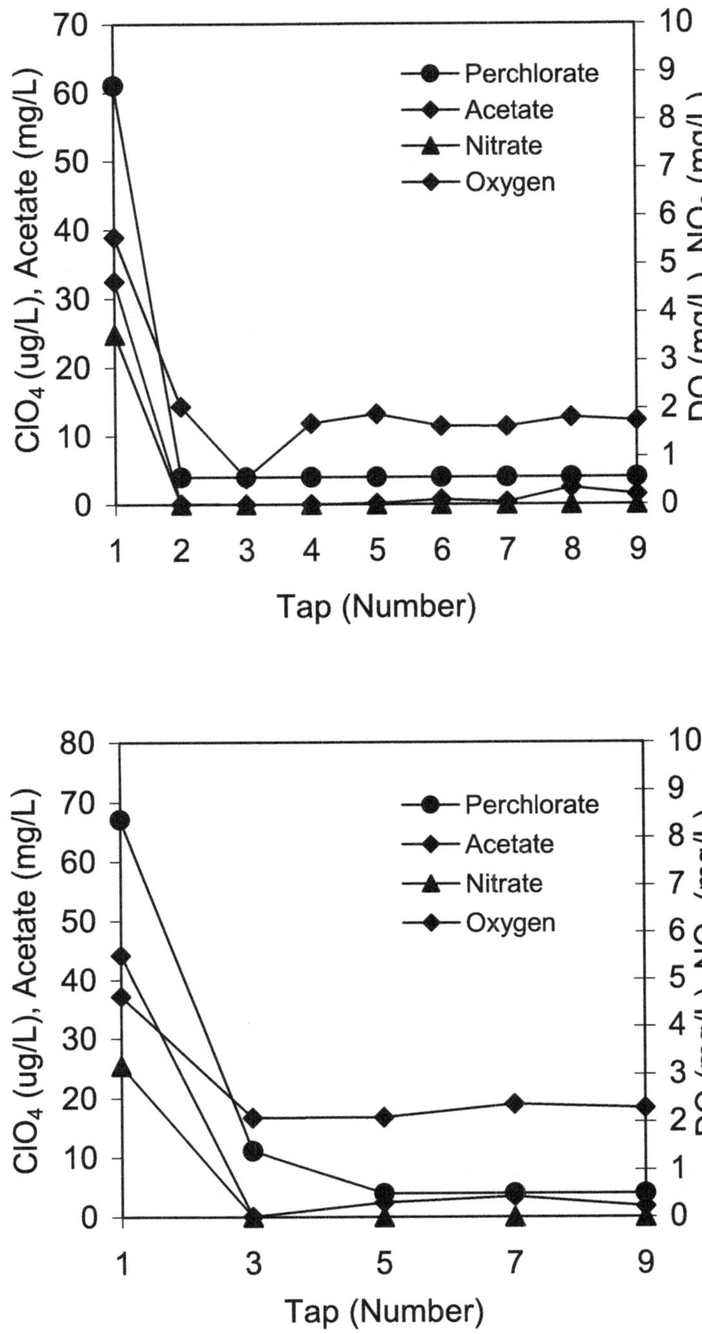


**Figure 7.6 Profile data from taps of the reactor at 1 gpm (upper) and 2 gpm (lower)**

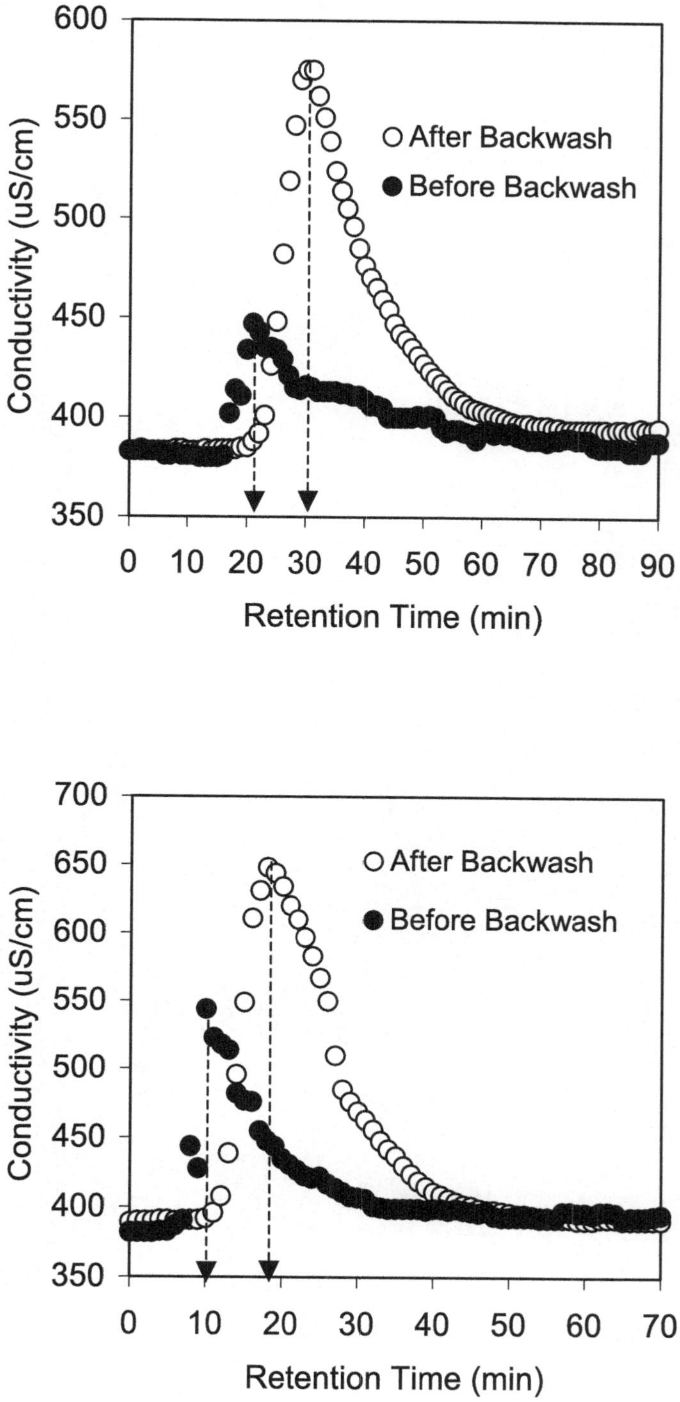


**Figure 7.7 Retention time test at 1 gpm (upper) and 2 gpm (lower)**

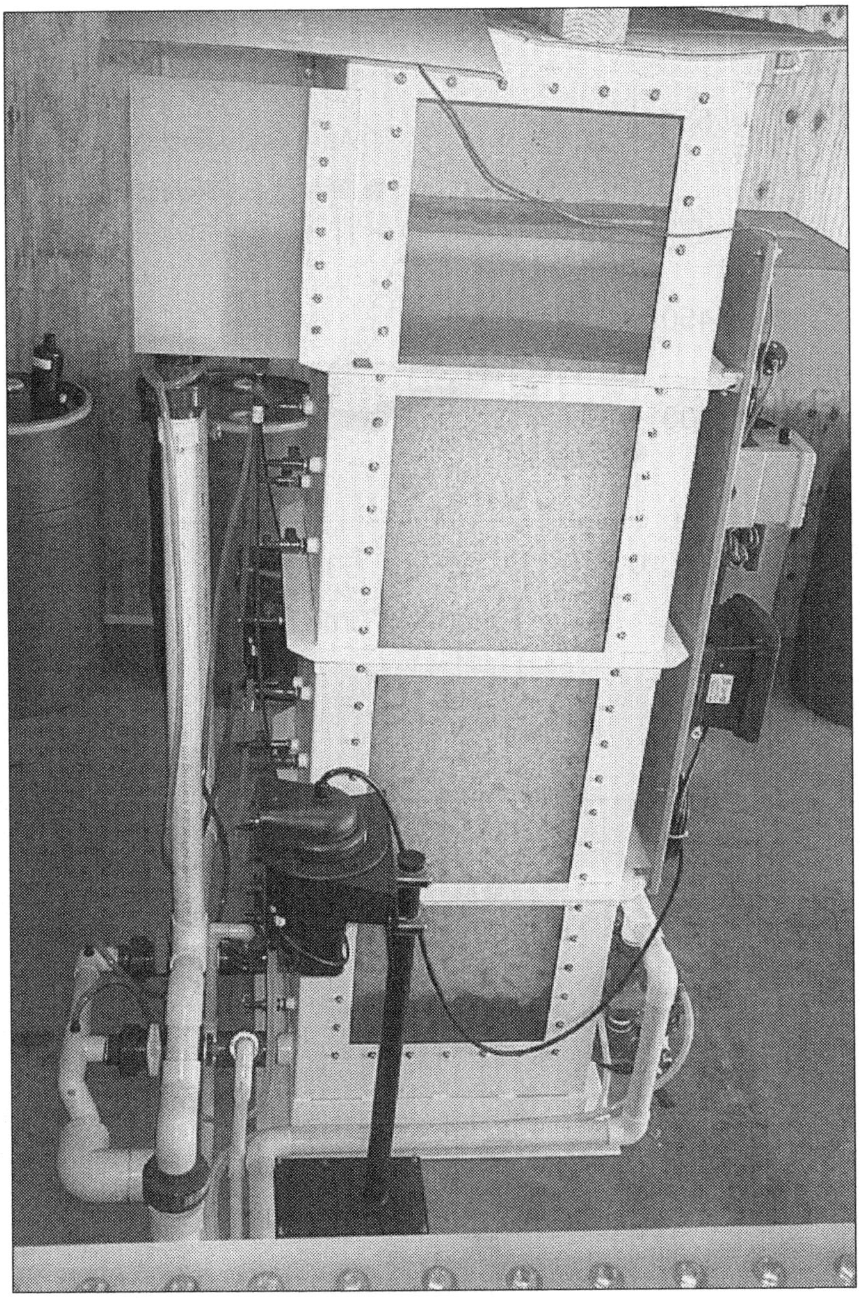

**Figure 7.8 Pilot-scale reactor with plastic media**

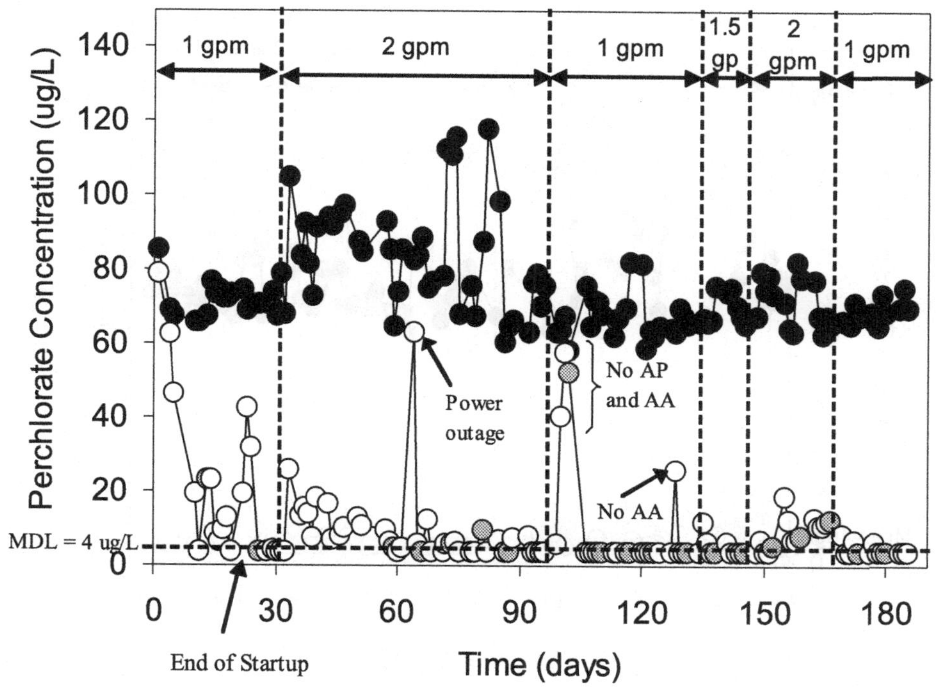


**Figure 7.9 Perchlorate raw data from influent (closed circle) and effluent (open circle; gray circle for indicating backwash time) in the plastic media reactor**

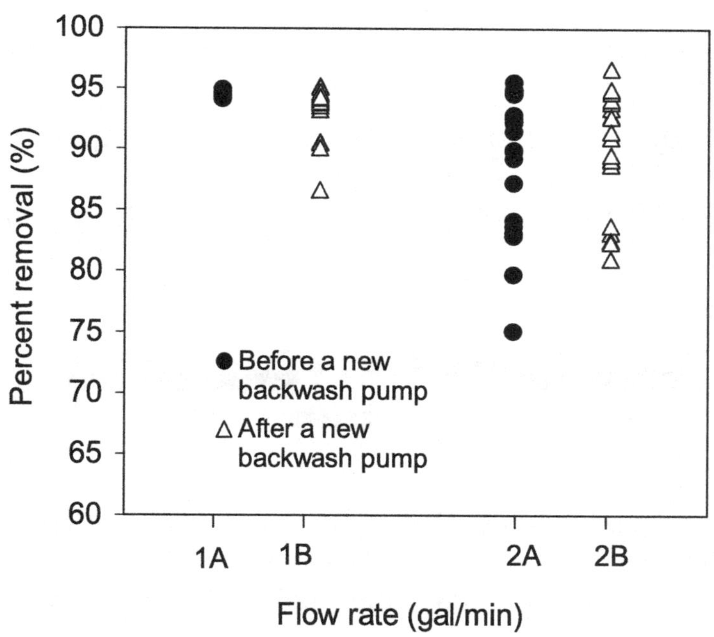


**Figure 7.10 Comparison of perchlorate removal before and after using a new backwash pump depending on flow rates**

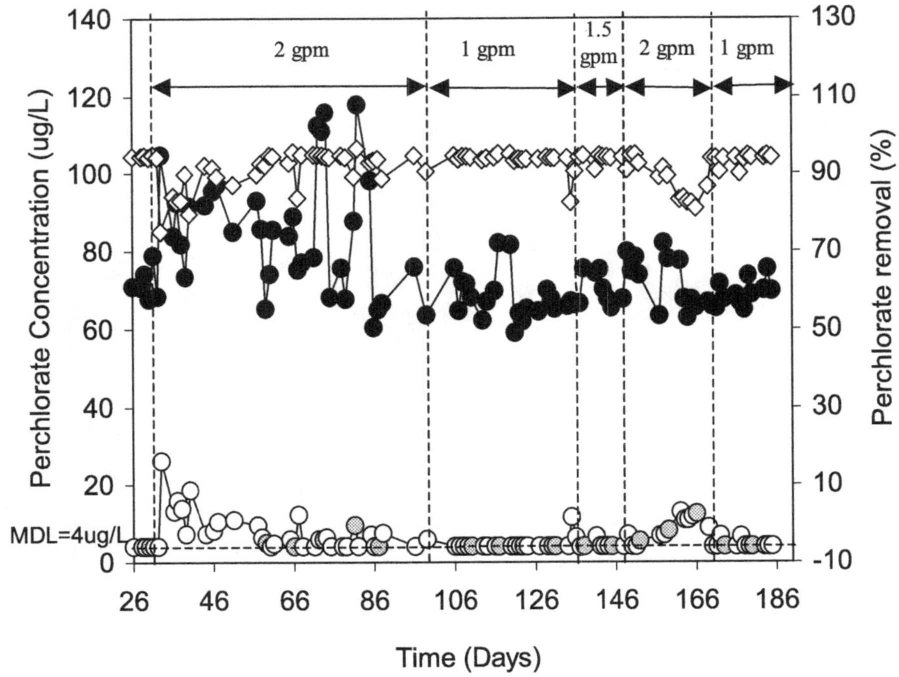


**Figure 7.11 Data for days of acceptable operation (no mechanical failures) of perchlorate from influent (closed circle) and effluent (open circle; gray circle for backwash time) in plastic media reactor (open diamond: Perchlorate removal)**

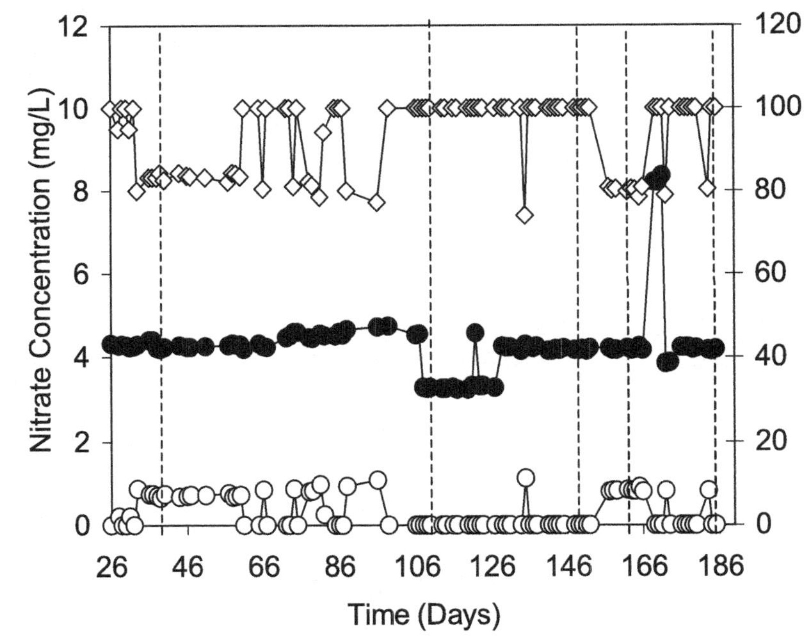


**Figure 7.12 Data for days of acceptable operation (no mechanical failures) of nitrate from influent (closed circle) and effluent (open circle) in plastic media reactor (open diamond for nitrate removal)**

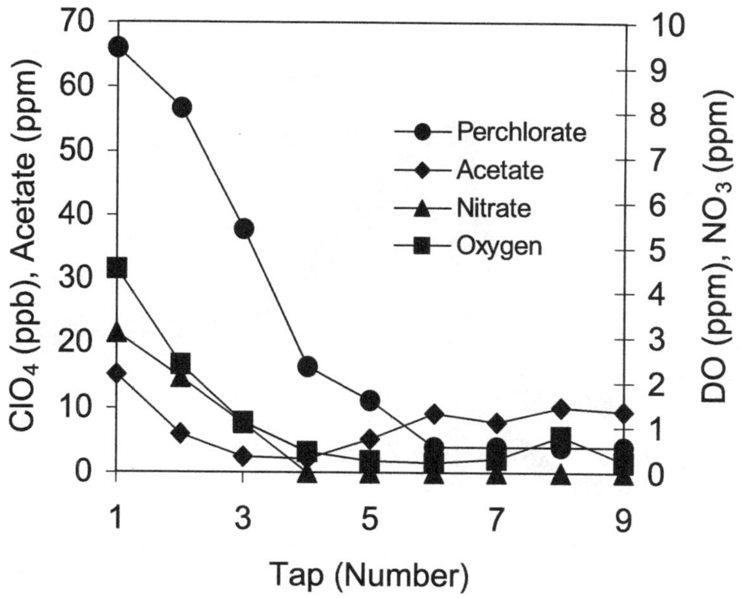


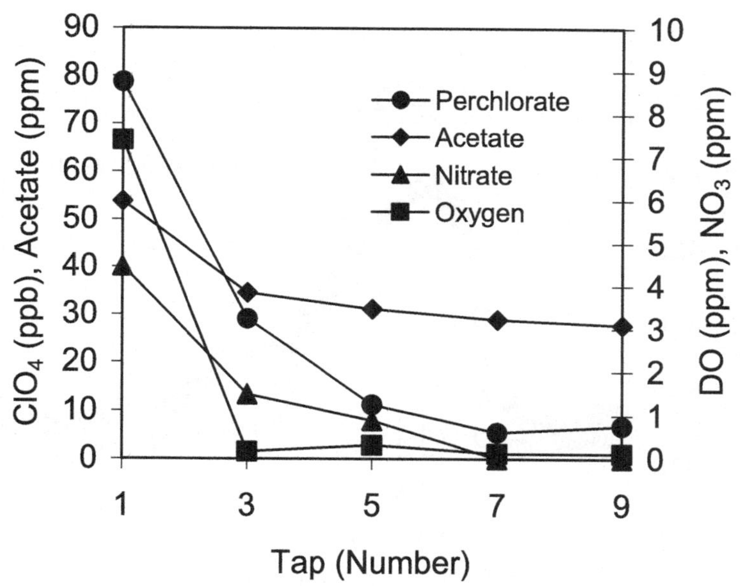


**Figure 7.13 Profile data from taps of the reactor at 1 gpm (upper) and 2 gpm (lower)**

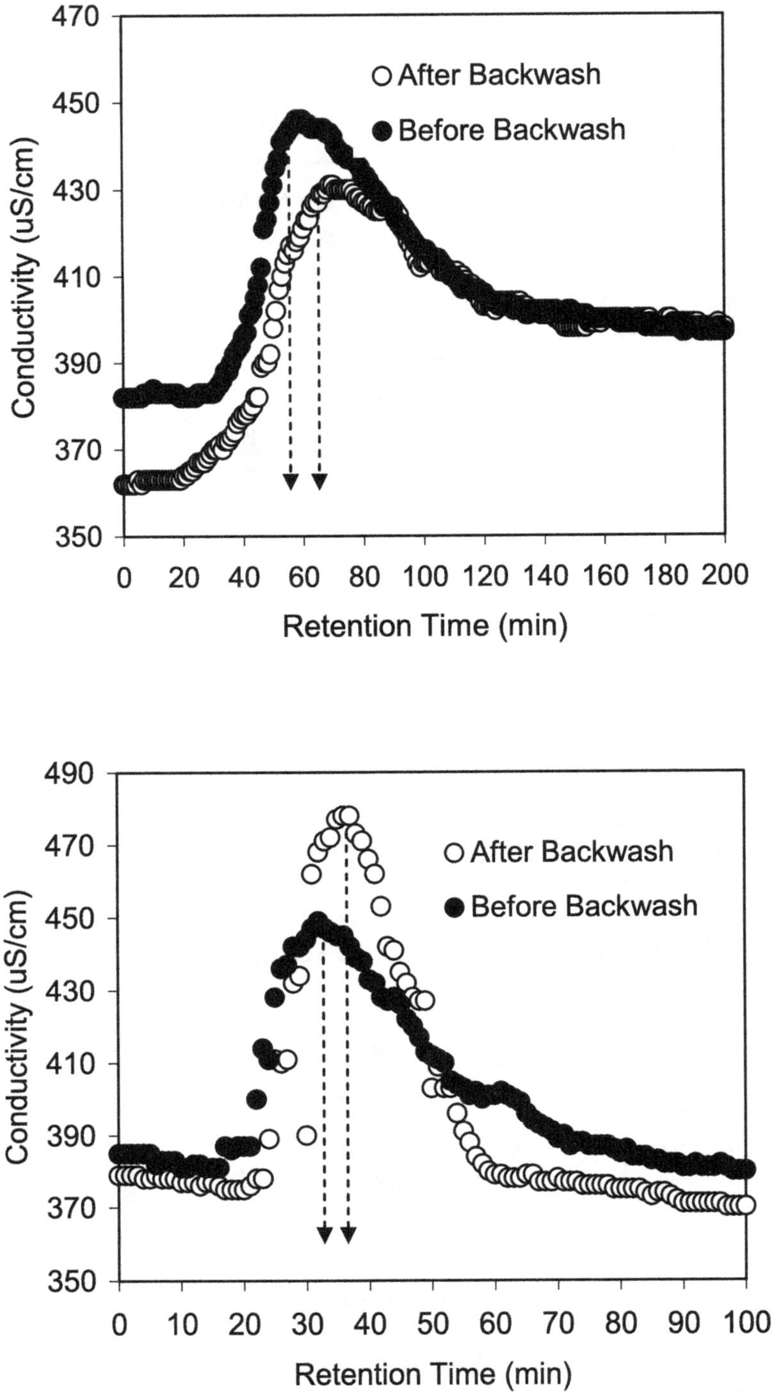


**Figure 7.14 Retention time test at 1 gpm (upper) and 2 gpm (lower)**

# CHAPTER 8
# ENGINEERING CONSIDERATIONS AND COST ESTIMATES

## INTRODUCTION

This chapter presents regulatory and engineering considerations affecting design of a full-scale biological perchlorate treatment process for drinking water treatment. The chapter is organized into the following sections:

- Drinking Water Regulations
- General Design Parameters
- Conceptual Full-Scale Design
- Conceptual Cost Estimate

## DRINKING WATER REGULATIONS

The EPA has not, to date, set a regulated maximum contaminant level (MCL) for perchlorate. The California Department of Health Services (DHS) developed an analytical method that reduced the detection limit of perchlorate to 4 μg/L. California DHS is using the 4-μg/L detection limit as the action level for perchlorate. A draft Toxicological Review and Risk Characterization (US EPA 2002) proposes an MCL of 1 μg/L for adults and 0.3 μg/L for children.

In addition to meeting the perchlorate action level, a groundwater source will need to comply with all other drinking water regulations under EPA's Safe Water Drinking Act (SWDA). This includes specific current regulations such as the Total Coliform Rule (TCR), the Lead and Copper Rule (LCR), and the Surface Water Treatment Rule (SWTR). Additionally, in the near future, all groundwater sources will need to comply with the Groundwater Rule (GWR) that was proposed by EPA in May 2000 and finalized in November 2000.

The TCR limits the presence of coliform bacteria in distribution systems. To comply with the TCR, many utilities with groundwater sources add a disinfectant such as chlorine to prevent pathogenic microorganisms from proliferating in distribution systems. Most treatment techniques to meet LCR requirements involve inhibitors, sequestering agents, or pH adjustment. For some groundwater systems, no treatment is needed for compliance with the LCR.

The proposed GWR includes the following basic requirements:

1. System sanitary surveys conducted by the State and identification of significant deficiencies;
2. Hydrogeologic sensitivity assessments for undisinfected systems;
3. Source water microbial monitoring by systems that do not disinfect and draw from hydrogeologically sensitive aquifers or have detected fecal indicators within the system's distribution system;
4. Corrective action by any system with significant deficiencies or positive microbial samples indicating fecal contamination; and
5. Compliance monitoring for systems, which disinfect to ensure that they reliably achieve 4-log (99.99 percent) inactivation or removal of viruses.

The SWTR specifically applies only to groundwater sources under the influence of a surface water source, such as a nearby stream or river. These groundwater sources must be treated as a surface water source and be filtered and adequately disinfected. For a groundwater that is subject to a biological treatment process, it will be necessary to filter the water downstream of any bioreactor process to capture solids that carry-over from the process and to remove excess nutrients added. The regulatory agency that has jurisdiction over the source water will most likely make this determination.

Due to the nature of the biological process developed by this project, the resulting process waters will require a disinfection step or steps (multi-barrier protection) to ensure the water meets applicable regulatory requirements. Finally, additional requirements (i.e. heightened source protection and monitoring, increased post-treatment sampling, additional failure response triggers) may be required by the primary regulatory agency. This is the case in California according to a Department of Health Services Policy Guidance (Reference Memorandum 97-005) for Direct Use of Extremely Impaired Sources.

## GENERAL DESIGN PARAMETERS

The anaerobic fixed bed filter (AFBF) is a packed-bed bioreactor. The filter operates in an up-flow mode and nutrients and an electron donor (i.e., acetic acid) are added to the influent to promote biological perchlorate reduction.

The pilot plant at the Texas Street Well Facility in Redlands consisted of two side-by-side AFBFs, one containing sand media and the other containing plastic media. The sand media was supported by a screen and a thin layer of gravel and the plastic media was contained with screens on the top and bottom. A screen on the top of the plastic media was necessary because the plastic had a specific gravity less than one and floated in water. A full-scale system would employ plastic media with a specific gravity greater than one and obviate the need for a top screen. Additional chemical tanks, mixers, process pumps, and control systems were required to facilitate the start-up, treatment and backwash of the AFBFs. Once biomass was established on the media, it was resilient to short disruptions of the system and only needed periodic backwashing to remove excess biomass. The pilot testing was performed in a garage at the Redlands Texas Street Well Facility.

The AFBF pilot units were operated in the following manner:

1. **Startup** – The startup phase established a perchlorate-respiring biofilm on the packing media. The length of this start-up was 8 days. During startup, perchlorate-respiring bacteria (strain KJ), acetate (an electron donor and carbon source required for perchlorate reduction) and nutrients (ammonium phosphate [$NH_4H_2PO_4$]) were mixed with groundwater containing perchlorate. The amended groundwater was pumped upward through the AFBF and was recirculated until the biofilm was established on the packing media. Once-through flow of groundwater was then initiated.
2. **Treatment** – During the treatment phase acetate and ammonium phosphate were added to the groundwater prior to the AFBF. The mixture was pumped upflow through the reactor. The biofilm in the bioreactor oxidized acetate and reduced dissolved oxygen, nitrate, and perchlorate. Acetate, nitrogen, and phosphorous were also assimilated by the growing bacteria in the biofilm. Periodic measurements of perchlorate, ammonia, phosphate, nitrate, dissolved oxygen (DO), temperature and pH measurements on

influent and effluent samples were conducted. AFBF effluent contained residual acetate, ammonia and phosphate.

3. **Backwashing** – The headloss increased over time as the biomass in the AFBF increased. Channeling and/or short-circuiting, especially in the sand media AFBF was also observed. Periodically (about once per week) the AFBFs were backwashed by increasing the groundwater flow and by adding air to the bottom of the AFBF to help scour the media. Backwash flow was typically 5 to 20 gpm/ft$^2$ and air scour flow was typically 1.5 to 3.0 scfm/ft$^2$. Backwashing and removal of excess biomass was conducted to limit backpressure, channeling, and short-circuiting. Following the backwash cycle the AFBF was put back in service (i.e., treatment mode).

The treatment and backwash conditions were varied during the pilot testing to measure the effects on treatment efficiency and to optimize operation for a full-scale plant. Table 8.1 lists the general design parameters determined by pilot testing.

Pilot testing was conducted in part to determine the general basis for engineering design of a full-scale plant. The following is a list some of the key findings from the pilot testing:

1. The biofilm established in the pilot-scale AFBF was very resilient. On several occasions the acetate and/or nutrient feed stopped for periods of hours to several days. Once the feed was restarted, the bacteria appeared to re-initiate perchlorate reduction within one day. Exposure to oxygen during the air scour did not appear to inhibit perchlorate reduction even though the bacteria have been shown in the laboratory to be sensitive to oxygen. Analysis of perchlorate reduction following these events was usually conducted the following day. The recovery time may have been less than one day, but data are not available to make a definitive conclusion.
2. Removal of perchlorate to less than 4 mg/L required an EBCT of 37 minutes for the sand media and 70 minutes for the plastic media. The EBCT for the sand media was based on a surface loading rate of 0.75 gpm/ft$^2$ and a media depth of 44 in. The EBCT for the plastic media was based on a surface loading rate of 0.5 gpm/ft$^2$ and a media depth of 56 in. The sand AFBF required less empty bed contact time (EBCT) than the plastic AFBF. This difference is thought to be caused by the significantly larger surface area per unit volume of the 1-mm sand media than the 1-¼" plastic media. The surface area of the 1-¼" Jaeger Tripack® was 70 ft$^2$/ft$^3$. The surface area of the 1-mm sand was estimated to be on the order of 1,800 ft$^2$/ft$^3$. However, much of the sand surface area was not available because of inter-particle contact and particle clumping. A smaller plastic media with commensurately larger surface area per unit volume theoretically will decrease the required EBCT.
3. DO and nitrate in the source water must be removed in the AFBF before perchlorate is removed. The influent concentrations of DO and nitrate were much greater than that of perchlorate and resulted in the high EBCT. Additionally, the high concentrations of DO and nitrate governed the required influent concentration of acetate. Therefore the source water DO and nitrate concentrations significantly govern the loading, sizing and operation of the AFBF. An alternative design, though not evaluated here, could involve biological pretreatment of groundwater for removal of DO and nitrate followed by an AFBF for perchlorate removal. Pretreatment could involve a variety of

bioreactor configurations. This two stage process may be more efficient and cost effective than the currently envisioned single stage process.

4. The headloss through the sand AFBF was much higher than that of the plastic media. Additionally, channeling and/or short-circuiting were more problematic with the sand media. Backwashing the sand media was problematic. Clumps of sand and biomass formed in the sand AFBF and could not be broken up by backwashing or air scouring. For this reason sand, should not be considered for a full-scale plant unless a proven means for breaking the clumps apart (e.g., a mechanical mixer) is provided. Air scour and backwash of the plastic media was easily accomplished and resulted in relatively uniform removal of excess biomass from the packing. Complete perchlorate removal was observed following backwash based on samples collected the next day. It is estimated that one to two AFBF volumes are required to reinstate perchlorate removal activity following an air scour and backwash.
5. The average concentrations of acetate, ammonia and phosphate that were consumed were 28 mg acetate/L (50 mg/L influent concentration), 0.8 mg-N/L (4.4 mg/L influent concentration), and 0.7 mg-P/L (11 mg-P/L influent concentration). The influent concentrations of these constituents should be reduced to minimize the residual amount in the effluent that has to be treated by subsequent processes. The average concentrations of DO, nitrate, and perchlorate that were consumed were 8.9 mg/L DO, 4.3 mg-N/L nitrate, and 7.5 μg/L perchlorate. These consumption data translate to 3700 μM electron equivalents of electron donor consumed and 2700 μM electron equivalents of total electron acceptors consumed. Consumption of the electron donor acetate was thus about 40% greater than theoretical. Acetate consumed for DO and nitrate reduction represented 99.8% of total acetate consumed based on an electron balance supporting the conclusion that AFBF design at Redlands is controlled primarily by DO and nitrate removal. In the pilot testing the acetate and was supplied in excess at 50 mg/L or a total of 6700 μM electron equivalents (80% greater than actually consumed). This feed could be reduced to 40 mg/L (40% greater than actually consumed) to produce a more efficient process.

## CONCEPTUAL FULL-SCALE DESIGN

The AFBF process design is dependent on site-specific water quality parameters. This section presents a conceptual process design for treating perchlorate at the Texas Street Well Facility in Redlands, California. The process was scaled-up based on results of pilot testing at the site.

The existing Texas Street Well Facility pumps ground water through granular activated carbon (GAC) units to remove VOC's. The water is then chlorinated to prevent biological growth in the distribution system and sent to a 1 million gallon water storage reservoir. The pressure of the water leaving the storage reservoir is boosted and then sent into the distribution system. Figure 8.1 shows schematic of the current configuration.

The existing GAC units are not able to effectively remove perchlorate because of the fast break-through time of perchlorate, thus requiring the addition of a perchlorate removal process. The proposed perchlorate removal system consists of three main components, the anaerobic fixed bed filter (AFBF), high rate granular filtration and disinfection. The AFBF system removes perchlorate but leaves residual nutrients and bacteria in the water. High rate granular filtration will provide biological removal of the remaining residual nutrients and provide filtration to remove suspended solids to meet turbidity requirements. The disinfection will consist of an ultraviolet (UV) system to inactivate the existing bacteria and provide multi-barrier protection in combination with the existing chlorination system.

The effluent from the AFBF is aerated prior to entering the high rate granular filter to allow for aerobic biological removal of the remaining nutrients. Because of this aeration step and the potential for stripping of VOCs from groundwater into the air, the GAC units will be used to remove VOCs prior to the perchlorate removal process. Normal operation will involve breakthrough of perchlorate through the GAC units followed by removal in the AFBF units.

The data obtained from the pilot plant was used to design a conceptual process to treat the 10-mgd capacity of the Texas Street Well Facility. The conceptual process consists of AFBF units to remove perchlorate, high-rate granular filters to remove residual nutrients and suspended solids, and a UV system to disinfect the water in combination with the existing chlorination system. The conceptual process schematic is shown in Figure 8.2. A proposed site plan showing existing and proposed process equipment is shown in Figure 8.3.

**AFBF System Description**

The AFBF is proposed to consist of a series of steel reactor tanks fed in parallel. Eleven tanks plus one spare tank are included. The 10-mgd flow will be directed to 11 of the 12 tanks at any one time while the remaining tank is being backwashed. The reactors will be fed from the bottom to create an upflow pattern through the packing media. The proposed plastic media is 3/8" diameter Biolox 10® manufactured by Rauschert or a similar high-surface area plastic packing with a specific gravity greater than 1.1. This plastic media has a higher surface to volume ratio than the 1-¼" media used in the pilot plant thus allowing for a shorter EBCT and smaller treatment area footprint. The individual treatment tanks may include internal baffles to divide the tank into 4 quarters and help minimize short-circuiting in each tank. Table 8.2 lists the proposed design criteria for each AFBF unit.

The AFBFs will require additional support processes for startup, nutrient addition and backwashing. Figure 8.4 illustrates the process schematic for a single anaerobic fixed bed filter. The additional support processes for start-up are not shown.

**High-Rate Granular Filters Description**

The effluent from the AFBF contains turbidity, excess acetate, ammonia, phosphate and bacteria. All of these constituents except phosphate will be removed by downstream polishing processes. It assumed that the low levels of phosphate would not violate any drinking water standards and may provide some corrosion protection to the distribution system. In addition, residual ammonia levels in the AFBF effluent will be less than current nitrate levels in Redlands groundwater. The AFBF units effectively remove nitrate. Thus nitrogen removal downstream of the AFBFs may be unnecessary.

The effluent from the AFBF will be sent through biologically active high-rate granular filter. Aeration of the filter influent will be conducted to create aerobic conditions for biological removal of acetate and ammonia in the filter. Ammonia will be assimilated into new bacteria in the biofilm and/or nitrified to nitrate.

The high-rate granular filter will also remove turbidity from the AFBF effluent. Biological growth in the AFBF will continually slough off and increase the turbidity of the effluent. The turbidity of the pilot plant effluent ranged from 2 to 4 NTU. The maximum allowed turbidity of the water would depend on the classification of this water, which is uncertain, however it is assumed that a final turbidity should be < 0.3 NTU. The backwash water from the AFBFs will be equalized in a basin prior being sent to the high-rate granular filters. Pilot testing indicated that this backwash water was significantly more turbid than normal AFBF effluent. The project team assumes the effectiveness of downstream filtration and disinfection processes selected will need to be demonstrated using pilot- or full-scale systems. A high-rate granular filter is proposed to minimize the footprint of the filter because of space constraints at the site. Table 8.3 contains the general design criteria for the filter.

### Ultraviolet Disinfection Description

The high-rate filters will be backwashed with water from the existing one million gallon storage reservoir. It is assumed that the backwash water could be sent to the sanitary sewer for disposal after the flows have been equalized in a basin to prevent overloading of the local sewer system.

UV disinfection will be provided to inactivate unfiltered bacteria in the high rate granular filter effluent. The proposed UV system and the existing chlorination system will provide two barriers to prevent the bacteria from the bioreactor from entering the distribution system. The UV system could also be modified, if needed, to provide additional oxidation of acetate and ammonia with the addition of hydrogen peroxide ahead of the UV reactors. In the presence of UV light, hydrogen peroxide breaks down to provide hydroxyl radicals. These hydroxyl radicals are extremely powerful oxidants; capable of oxidizing both acetate and ammonia to carbon dioxide, water, and nitrate.

## CONCEPTUAL COST ESTIMATE

The treatment of water at the Redlands Site requires the removal of perchlorate and additional secondary treatment to meet drinking water regulations. The conceptual cost estimate represents an opinion of cost to treat water at this site and water quality and is presented in Table 8.4. The AFBF process is cursorily compared to fluidized bed bioreactor (FBBR) and ion exchange processes. The FBBR estimate is based on reasonable assumptions but is not necessarily indicative of actual costs for a proprietary process available on the market. The ion exchange estimates are for Calgon ISEP™ and ISEP+™ processes. ISEP™ is a continuous ion exchange process. ISEP+™ is an optimzed continuous ion exchange process combined with perchlorate and nitrate destruction and brine reuse. The capital and operating and maintenance costs for 7,800 gallon per minute ISEP™ and optimized ISEP+™ (i.e., completed optimization) systems were obtained from Calgon (Tramposch 2001).

The Redlands Site contains very little space for additional process equipment. It is assumed for this cost estimate that the equipment will fit on this site, however a detailed design may reveal that additional land is required. The high rate filter has not been pilot tested and many assumptions have been made to the required size and operation.

The estimated capital cost for the additional perchlorate removal at the Redlands Site using the AFBF process is $0.91 per gpd of capacity. It should be noted this cost is in addition to the existing capital expenditures at the site for the wells, reservoir and GAC units. The capital cost of a FBBR process is less ($0.76/gpd) because the fluidized bed requires fewer reactor tanks. The capital cost for ISEP™ ($0.86/gpd) is similar to that for AFBF and the capital cost for ISEP+™ ($1.47/gpd) is greater. The ion exchange process has the advantage that downstream polishing processes are not required as for the biological processes.

The operating cost for the perchlorate removal process consists primarily of the chemical costs for the acetate and ammonium phosphate, the power costs for the pumps and cost of additional staff to run the facility. Additional costs of repairs, equipment replacement and testing have not been included. The AFBF process has a lower operating cost than the fluidized bed because pumping is not required to keep the media in the reactors fluidized. The net present value costs for the AFBF and FBBR systems are similar and are less than the ion exchange processes. The O&M costs for the AFBF and FBBR systems are predominately controlled by the chemical costs and the need to remove dissolved oxygen and nitrate. Quotations were obtained for food-grade acetic acid and ammonium phosphate which exceed NSF standard 60. The O&M costs for technical-grade chemicals would be about $7 less per acre-foot. O&M costs for water containing lower dissolved oxygen and nitrate concentrations would be substantially less. The O&M costs of the ion exchange are significantly greater, include regeneration costs, and assume in the case of ISEP™ that brine can be disposed to the sanitary sewer.

The total costs account for the capital and net present value of the O&M costs. The AFBF and FBBR systems had the lowest total cost and the ISEP system had the greatest total cost.

**Table 8.1**
**General design parameters based on pilot testing**

| Parameter | Unit | Range |
|---|---|---|
| Surface loading rate | gpm/ft$^2$ | 0.5 to 0.75 |
| Media depth | ft | 3.7 to 4.7 |
| Empty bed contact time (ebct) | min | 35 to 70 |
| Backwash rate | gpm/ft$^2$ | 5 to 20 |
| Air scour | scfm/ft$^2$ | 1.5 to 3.0 |
| Influent acetate | mg/l | 50 |
| Effluent acetate | mg/l | 22 |
| Influent ammonia | mg-N/l | 4.4 |
| Effluent ammonia | mg-N/l | 3.6 |
| Influent phosphate | mg-P/l | 11.4 |
| Effluent phosphate | mg-P/l | 10.7 |

**Table 8.2**
**Design criteria for full-scale AFBF units**

| Parameter | Unit | Quantity |
|---|---|---|
| **Water Quality Conditions of the Texas Street Well** | | |
| Perchlorate – influent | μg/l | 75 avg. |
| Perchlorate – effluent | μg/l | < 4.0 |
| DO – influent | mg/l | 8.0 – 10.0 |
| DO – effluent | mg/l | 0 – 0.5 |
| Nitrate – influent | mg/l as N | 4.0 – 4.5 |
| Nitrate – effluent | mg/l as N | 0 |
| pH – influent | - | ~ 6.8 |
| pH – effluent | - | ~ 6.8 |
| Turbidity - effluent | NTU | 2 – 3 |
| **Chemical Addition** | | |
| Acetate – added to influent | mg/l | 40 |
| Acetate – remaining in effluent | mg/l | 5 – 10 |
| Ammonia – added to influent | mg-N/l | 1.0 |
| Ammonia – remaining in effluent | mg-N/l | 0 – 0.5 |
| Phosphate – added to influent | mg-P/l | 2.2 |
| Phosphate – remaining in effluent | mg-P/l | 1.0 – 1.5 |
| **Treatment Reactor Configuration** | | |
| Process flow rate | mgd | 10 |
| Media type | - | 3/8" Biolox by Rauschert, 195 $ft^2/ft^3$, specific gravity = 1.15 |
| EBCT | min. | 35 |
| Media height | feet | 15 |
| Surface loading rate | $gpm/ft^2$ | 3.2 |
| Total surface area | $ft^2$ | 2,170 (for 11 operational tanks) |
| Reactor configuration | - | 12 tanks (11 operational plus 1 spare), 16-ft diameter |
| Flow per tank | gpm | 630 |
| **Backwash Cycle** | | |
| Backwash rate | $gpm/ft^2$ | 20 |
| Backwash flow per tank | gpm | 4000 |
| Backwash air scour | $scfm/ft^2$ | 3 |
| Air flow per tank | scfm | 600 |
| Backwash time | min | 15-30 |
| Backwash frequency | per week | 1 |

**Table 8.3**
**Design criteria for high-rate granular filter units**

| Parameter | Unit | Quantity |
|---|---|---|
| Process flow rate | mgd | 10 |
| Surface loading rate | $gpm/ft^2$ | 10 |
| Media depth | ft | 6 |
| Total bed surface area | $ft^2$ | 700 |
| Number of beds | - | 5 (one redundant for backwashing) |
| Surface area per bed | $ft^2$ | 175 |

**Table 8.4**
**Comparison of conceptual costs for AFBF, FBBR, and ion exchange processes**

| Costs | Anaerobic fixed bed filter | Fluidized bed reactor | ISEP™ | ISEP+™ completed optimization |
|---|---|---|---|---|
| **Capital Costs** | | | | |
| Perchlorate removal | $2,752,466 | $1,909,520 | $3,826,087 | $6,869,565 |
| Filtration / disinfection | $1,905,835 | $1,905,835 | N/A | N/A |
| Booster pumps | $605,059 | $605,059 | $605,059 | $605,059 |
| Subtotal | $5,263,361 | $4,420,414 | $4,431,146 | $7,474,624 |
| | | | | |
| Contractor overhead and profit (15%) | $789,504 | $663,062 | $664,672 | $1,121,194 |
| Subtotal | $6,052,865 | $5,083,476 | $5,095,818 | $8,595,818 |
| | | | | |
| Contingencies (25%) | $1,513,216 | $1,270,869 | $1,273,954 | $2,148,954 |
| Subtotal | $7,566,081 | $6,354,345 | $6,369,772 | $10,744,772 |
| | | | | |
| Engineering design (10%) | $756,608 | $635,434 | $636,977 | $1,074,477 |
| Construction management (5%) | $378,304 | $317,717 | $318,489 | $537,239 |
| Legal and administration (5%) | $378,304 | $317,717 | $318,489 | $537,239 |
| Engineering total | $1,513,216 | $1,270,869 | $1,273,954 | $2,148,954 |
| | | | | |
| **Total Construction & Engineering Costs** | **$9,100,000** | **$7,600,000** | **$7,600,000** | **$12,900,000** |
| | | | | |
| **Operation and Maintenance** | | | | |
| Cost (yearly) | | | | |
| Power cost | $172,245 | $246,368 | $379,807 | $343,062 |
| Chemicals, supplies, waste, etc. | $752,553 | $752,553 | $1,145,070 | $944,609 |
| Staffing cost | $161,000 | $161,000 | $193,031 | $159,238 |
| Subtotal O&M | $1,085,798 | $1,159,921 | $1,717,909 | $1,446,909 |
| **O&M Net Present Worth (20 yr, 5%)** | **$13,500,000** | **$14,500,000** | **$21,400,000** | **$18,000,000** |
| | | | | |
| **Total Cost** | **$22,600,000** | **$22,100,000** | **$29,000,000** | **$30,900,000** |
| | | | | |
| **Annual O&M Cost / Acre Ft** | **$97** | **$104** | **$153** | **$129** |

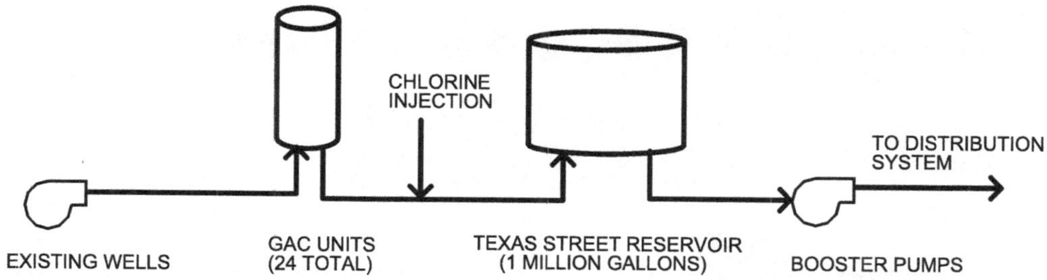


**Figure 8.1 Existing Texas Street Well Facility process schematic**

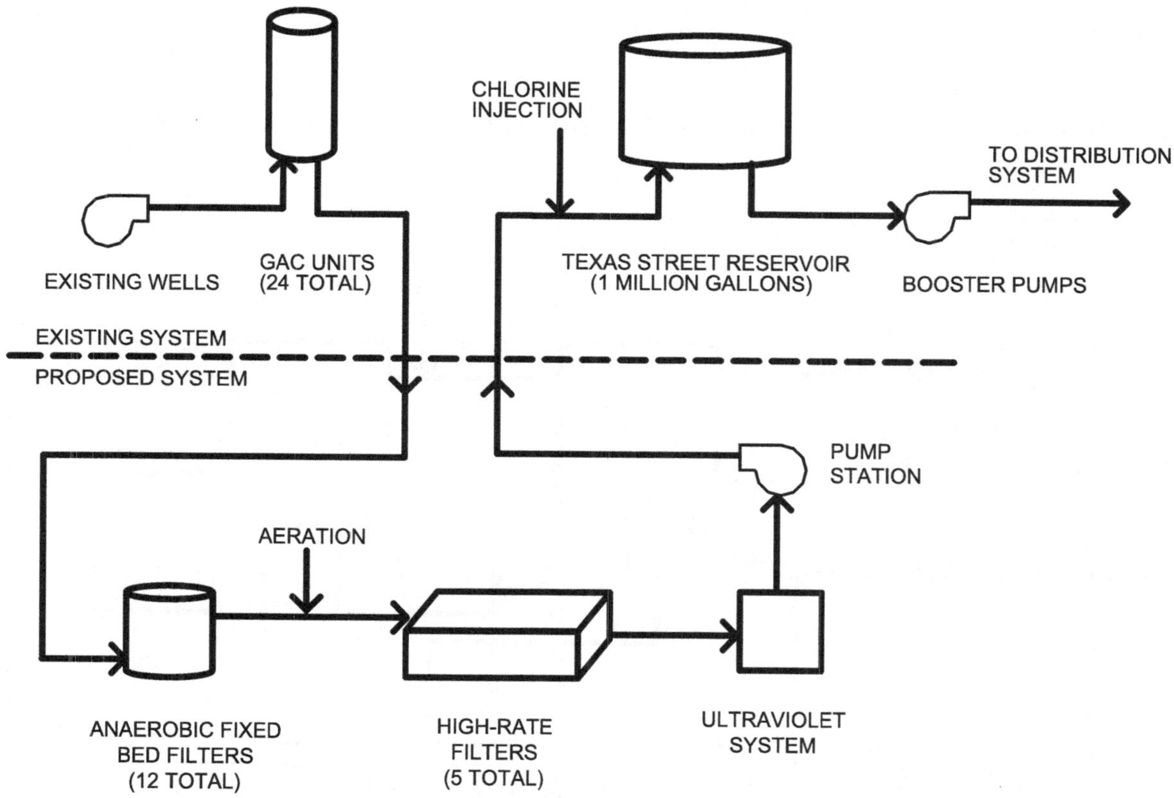


**Figure 8.2 Proposed Texas Street Well Facility conceptual process schematic**

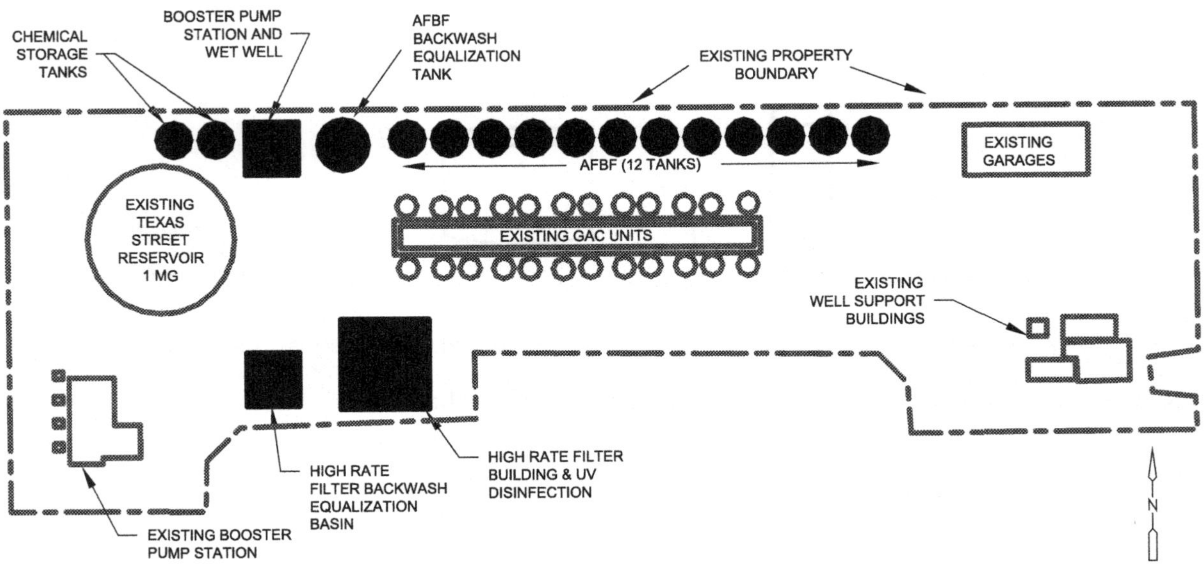


**Figure 8.3 Site plan of Texas Street Well Facility with proposed process facilities**

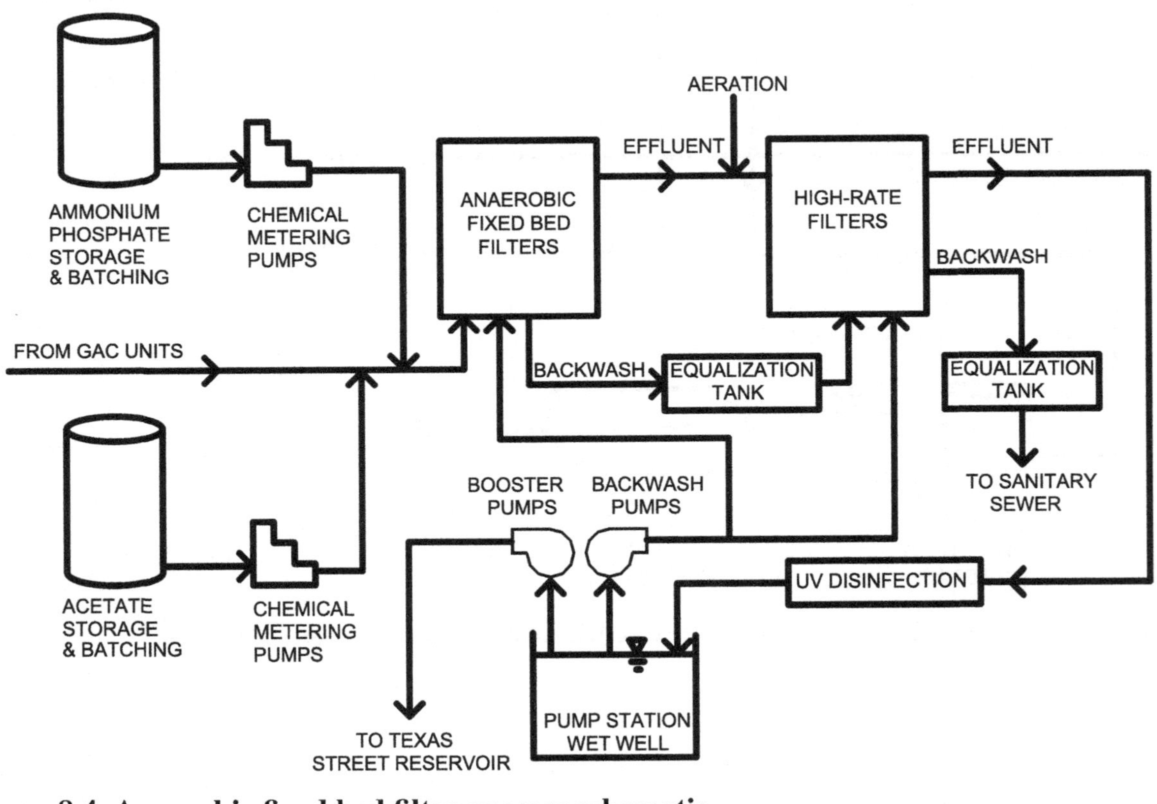


**Figure 8.4 Anaerobic fixed bed filter process schematic**

# CHAPTER 9
# CONCLUSIONS

It was determined from bench-scale tests that all three systems could be used to successfully remove perchlorate to levels necessary for drinking water. The perchlorate removal rates in the three bioreactors were a function of hydraulic loading rates, substrate and the presence of nitrate. Only one reactor type could be selected for pilot scale testing, and the acetate-fed bioreactor design was chosen. The following conclusions are drawn from these studies for each reactor type and from pilot plant tests.

**Packed-Bed Acetate-Fed Bioreactor (PSU-$O_4$ System)**

1. The acetate-fed packed bed bioreactor was recommended as the system for future field tests. This system, referred to here as the PSU-$O_4$ system, is a type of anaerobic fixed bed bioreactor.
2. The optimum support medium among those examined (sand, glass beads, granular activated carbon) in the laboratory for biofilm growth was sand. Perchlorate desorption from GAC following a high-perchlorate concentration dose during a start up period contributed to high perchlorate concentrations in the reactor effluent. Glass beads failed to hold sufficient biofilm to achieve high perchlorate degradation rates.
3. A bioreactor inoculated with a pure culture (isolate KJ) provided higher perchlorate removal rates, and therefore a lower detention time, than reactors inoculated with mixed cultures. With mixed cultures, perchlorate was reduced in sand column bioreactors from 20 mg/L to below the detection limit (<4 μg/L) at detention times as low as 18 minutes. With a pure culture, perchlorate was removed from concentrations of ~26 mg/L to non-detectable levels at a detention time of < 1 min (50 s; EBCT <2.1 min). Based on a total of 147 samples taken over 83 days, perchlorate concentrations were not detectable in 84.4% of effluent samples, and all the samples contained perchlorate concentrations less than 146 μg/L for the high-perchlorate concentration feed.
4. Chloride measurements confirmed that all of the perchlorate was converted to chloride. The average molar ratio of chloride produced to perchlorate reduced was 1.0 for 91 samples taken during the 83-day sampling period.
5. Perchlorate-contaminated water from the City of Redlands was successfully treated in the pure culture bioreactor despite the presence of nitrate. This Redlands water contained approximately 65 μg/L of perchlorate and 20 mg/L of nitrate. Perchlorate was removed to below detectable levels. Nitrate was removed from a concentration of approximately 20 mg/L to effluent concentrations in the range of 9.3 to 11.3 mg/L during the first 6 days, to 830 μg/L after 12 days, and to non-detectable concentrations (180 μg/L) after day 20 for all samples except one.
6. An economic and engineering analysis indicated that the acetate-fed packed bed bioreactor was the least-expensive, most reliable system. The packed-bed sand reactor achieved the highest perchlorate removal rates of the three systems and was operated.

### Unsaturated-Flow Hydrogen Gas Bioreactor

1. A hydrogen gas reactor was tested at bench scale and determined to have perchlorate removal rates suitable for scale up. The biofilm reactor, operated in unsaturated-flow mode, was fed water containing $ClO_4^-$ and a gas mixture of $H_2$ (5%) and $CO_2$. This acetate-fed packed-bed reactor was recommended over the hydrogen-gas reactor due to the more conventional nature of the packed bed bioreactor and a smaller tank size requirement.
2. Over a 140 day period at an influent perchlorate concentration of 740 μg/L, and at a constant hydraulic loading rate of 0.45 cm/min, 39±9% of the perchlorate was removed in the reactor at detention times of 1.1 to 1.3 min producing an average removal rate of perchlorate of 230 μg $L^{-1}$ $min^{-1}$. This perchlorate removal rate was comparable to acetate-fed bioreactor. Perchlorate removal was not complete in this reactor due to the short length of the reactor. It was assumed that perchlorate could be completely removed using a longer column, but this was not tested in the laboratory.
3. Nitrate slightly decreased the rate of perchlorate removal in the hydrogen gas bioreactor. At a perchlorate concentration of 826 μg/L, perchlorate was reduced by 192 ± 51 μg/L in the absence of nitrate, and decreased to 164 ± 56 μg/L in the presence of nitrate. The nitrate removal rate under these conditions was 1.5 mg-$NO_3^-$/min. At a perchlorate concentration of 79 μg/L, the average removal rate of perchlorate in the absence of nitrate was 15.2 ±3.6 μg/L, and was 8.7 ±1.6 μg/L in the presence of 25.3 mg/L of nitrate.

### Membrane Bioreactor

1. The diffusion coefficient for perchlorate through the three membranes tested, Memcor BTS-55, Memcor PVDF and Millipore FGLP, were found to be $6.64 \times 10^{-6}$, $3.75 \times 10^{-6}$, $6.67 \times 10^{-6}$ $cm^2$/sec, respectively. These diffusion coefficients are much smaller than that of perchlorate in water in the absence of a membrane ($1.53 \times 10^{-5}$ $cm^2$/sec) as calculated by the Wilke-Chang method. These results confirm the initial hypothesis of this study that perchlorate would migrate through semi-permeable membrane solely by diffusion.
2. The diffusion coefficient for nitrate through the Memcor BTS-55 membrane was determined to be $4.74 \times 10^{-6}$ $cm^2$/sec. This data is very close to the value previously reported in the literature for a similar membrane type. However, this diffusion coefficient is much smaller than that of nitrate in water ($2.12 \times 10^{-5}$ $cm^2$/sec).
3. The diffusion coefficient for sulfate through the Memcor BTS-55 membrane was found to be $3.79 \times 10^{-6}$ $cm^2$/sec. This coefficient is much smaller than that of sulfate in water ($1.47 \times 10^{-5}$ $cm^2$/sec).
4. When comparing the diffusion coefficients of perchlorate, nitrate, and sulfate, as expected from the values of their molar volumes, the diffusivity in water is from the fastest to the slowest: nitrate > perchlorate > sulfate. However, the diffusivity through the BTS-55 membrane follows: perchlorate > nitrate > sulfate. Thus, through the BTS-55 membrane, the diffusivity of nitrate is higher than that of sulfate, as expected from their diffusivities in water. However, perchlorate diffusivity is larger than that of nitrate and sulfate. Thus, sulfate and nitrate contained in perchlorate-contaminated waters

may not reach the biofilm (located on the BR side of the reactor) as fast as perchlorate and their effects on perchlorate biodegradation may be minimized.

5. Enrichment of a perchlorate-reducing microbial culture, from returned activated sludge taken from the Clark County Wastewater Treatment Plant, was achieved quickly and easily by acclimating the culture anaerobically to perchlorate.
6. With time and after the enrichment culture (BALI) has been used in several bioreactor tests, the culture developed a red coloration. Microscopic examination and Gram staining testing revealed that BALI is a mixed culture containing gram-positive and gram-negative microorganisms. Gram-negative microorganisms were dominant with the majority being motile rods, but also present were gram-negative microorganisms with a helical morphology. Identification by 6S rRNA analysis of three microbial isolates from the mixed culture revealed that all three isolates are *Dechlorisoma* sp. with 98% match.
7. The development of biofilms on the surface of the membranes was accomplished in a short period by feeding the culture with a high carbon (lactate) to perchlorate ratio (approximately 5:1).
8. Biodegradation of perchlorate through a FGLP membrane- immobilized biofilm was very slow because of the poor perchlorate diffusion through the thick biofilm developed on the surface of this membrane. This happened because this membrane has a plastic backing that makes microbial attachment easier. It was concluded from the results that the FGLP membrane is not suitable for immobilizing biofilm for perchlorate biodegradation.
9. Biodegradation of perchlorate through a BTS-55 membrane-immobilized biofilm was rapid and the concentration of perchlorate in the BR reactor was very small. The lactate in the BR reactor decreased proportionally to perchlorate in the DR reactor. The molar ratios of perchlorate biodegraded to that of chloride formed were 0.82, 1.03 and 0.86 for three testing cycles. Theoretically, each mole of perchlorate would generate one mole of chloride, thus these ratios are approximately stoichiometric considering the larger size of the reactor used.
10. For the BTS-55 membrane, the thickness of the biofilm during the first testing cycle was found to be 24.7 μm. Perchlorate biodegradation rates were found to be 0.25, 0.11, and 0.22 moles $ClO_4^-$ biodegraded /day/$m^2$ of membrane area for the three testing cycles. The difference in biodegradation rates in the three cycles may be explained by the change in biofilm thickness during the three testing cycles.
11. During the BTS-55 testing, chloride concentrations changed significantly from one reactor to another. This was attributed to diffusion of chloride through the membrane due to its small size and to movement of water from the DR reactor (smaller concentration of ions) to the BR reactor (higher concentration of ions) by osmotic pressure.
12. Biodegradation of perchlorate through a PVDF membrane-immobilized biofilm was found faster than that of the FGLP membrane, but slower than that of the BTS-55 membrane. The biodegradation rate (0.06 moles of perchlorate per day per $m^2$ of membrane area) was much smaller than the values found for the BTS-55 membrane (0.11 to 0.25 moles of perchlorate per day per $m^2$ of membrane area). The PVDF membrane has the smallest diffusion coefficient from all three membranes tested. Therefore, the lower biodegradation rate is the result of low diffusion of perchlorate through the membrane. The perchlorate to chloride ratio for this test was found to be 0.99.

13. The results of the kinetic studies on perchlorate biodegradation showed that the BALI culture required a minimum organic carbon to perchlorate ratio for perchlorate biodegradation to occur at acceptable rates. Under lactate-limited conditions, no microbial growth was observed, except in bottles in which the lactate to perchlorate ratio was equal to three. For all perchlorate-limited bottles, microbial growth was observed. The microbial growth was coupled with a decrease in perchlorate and lactate concentrations and a stoichiometric increase in chloride concentration. The results further indicated that a lactate to perchlorate ratio of at least three is needed for perchlorate biodegradation to occur at acceptable rates.
14. The results of batch tests using the BALI culture on the influence of nitrate on perchlorate biodegradation showed that the rate of nitrate biodegradation was faster than that of perchlorate. Thus, nitrate negatively affects perchlorate biodegradation.
15. The results of sulfate interference on perchlorate biodegradation in batch testing showed that perchlorate biodegradation occurred before sulfate reduction. Sulfate biodegradation, after almost all the perchlorate had been reduced, was very slow.
16. The results of the interference of sulfate on perchlorate biodegradation in BTS-55 membrane-immobilized biofilm reactors showed that the higher the initial perchlorate concentration, the higher the perchlorate biodegradation rate, independent of the initial sulfate concentration. The sulfate biodegradation rates, in the presence of perchlorate, were much smaller (or null) than those of perchlorate. In the absence of perchlorate, however, the sulfate biodegradation rates were larger than those in the presence of perchlorate, except for the case in which the perchlorate and sulfate concentrations were about 50 mg/L. The reason for that is not clear from the data available. It can be concluded that sulfate does not affect perchlorate biodegradation.
17. The studies on the interference of salinity on perchlorate biodegradation showed that as the salinity was increased, the growth rate of the culture decreased. At a salinity of 0.5% (5,000 mg/L as NaCl) there was a 27% reduction in the growth rate and at 4% salinity no growth was observed. However, this culture was not acclimated for highly saline conditions but for potable waters which typically have total dissolved solids (TDS) values between 25 and 1000 mg/L (conductivity between 50 and 2000 μmhos/cm). Thus, when treatment of more saline waters or ion exchange brines (6 to 8% salinity) is necessary, cultures specifically adapted for these conditions will need to be developed.

**Pilot Scale Tests With the Acetate-Fed Fixed Bed Bioreactor**

1. It was found that the sand media column could remove perchlorate from ~75 ppb to the detection level (<4 ppb) at hydraulic loading rates of 1 to 2 gpm/ft$^2$ when we used a regular backwashing (weekly) cycle to clean and redistribute sand in the column. Salt tracer experiments indicated that the detention time for the cleaned medium was between 30 minutes (1 gpm/ft$^2$) and 18 minutes (2 gpm/ft$^2$). Prior to backwashing, the detention time was 30-45% lower than after backwashing.
2. The plastic media bioreactor produced the most consistent perchlorate removals at 1 gpm/ft$^2$. Perchlorate removal below the detection level was possible at hydraulic loading rates of 1 or 1.5 gpm/ft$^2$ and was less sensitive to the backwashing frequency than the sand medium likely as a higher porosity of the plastic medium. The detention

times at a hydraulic loading rate of 1 gpm/ft$^2$ ranged from 60 minutes before back-washing to 70 minutes after backwashing.

3. The fixed bed reactor cost is comparable to the cost of a fluidized bed system, and is less expensive than the ion exchange system. The total estimated cost for perchlorate removal at the Redlands Site using the PSU-$O_4$ process based on a fixed bed is $22.6 million. It should be noted this cost is in addition to the existing capital expenditures at the site for the wells, reservoir and GAC units. The total cost of a fluidized bed reactor is approximately the same ($22.1 million) at this site because the fluidized bed requires fewer reactor tanks but has higher operating costs due to recycle. The cost for an ionexchange system (ISEP+™) is much larger at $30.9 million. The ion exchange process has the advantage that downstream polishing processes are not required as for the biological processes.
4. The plastic media reactor is recommended for full scale systems based on consistent operation, ease of backwashing the system, and lower back pressure compared to the sand medium reactor.

## RECOMMENDATIONS

Based on the laboratory, engineering, and economic analysis performed during this study, it is recommended that a pilot-scale system based on the acetate-fed, packed-bed reactor (PSU-$O_4$ system) be constructed and tested at the Crafton-Redlands site in Redlands, CA, as part of Phase 2 of this project. To ensure long-term performance of the system, the system will have the capability for temporary fluidization of media (sand) or backwashing (plastic random media), depending on the system, to reduce the potential for reactor plugging due to biological growth. Such backwashing will be needed during long-term operation to prevent excessive biofilm accumulation leading to reactor plugging. There will also be a separate regeneration tank that can be inoculated with perchlorate-reducing microorganisms (PRMs) for reactor bioaugmentation for system start up or perchlorate-reducing biofilm community revival.

# APPENDIX A
# QAPP

# 1.0 PROJECT DESCRIPTION

## 1.1 GENERAL OVERVIEW

Perchlorate has recently been detected in several surface waters and groundwater wells used to supply drinking water at concentrations above the detection limit (4 ppb) to 0.37%. The California Department of Health Services (CDHS), based on EPA work, has established a provisional action level of 18 ppb for drinking water. This relatively low concentration was established because of perchlorate's interference with iodine in the production of hormones in the thyroid. The presence of perchlorate at these high concentrations in the environment, coupled with a very low drinking water standard, has created a national water contamination crisis in the US potentially affecting 12 million people. Perchlorate is readily biodegradable, and under proper conditions, can be reduced to non-detectable levels by fixed and suspended cultures of microorganisms.

Currently, there is no EPA established Maximum Contaminant Level (MCL) or Maximum Contaminant Level Goal (MCLG) for perchlorate. However, based on the guidelines established by CDHS, for the purposes of this research, the perchlorate MCLG will be considered to be <4 ppb (defined as less than the minimum detection level) and the MCL as 18 ppb.

In Phase I of the proposed research, the potential for using anaerobic, fixed-film bioreactors to treat water contaminated with perchlorate concentrations ranging from 100 to 0.1 mg/L to concentrations lower than those necessary to meet the anticipated drinking water MCL of 18 ppb will be investigated. This is a basic research project; therefore, the Quality Assurance Project Plan (QAPP) that is presented was prepared using the Category IV QAPP guidelines.

Based on the results of Phase I activities, and on engineering and economic analyses, one of the three proposed treatment systems will be recommended for further testing (Phase II). This second phase will involve on-site implementation of a specific treatment technology. It is not certain whether the reactor recommended by this project, or another project funded through the Awwa Research Foundation (AwwaRF), will be selected for Phase II. Therefore, since the Phase II testing of the proposed reactor system will be covered under a separate project, the presented QAPP covers only Phase I (basic research) and not field studies that will be conducted under a new grant for Phase II. Near the completion of Phase I, a separate Category III QAPP will need to be completed to discuss Phase II activities.

The research described here is a feasibility study intended to establish the most economically efficient method (from the methods shown to be technically feasible) to remove perchlorate from potential drinking water sources. Calculation of incorrect removal rates of perchlorate would result in the incorrect sizing of a reactor to be used during Phase II of research, but this QAPP is designed to avoid these errors. Specific design factors recommended during Phase I research will be tested, and refined, during subsequent Phase II work.

## 1.2 THE PROCESS

Bench scale experiments will be conducted on three different fixed-film biological treatment processes to determine their feasibility for being scaled up to treat large quantities of drinking water. These treatment systems are: a packed bed (slow sand filter) amended with soluble microbial carbon sources (acetate, methanol, or ethanol); a hydrogen gas fed four-phase (hydrogen gas, water, biofilm, and support media), unsaturated trickle-type packed column; a membrane-bound biofilm reactor.

Two different types of water sources will be used throughout the project. For most initial screening work, it will be necessary to control the water quality so that reactor performance can be compared (i.e. there is no variation in the water quality fed to the different reactors). Therefore, an artificial groundwater will be used in most studies conducted in Phase I. This water will be constructed using ultrapure/low ionic strength water (~0.01 mM, produced from a Millipore Academic Q water treatment system) amended with trace metals and ionic species to mimic the groundwater at the Redlands, CA, site in conductivity, major ionic species (i.e. sulfate, nitrate, phosphate), pH, and dissolved organic carbon (DOC).

Near the completion of Phase I studies, when reactor designs are optimized, a final set of experiments will be conducted using: (a) water samples from groundwater wells that have been turned off in the Redlands area (due to perchlorate contamination); (b) water from Lake Mead (spiked with perchlorate). These water samples will be used to verify the general applicability of our findings obtained using the artificial groundwater. The specific number and quantity of water samples used from Lake Mead or the Redlands area will be determined later as the research project progresses. The specific number of samples used will be a function of the number of reactor types and conditions that prove to be technically capable of perchlorate removal. There is no impact of these different water samples on the QAPP because the same reactor sampling and analytical procedures will be applied to all water samples.

As a common starting point in our studies, we will use water containing perchlorate at 1000 ppb, with an experimental influent range of 10 to $10^5$ ppb, and set as a goal a reduction to the MCLG (nondetectable, or <2.5 ppb). As indicated, there currently is no EPA established MCL or MCLG for perchlorate. Because the CDHS acceptable perchlorate level in drinking water sources is 18 ppb, biological perchlorate removal will be deemed successful if reactor effluent concentrations fall below the 18 ppb concentration. Regardless of reactor type, influent and effluent concentrations of perchlorate will be measured to determine to extent of perchlorate removal.

## 1.3 STATEMENT OF PROJECT OBJECTIVES

This proposal is targeted to investigate removal of perchlorate from water, leaving it suitable to be used as drinking water. The specific objectives of the proposed research grant will be to:

- Modify a sand filter to treat perchlorate-contaminated water to drinking water standards
- Design and test a gas (hydrogen) phase fixed-film bioreactor for efficiency of perchlorate removal
- Test a membrane-bound hydrogen fed biofilm reactor
- Make an economic comparison of these reactors if they were to be scaled up to capacities sufficient for treatment of large volumes of water (millions of gallons per day).

With information gained in this proposal, we will estimate the costs of treating waters using the reactors and feed substrates that successfully remove perchlorate down to drinking water levels (both <2.5 ppb and <18 ppb). Based on the engineering and economic analysis, one of these treatment systems will be recommended for further testing in Phase II at the Crafton-Redlands site in Redlands, CA.

## 1.4 SCHEDULE

This project will be conducted over a two-year period to begin in August of 1998. Each year is divided into three sections in order to be compatible with the preparation of reports to AwwaRF every four months. In the first year, water samples from the two sites (Redlands and Nevada) would be collected and analyzed, and the three reactors built and tested using base conditions of the artificial groundwater. In the second year, the reactors would be optimized for performance by varying input loads (such as concentration of applied substrate, nutrients, etc.). The second year's work would end with an economic analysis (by CDM) and a final report submitted to AwwaRF. Research at Penn State University will be conducted on the sand filter and the gas phase fixed-film bioreactor. Research at UNLV would proceed along the same time line, except that all work would be done on the membrane bioreactor. Camp, Dresser and McKee (CDM) Consulting Engineering will be responsible for working with the City of Redlands in obtaining water samples, running conventional water treatment analyses, and in the second year conducting an economic evaluation of the treatment systems' costs and projected costs for scale up of the systems.

During all phases of research, Frank Blaha, Project Manager, will be the point of reference at AwwaRF, the organization responsible for administering funds from the East Valley Water District.

## 1.5 PROJECT ORGANIZATION AND RESPONSIBILITIES

This project will involve the following researchers:

- Bruce Logan, Ph.D. Kappe Professor of Environmental Engineering, Penn State University. Email: blogan@psu.edu. Telephone: 814-863-7908.
- Jacimaria Batista, Ph.D. Assistant Professor, Department of Civil and EnvironmentalEngineering, University of Nevada, Las Vegas. 702-895-1585). Email: jaci@ce.unlv.edu. Telephone: 702-895-1585.
- Steve Price, Camp, Dresser and McKee Consulting Engineering, 100 Pringle Ave, Suite 300, Walnut Creek, CA, 94596. Email: prices@cdm.com. Telephone: 923-296-8056.

## 2.0 QUALITY ASSURANCE OBJECTIVES

## 2.1 EXPERIMENTAL DESIGN

The critical measurement that will be used in the evaluation of reactor performance is effluent perchlorate ($ClO_4^-$) concentrations. It is hypothesized that no intermediates will accumulate in solution during the reduction of $ClO_4^-$ to chloride ($Cl^-$). However, potential intermediates are chlorate ($ClO_3^-$) and chlorite ($ClO_2^-$). Theoretically, concentrations of $Cl^-$ in the treated water should be equal to the quantity of $ClO_4^-$ removed (raw water $ClO_4^-$ concentration - treated water $ClO_4^-$ concentration ).

Noncritical measurements will be made in an attempt to optimize reactor removal efficiency. Noncritical measurements include the reactor influent and effluent concentrations of: potential perchlorate degradation products, including $ClO_3^-$, $ClO_2^-$, and $Cl^-$; other potential contaminants found in water at sites in California, including TCE and PCE; energy and/or carbon sources, including acetate, methanol, ethanol (added in reactor liquid feed) and hydrogen gas (the electron donor for microorganisms in the gas phase fixed-film bioreactor); and other inorganic ions that can serve as competing electron acceptors, including nitrate ($NO_3^-$), and sulfate ($SO_4^{2-}$). Other noncritical measurements that can affect bioreactor performance, and will therefore be monitored include: pH, dissolved oxygen (DO), total organic carbon (TOC), conductivity, and oxidation-reduction potential (ORP). Table A.1 summarizes the above parameters.

Noncritical parameters, such as $NO_3^-$, $SO_4^{2-}$, DO, and TOC will be monitored for two reasons. First, the artificial groundwater that will be used in most of our studies should be similar in quality to the actual groundwater in the Redlands area. By monitoring these parameters we can ensure that the artificial groundwater is representative of the Redlands groundwater. Second, these parameters can affect $ClO_4^-$ removal efficiency. For instance, $ClO_4^-$ reducing microorganisms require very low DO levels. A possible reason for lower $ClO_4^-$ reduction could be that the DO levels in the reactor are too high. Also, $NO_3^-$ and $SO_4^{2-}$ may serve as alternate electron acceptors in microorganisms. Therefore, a lower $ClO_4^-$ reduction could indicate the use of one of these terminal electron acceptors in preference to $ClO_4^-$.

Acetate, methanol, ethanol, and hydrogen gas are important noncritical parameters because they (acetate, methanol, or ethanol) serve as a carbon source and/or electron donor (hydrogen gas) to the microorganisms that will degrade $ClO_4^-$. If it is found that influent and effluent concentrations of a specific carbon source (e.g., acetate) are equal, then no $ClO_4^-$ removal will have occurred.

A comparison of the influent and effluent concentrations of the microbial carbon source (acetate, methanol, or ethanol) will allow a determination of the proper feed quantities to enable the reactor to function most efficiently. These proper feed quantities, for the individual carbon sources, will be used in the cost analysis of reactor operation.

In the second year of study, we will introduce co-contaminants (such as TCE and PCE) at 10 mg/L levels in order to study the effect of the potential degradation of these chemicals in the highly reducing environment achieved during $ClO_4^-$ reduction (~-300 mV; Bliven 1996).

Reactors will be sampled for effluent $ClO_4^-$ concentrations (single sample per detention time) until steady state conditions are achieved. Quantitatively, steady state will be defined as being achieved when effluent $ClO_4^-$ concentrations measured over three detention times are within $\pm$ 1 standard deviation. Upon reaching steady state conditions, samples will be taken from the reactor and these samples will be those that are reported for reactor steady state conditions. A total of three reactor samples will be taken, one sample taken per detention time. Perchlorate samples (for each of three steady state detention times) will be taken in replicate, and each sample analyzed as duplicate. All other noncritical parameter samples (for each of three steady state detention times) will be taken singularly and will be analyzed as duplicate.

The measurements (both critical and noncritical) will be made in the laboratories of Dr. Bruce Logan and Dr. Jacimaria Batista at The Pennsylvania State University (University Park, Pennsylvania) and The University of Nevada, Las Vegas (Las Vegas, Nevada), respectively.

## 2.2 DETERMINING QA OBJECTIVES

The purpose of establishing QA objectives for this research is to ensure that the data generated throughout the project are of acceptable quality so that the project's technical objectives are achieved.

## 2.3 QUANTITATIVE QA OBJECTIVES: ACCURACY, PRECISION, METHOD DETECTION LIMIT, AND COMPLETENESS

### 2.3.1 Accuracy

The accuracy of effluent $ClO_4^-$ concentration measurements will be determined based on calibration techniques outlined by the equipment manufacturer. The validity of the constructed calibration curves will be determined by performing a linear least squares analysis. Water samples received will be spiked with a known concentration of $ClO_4^-$ to determine the accuracy of laboratory techniques.

Accuracy of in-house $ClO_4^-$ measurements will also be determined monthly by analyzing blind samples that have been prepared by Dr. Jaci Batista (if determining accuracy of laboratory procedures conducted at Penn State), or prepared by Dr. Bruce Logan (if determining accuracy of laboratory procedures conducted at UNLV).

### 2.3.2 Precision

Samples used for $ClO_4^-$ concentration determination will be reported as the average of duplicate analyses of each replicate sample. Precision of these multiple analyses will be measured by calculation of the relative percent deviation (RPD).

### 2.3.3 Minimum Detection Level

The MDL for $ClO_4^-$ method will be computed by injecting 7 samples of 5 ppb into the IC unit and computing the peak area given by the 7 injections. The standard deviation of the 7 corresponding peak areas will be calculated and subsequently, the MDL will be calculated as:

$$MDL = (SD * t * C)/Am$$

where

SD = the standard deviation of the 7 areas recorded
t = student distribution for n=7
C = 5 ppb
Am = the mean area of the 7 peak areas measured

The above method for determining the $ClO_4^-$ method follows Standard Method 1030-E (APHA, AWWA, and WEF 1995).

Dionex defines the MDL of its instruments as 2.5 ppb; however, researchers have been able to obtain lower concentrations (Batista 1998). Therefore, an in-house MDL will be determined, and as long as it $\leq$ 2.5 ppb, it will be accepted as the MDL.

### 2.3.4 Completeness

Completeness is a measurement of the amount of valid data obtained compared to the total amount of data collected. The degree of completeness is the number of samples with acceptable data divided by the total number of samples collected and tested, and multiplied by 100.

Table A.2 summarizes the quantitative QA objectives for the critical parameter $ClO_4^-$ concentration. Perchlorate concentrations will be measured in the untreated reactor influent water and treated reactor effluent water.

Average effluent $ClO_4^-$ concentrations will have to satisfy the QA objectives outlined in Table A.2. If the QA objectives are not met, adjustments will be made to the experimental conditions and measurements will be conducted again.

## 2.4 QUALITATIVE QA OBJECTIVES: COMPARABILITY AND REPRESENTATIVENESS

### 2.4.1 Comparability

Comparability is a qualitative parameter that expresses the degree of confidence that data are equivalent for a specific parameter or group of parameters. Comparability of the data obtained here to other data sets will be achieved by strict observance of typical analytical procedures (Section 3.3). Accurate and prompt tabulation of data will also aid comparability.

### 2.4.2 Representativeness

The degree to which sample data accurately and precisely represent the characteristics of a population, parameter variation at a sampling point, or an environmental condition is defined here as sample representativeness. Consistent sampling procedures and frequent sampling throughout the operation of the reactor will achieve representativeness and comparability.

## 2.5 OTHER QA OBJECTIVES

No further QA objectives are expected.

## 2.6 WHAT IF QA OBJECTIVES ARE NOT MET?

A lack of precision and accuracy will be corrected by employing corrective actions. Examples of possible corrective actions include reanalyzing samples, checking equipment for possible malfunctioning, checking mathematical calculations, or modifying sampling method/bioreactor design.

Data will be promptly evaluated to determine whether individual QA objectives are being achieved so corrective action can be immediately taken if necessary (corrective action will be noted in log book).

If particular QA objectives are not met, data will be noted and a discussion will made in each progress report, as well as in the final report, relating the significance of the particular QA objectives not being achieved.

## 3.0 SAMPLING AND ANALYTICAL PROCEDURES

### 3.1 SAMPLING PROCEDURES

The exact design of the individual reactors has not yet been determined as their design is a goal of this research. Therefore, exact sampling locations cannot yet be specified, but at a minimum, reactor influent and effluent conditions will be monitored. It is anticipated that 2-3 other sampling ports will also be located throughout the column depths (applies to modified sand reactor and gas-phase fixed-film reactor). Influent feed and effluent ports will most likely be the only sampling locations on the membrane bound reactor. When reactor design is finalized, sampling procedures will be developed to allow a representative and contaminant free sample to be obtained.

Table A.3 summarizes all analysis parameters (both critical and noncritical) that will be measured, and the liquid volumes that will be necessary for each analytical procedure.

Reactor effluent samples will be collected in glass or HDPE containers, while TCE and PCE samples will be collected in crimp-sealed vials (to allow for zero headspace in sample bottle). If the sample to be collected is to be used in the determination of TOC or DOC, a glass sample container that has been previously fired in a muffle furnace (to remove residual organic material) will be used. HDPE sample containers will undergo triplicate rinses in MilliQ water (Section 1.2) and will be reused.

Samples will be analyzed as soon as possible after collection, but holding times will be less than seven days. Samples that are not analyzed immediately will be stored on ice.

All water samples from non-laboratory sites (water from Lake Mead or groundwater from Redlands, CA) will be shipped in 55-gallon drums and stored in a refrigerator at 4°C. The drums will be constructed of plasticizer-free plastic, and a comparison of water sample DOC will be made before and after sample shipment.

Smaller samples will be shipped using overnight mail, and preserved with dry ice to keep them cool during shipping. Water samples will be held in polycarbonate carboys (made of plasticizer-free plastic). Perchlorate is non-volatile and non-adsorbing, and therefore we do not expect any degradation of samples during shipping. Soil samples used to develop a $ClO_4^-$ degrading consortium will be placed in 5 gallon carboys. Cooling these samples is essential to preserve the microbial community in the samples. Samples will be marked to indicate: person taking samples, date, time, contents (water), location sample obtained, conditions (temperature), and date received.

Details of sampling procedures and results of all analyses will be kept in hand-written, bound laboratory notebooks that will be available for inspection. All analytical results will be calculated using appropriate equations and reported in appropriate concentration units (mg/L or ppb). Reduced data will be transferred to spreadsheets, and will be periodically reviewed by the analyst and principal investigator. All data classified as outliers (data that does not satisfy QA objectives) will be immediately examined in detail and corrective actions taken.

### 3.2 PROCESS MEASUREMENTS

Field process measurements are not a component of the Phase I research project.

## 3.3 ANALYTICAL PROCEDURES AND CALIBRATION

Water samples from the Redlands and Nevada sites will be analyzed for $Cl^-$, TCE, PCE, $NO_3^-$, $SO_4^{2}$, pH, and conductivity by an EPA certified commercial water laboratory.

Sands used in columns will be analyzed for size distribution and organic carbon. Sand grain sizes will be measured by sieving, followed by image analysis of particle sizes (Galai Image analysis system) to determine the particle size distribution within size classes. Organic carbon will be measured through combustion by a commercial laboratory.

Perchlorate concentrations above 1 mg/L can be analyzed using an ion specific probe (Orion Research Inc.), and verified by ion chromatography (Dionex DX-100, DX-500 at Penn State; DX-120 at UNLV). The perchlorate specific probe will only be used as a monitoring tool to gauge the performance of a reactor (i.e. prior to steady state reactor operation). All final data that will be reported on steady state reactor performance will be measured using an IC. For verification of perchlorate probe measurements at >1mg/L concentrations, and analysis of samples at concentrations of 2.5 ppb (detection limit) to 1 mg/L or 20 mg/L (see below), we will use the perchlorate IC method developed by Dionex. The Dionex method is included as Appendix C.

### 3.3.1 EPA-Approved or Other Validated Standard Methods

Sample concentrations of the following noncritical measurements will be conducted according to Standard Methods for the Examination of Water and Wastewater (1995): $Cl^-$, acetate, TCE/PCE, conductivity, ORP, reactor headspace hydrogen gas ($H_{2(gas)}$), pH, DO, TOC, $NO_3^-$, and $SO_4^{2-}$. Methanol/ethanol aqueous concentrations, $ClO_2^-$ , and $ClO_3^-$ concentrations (noncritical measurements) will be measured according to USEPA Methods.

TOC measurements for all water samples will be determined by a Shimadzu TOC 5000A, which uses a platinum catalyst combustion method. Prepared standards will be spiked with ethanol or methanol to determine if the analytical method is sensitive to these noncritical analytes. It will be assumed that any volatile organic compound (e.g. TCE, PCE) will be sparged by the nitrogen flow, and therefore will not be included in the TOC measurement.

Sample $ClO_4^-$ concentrations (critical measurement) will be measured according to the Dionex method (Appendix C). The Dionex method is the same as the CA Department of Health Method, except different eluents are used (sodium hydroxide in Dionex method, para-cyanophenoxide in CA Dept Health method). Prior to its use in the IC unit, the sodium hydroxide eluent will be vacuumed to de-gas the solution. A 1-L solution of sodium hydroxide can be used for up to a month (Dionex 1998), but will likely be used much more rapidly. Perchlorate can be measured down to concentrations of 4 ppb using a Dionex DX-500 with an AS-11 column, with a 100 mM NaOH solution, sparged with helium gas, as eluent (Wirt et al. 1998). For concentrations up to 1000 ppb, a 1 mL injection is used; at higher concentrations, a 25 uL injection is used. Concentrations larger than 20 mg/L are diluted manually with ultrapure water.

Chloride and acetate (noncritical measurements) are similarly measured on the IC, except the necessary eluent is 0.3 mM $NaHCO_3$±0.9 mM $Na_2CO_3$ and only the 25 uL injection loop is used. An AS-4 column will be used in the determination of $Cl^-$ and acetate by IC.

In all IC measurements, the liquid sample will be filtered using 0.25 μm polycarbonate filters before injection into the IC.

Table A.4 summarizes these analytical methods.

### 3.3.2 Nonstandard or Modified Methods

Perchlorate will be measured at concentrations above 1 mg/L using a $ClO_4^-$ specific probe (Orion Research Inc.). Interferences by other anions are known, and include for example at the 10 mg/L concentration: $I^-$, 25 mg/L; $ClO_3^-$, 166 mg/L; $NO_3^-$, 310 mg/L; $SO_4^{-2}$, 19,200 mg/L. In order to assure that other anions do not interfere with the values obtained using the $ClO_4^-$ probe, we will verify probe results using the Dionex IC method for any new water sample received (i.e. water from Lake Mead or groundwater from Redlands, CA). It should be emphasized that the $ClO_4^-$ probe will only be used for determining $ClO_4^-$ concentrations before a reactor reaches steady state. All $ClO_4^-$ data that will appear in quarterly and the final report will be measured using the Dionex method.

### 3.3.3 Calibration Procedure and Frequency

Our ion chromatograph (IC) has an autosampler. The following calibration procedure will be used before each day's sample analysis. Each step represents a run through the IC unit.

1. Deionized water sample
2. Deionized water sample
3. Calibration standard-Level 1- 5 ppb $ClO_4^-$
4. Calibration standard-Level 2- 10 ppb $ClO_4^-$
5. Calibration standard-Level 3- 15 ppb $ClO_4^-$
6. Calibration standard-Level 4- 30 ppb $ClO_4^-$
7. Calibration standard-Level 5- 60 ppb $ClO_4^-$
8. Calibration standard-Level 6- 80 ppb $ClO_4^-$
9. Calibration standard-Level 7- 100 ppb $ClO_4^-$
10. Deionized water sample
11. Deionized water sample

After the calibration procedure is complete and a $R^2$ value (from linear least squares analysis) of 0.997 or greater is achieved, the following sampling procedure will be used.

1. Run samples
2. Run Quality Control (QC) and mid-standard check

This procedure will be repeated provided that the QC and mid-standard check have accurate results. A QC sample is a sample prepared by a member of the research group, different from the person who prepared the calibration standards. A mid-standard check consists of injecting one of the standards after every 20 samples, to check the accuracy of the calibration.

If the $R^2$ value of the $ClO_4^-$ standard curve is not $\geq$0.997, the calibration standards will be reanalyzed, and a new calibration curve constructed. If the criterion of $R^2 \geq$0.997 is again not met, new standards will be prepared and the process repeated. The IC will also be examined to make sure that it is functioning properly.

The $ClO_4^-$ ion specific probe will be calibrated daily using four calibration standards at concentrations of 5, 10, 25, and 50 mg/L. If it is expected that the sample concentration will be above 50 mg/L, the sample will be diluted.

Calibration of other analytical equipment used for the determination of noncritical parameters will follow procedures described by equipment manufacturer. For example, DO probe will be air calibrated, pH meter will be calibrated against two known pH solutions, etc. Calibration standards will be prepared using the same methods as used to prepare the samples in order to reduce bias from the preparation method. Reagents used in the preparation of all standard solutions will be ≥99% pure.

The validity of the constructed calibration curves will be determined by performing a linear least squares analysis.

## 4.0 APPROACH TO QA/QC

## 4.1 INTERNAL QUALITY CONTROL

As stated previously, $ClO_4^-$ calibration standards and spiked water samples (from Lake Mead and groundwater from Redlands, CA) will be prepared using ≥99% pure standards, and the standards will be prepared using Milli-Q water (18 M cm DI water). A 7 point calibration curve will be generated on the IC (4 point for $ClO_4^-$ probe). Acceptable goodness of fit will require a $R^2$ ≥0.997 (for IC method; 0.97 for probe method). Duplicate sample analyses will agree in precision to ± 20% RPD. Each new water sample received (from Lake Mead or Redlands, CA), will be spiked with $ClO_4^-$, and will have a percent recovery of ± 20%. Prepared blind samples that are tested monthly (Section 2.3) will have a percent recovery of ± 20% as well. A QC sample and mid-standard check (Section 3.3.3) will also be run with each set of samples on IC check the accuracy of the calibration procedure.

## 4.2 CALCULATION OF DATA QUALITY INDICATORS

### 4.2.1 Precision

The calculation of relative percent difference (RPD) for duplicate measurements will be calculated as outlined in Category IV Prep Aids (Simes 1991).

### 4.2.2 Accuracy

The calculation of matrix spike percent recovery will be calculated as outlined in Category IV Prep Aids (Simes 1991).

### 4.2.3 Completeness

Completeness will be calculated as outlined in Category IV Prep Aids (Simes 1991).

### 4.2.4 Method Detection Limit (MDL)

The MDL for the determination of $ClO_4^-$ concentrations will be calculated as per Standard Method 1030-E (APHA, AWWA, and WEF 1995).

## 5.0 REFERENCES


Batista, J. (1998). Verbal communication.

Bliven, A.R. (1996). Chlorate respiring bacterial: isolation, identification, and effects on environmentally significant substrates. M.S. Thesis, Department of Chemical and Environmental Engineering, University of Arizona, Tucson, AZ.

Dionex Corporation (1998). Verbal communication.

Hewlett Packard (1987). *A Practical Primer to Gas Chromatography.*

Simes, G.F. (1991). Preparation Aids for the Development of Category IV Quality Assurance Project Plants. Risk Reduction Engineering Laboratory, Office of Research and Development, U.S. Environmental Protection Agency, Cincinnati, OH.

Standard Methods for the Examination of Water and Wastewater (1995). 19th ed., American Public Health Association, Washington, D.C.

U.S. EPA (1979). Methods for Chemical Analysis of Water and Wastes. U.S. Environmental Protection Agency, Washington, D.C.

Wirt, K., M. Laikhtman, J. Rohrer, and P.E. Jackson (1998). Low-level perchlorate analysis in drinking water and groundwater by ion chromatography. *Amer. Environ. Lab.* 10(3):1-5.

**Table A.1**
**Summary of critical and noncritical measurements**

| Measurement | Sample matrix | Parameter classification |
|---|---|---|
| $ClO_4^-$ | Raw water, treated water | Critical |
| $ClO_3^-$ | Raw water, treated water | Noncritical |
| $ClO_2^-$ | Raw water, treated water | Noncritical |
| $Cl^-$ | Raw water, treated water | Noncritical |
| TCE | Raw water, treated water | Noncritical |
| PCE | Raw water, treated water | Noncritical |
| Acetate | Raw water, treated water | Noncritical |
| $NO_3^-$ | Raw water, treated water | Noncritical |
| Methanol/ethanol | Raw water, treated water | Noncritical |
| $SO_4^{2-}$ | Raw water, treated water | Noncritical |
| pH | Raw water, treated water | Noncritical |
| DO | Raw water, treated water | Noncritical |
| TOC | Raw water, treated water | Noncritical |
| Conductivity | Raw water, treated water | Noncritical |
| Headspace hydrogen gas | Reactor headspace | Noncritical |
| ORP | Treated water | Noncritical |

**Table A.2**
**Summary of QA objectives for the critical parameter ($ClO_4^-$)**

| Critical measurement | Matrix | Method | Reporting units | MDL[a] | Precision (RPD)[b] | Accuracy[c] (% recovery) | Completeness[d] |
|---|---|---|---|---|---|---|---|
| Perchlorate | Untreated raw water | Ion chromatography | ppb | $\leq$2.5 | $\leq$ 20 | 80-120 | 95 |
| Perchlorate | Treated water | Ion chromatography | ppb | $\leq$2.5 | $\leq$ 20 | 80-120 | 95 |

(a) Minimum detection level
(b) Given as RPD for laboratory replicate samples
(c) As percent recovery of matrix spike
(d) Based on the number of valid measurements, compared to the total number of measurements

**Table A.3**
**Summary of sample parameters**

| Analysis parameter | Container size | Sample quantity required for analysis |
|---|---|---|
| Conductivity | 20 mL | 5 mL |
| ORP | 20 mL | 5 mL |
| Hydrogen gas | 100 uL syringe | 100 uL syringe |
| $Cl^-$ | 20 mL | 5 mL[a] |
| pH | 20 mL | 5 mL |
| $NO_3^-$ | 20 mL | 5mL[a] |
| DO | 20 mL | 5 mL |
| TOC | 20 mL | 3 mL |
| $ClO_4^-$ | 20 mL | 5mL[a] |
| $ClO_3^-$ | 20 mL | 5mL[a] |
| $ClO_2^-$ | 20 mL | 5mL[a] |
| Acetate | 20 mL | 5mL[a] |
| $NO_3^-$ | 20 mL | 5mL[a] |
| Methanol/Ethanol | 20 mL | 5 mL |
| $SO_4^{2-}$ | 20 mL | 5mL[a] |
| TCE | 20 mL | 1 mL |
| PCE | 20 mL | 1 mL |

(a) All parameters that will be determined by ion chromatography require 5 mL volume, which can be analyzed once, or as a duplicate or a triplicate. Actual sample volume may be less than 5 mL, when sample requires dilution to bring expected concentration into range of ion chromatography machine detection.

**Table A.4**
**Method description of analytical techniques**

| Parameter | Method number | Method title | Source (reference) |
|---|---|---|---|
| $ClO_4^-$ | | Analysis of Low Concentrations of Perchlorate in Drinking Water and Ground Water by Ion Chromatography | Dionex |
| Methanol/ ethanol | USEPA Method 8015 | | USEPA (2) |
| $ClO_3^-$ | USEPA Method 300.1 | | USEPA (2) |
| $ClO_2^-$ | USEPA Method 300.1 | | USEPA (2) |
| $Cl^-$ | 4500-$Cl^-$ F | Chloride-Ion Chromatography Method | Standard Methods (1) |
| Acetate | 4110 | Determination of Anions By Ion Chromatography | Standard Methods (1) |
| TCE/PCE | 6232 B. | Liquid-Liquid Extraction Gas Chromatographic Method | Standard Methods (1) |
| Conductivity | 2510 B. | Conductivity-Laboratory Method | Standard Methods (1) |
| ORP | 2580 B. | Oxidation-Reduction Potential Measurement in Clean Water | Standard Methods (1) |
| $H_{2(gas)}$ | 2720 C. | Anaerobic Sludge Digester Gas Analysis-Gas Chromatographic Method | Standard Methods (1) |
| pH | 4500-$H^+$ | pH Value-Electrometric Method | Standard Methods (1) |
| DO | 4500-O | Oxygen (Dissolved)-Membrane Electrode Method | Standard Methods (1) |
| TOC | 5310 B. | Total Organic Carbon (TOC)-Combustion-Infrared Method | Standard Methods (1) |
| $NO_3^-$ | 4500-$NO_3^-$-C | Nitrogen (Nitrate)-Ion Chromatographic Method | Standard Methods (1) |
| $SO_4^{2-}$ | 4500-$SO_4^{2-}$-B | Sulfate-Ion Chromatographic Method | Standard Methods (1) |

(1) *Standard Methods for the Examination of Water and Wastewater* (APHA, AWWA, and WEF 1995).
(2) *Methods for Chemical Analysis of Water and Wastes* (US EPA 1979).

# APPENDIX B
# ADJUSTMENTS TO QAPP

## ADJUSTMENTS TO ORIGINAL QAPP

A preliminary notice of an award from AwwaRF was received well before the official grant started, and many experiments on chlorate and perchlorate reduction were already underway in our laboratory before we prepared our QAPP. As a result, there are several experiments that were conducted that we believe are valid, but did not strictly adhere to our QAPP. The QAPP submitted to EPA was based on our experience in conducting laboratory investigations, and was our best effort at anticipating laboratory protocols that would assure the quality of our data. Experiences gained since our original QAPP suggests that some modifications will allow us to make more rapid progress in our research efforts while not sacrificing the quality and accuracy of our chemical measurements. Below, we list statements were made in our original QAPP, and our new changes to the QAPP. For each change, we include our reasoning for each amendment.

1. *Original statement:* "All other noncritical parameter samples (for each of the three steady state detention times) will be taken singularly and will be analyzed as duplicate."

   *Replace with:* "All noncritical parameters will not be measured when a reactor reaches steady state. Only the applicable parameters will be measured."

   *Reasoning:* Not all of the parameters listed need to be taken for every experiment. The change indicates that only necessary parameters will be measured.

2. *Original statement:* "It is anticipated that 2-3 other sampling ports will also be located throughout the column depths" (This statement was made with reference to both the modified sand reactor and the gas-phase fixed-film reactor).

   *Replace with:* "Only influent and effluent sampling ports have been built into the gas-phase fixed-film reactor."

   *Reasoning:* The packed sand column can have (and does have) multiple sampling ports. However, the small size of the hydrogen-gas reactor and relatively large size of the packing material (3 mm glass beads) does not allow easy sampling throughout the column depth. Therefore, we can only sample influent and effluent samples to this column.

3. *Original statement:* "Accuracy of in-house perchlorate measurements will also be determined monthly by analyzing blind samples that have been prepared by Dr. Jaci Batista (if determining accuracy of laboratory procedures conducted at Penn State), or prepared by Dr. Bruce Logan (if determining accuracy of laboratory procedures conducted at UNLV)."

   *Replace with:* "Accuracy of in-house perchlorate measurements will also be determined every four months by analyzing blind samples that have been prepared by Dr. Jaci Batista (if determining accuracy of laboratory procedures conducted at Penn State), or prepared by Dr. Bruce Logan (if determining accuracy of laboratory procedures conducted at UNLV)."

*Reasoning:* A monthly exchange of samples would be extremely time consuming and we do not believe would add to our data accuracy. In fact, in the rapid start up of this project we have already neglected to exchange samples between our laboratories. This is not a detriment to the project, as reactors at Penn State and UNLV have only recently been built and are only in the initial start-up phase of operation. When detailed experiments are begun, and steady-state reactor samples are taken, blind samples will be run once each reporting period (every four months). These samples will be prepared by a graduate student at either Penn State or UNLV, and will be promptly sent to the respective university.

4. *Original statement:* "Two deionized water samples, followed by seven perchlorate calibration standards (ppb level), and two additional deionized water samples will be run before each day's sample analysis."

   *Replace with:* "One deionized water sample, followed by four perchlorate calibration standards (ppb level), and one additional deionized water sample as a blank will be run before each day's sample analysis."

   *Reasoning:* The large number (7) of calibration standards originally placed in our QAPP has not been found to be necessary for accurate reporting of perchlorate concentrations. We believe that our current method of 4 calibration standards is sufficient as all $R^2$ are quite high. When reactor studies advance past the initial start-up phase, quality control samples will be run monthly, and mid standard check samples will be run with each day's sample analysis.

# APPENDIX C
# PERCHLORATE ANALYTICAL PROCEDURE

DIONEX

# Application Note 12

# Analysis of Low Concentrations of Perchlorate in Drinking Water and Ground Water by Ion Chromatography

## INTRODUCTION

Perchlorate (as ammonium perchlorate), which is widely used in solid rocket propellants, has recently been found in drinking water wells in areas where aerospace materials and munitions have been manufactured and tested.[1] Perchlorate is a health concern, as it interferes with the production of thyroid hormones. Current data suggest that an exposure level range of 4 to 18 µg/L (ppb) is acceptable.[2] Although perchlorate is not yet regulated in the U.S. under the Federal Safe Drinking Water Act, the State of California requires remedial action for drinking water sources containing greater than 18 µg/L of perchlorate.

This Application Note details a new method developed to quantify low levels of perchlorate. A large loop injection (1000 µL) is used with an IonPac® AS11 column and suppressed conductivity detection to quantify perchlorate in drinking water down to approximately 2.5 µg/L.

## EQUIPMENT

Dionex DX-500 Ion Chromatography system consisting of:
- GP40 Gradient Pump
- CD20 Conductivity Detector
- AS40 Automated Sampler
- LC20 Chromatography Enclosure with a rear-loading valve

4-L Plastic bottle assemblies (two for external water mode)

PeakNet Chromatography Workstation

## REAGENTS AND STANDARDS

Deionized water (DI $H_2O$), Type I reagent grade, 18 MΩ-cm resistance or better

Sodium hydroxide, 50% (w/w) aqueous solution (Fisher Scientific or other)

Sodium perchlorate, 99% ACS reagent grade or better (Aldrich or other)

Potassium sulfate, 1000 mg/L aqueous solution (Ultra Scientific or other)

## CONDITIONS

| | |
|---|---|
| Columns: | IonPac AS11 Analytical, 4 x 250 mm (P/N 44076)<br>IonPac AG11 Guard, 4 x 50 mm (P/N 44078) |
| Eluent: | 100 mM Sodium hydroxide |
| Run Time: | 12 min |
| Flow Rate: | 1.0 mL/min |
| Sample Volume: | 1000 µL |
| Detection: | Suppressed conductivity, ASRS® (4 mm) AutoSuppression® external water mode |
| System Backpressure: | 600ñ900 psi (3.95ñ5.93 MPa) |
| Background Conductance: | 2ñ5 µS |

## PREPARATION OF SOLUTIONS AND REAGENTS

### Standard Solution

*Stock perchlorate standard solution (1000 mg/L)*

Dissolve 1.231 g of sodium perchlorate in 1000 mL of deionized water to prepare a 1000 mg/L standard. Standard is stable for at least one month when stored at 4 °C.

### Working Standard Solutions

Dilute 1000 mg/L standard solution as required with deionized water to prepare the appropriate working standards.

### Eluent Solution

*100.0 mM Sodium hydroxide*

Weigh 992.0 g of deionized water into an eluent bottle. Degas water for approximately 5 minutes. Carefully add 8.0 g of 50% sodium hydroxide directly to the bottle. Mix then quickly transfer the eluent bottle to the instrument and pressurize the bottle with helium at 8 psi (0.055 MPa).

## RESULTS AND DISCUSSION

For the best perfomance at low-ppb levels, it is critical that baseline noise be kept to a minimum. To minimize baseline noise, it is necessary to use the ASRS in external water mode rather than the recycle mode. An equilibrated system will produce a background conductance between 2ñ5 µS. Peak-to-peak noise is typically 10 nS and system backpressure is 600ñ900 psi (3.95ñ5.93 MPa). A system blank is determined by using deionized water as a sample. This blank establishes the baseline and confirms the lack of contamination in the system. The linear concentration range was determined to ensure accurate quantification of perchlorate in the 2.5ñ100 µg/L range. Figure 1 shows the results of a linearity study.

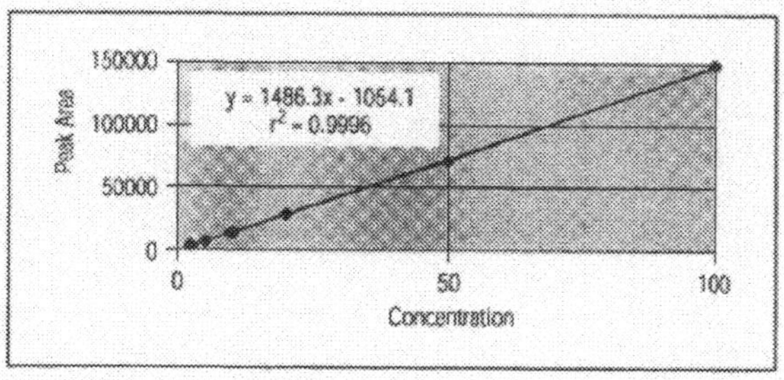


*Figure 1* *Perchlorate calibration*

This plot demonstrates that calibration of perchlorate is linear in the low-ppb range. Figure 2 shows a typical chromatogram of a 20 µg/L perchlorate standard. To determine the method detection limit (MDL), seven injections of the 2.5 µg/L perchlorate standard were made. Table 1 shows the results of a method detection limit study. The 1000 µL injection is large enough to achieve the desired detection limit without overloading the column. Note that this method is not intended for use with high (ppm) levels of perchlorate. The calculated MDL equals 254 ng/L (ppt).

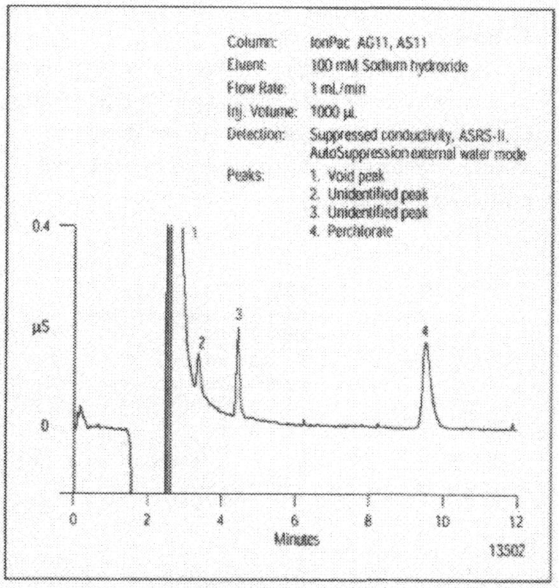


*Figure 2* *20 µg/L Perchlorate standard*

| Injection # | Area counts | Retention time (min) |
|---|---|---|
| 1 | 3391 | 9.48 |
| 2 | 3405 | 9.57 |
| 3 | 3504 | 9.50 |
| 4 | 3503 | 9.45 |
| 5 | 3435 | 9.47 |
| 6 | 3301 | 9.52 |
| 7 | 3315 | 9.43 |
| Average | 3408 | 9.49 |
| SD | 81 | 0.05 |
| RSD | 2.38 | 0.49 |

MDL=254 ng/L (ppt). MDL=SD*$t_{s,99}$ where $t_{s,99}$=3.14 for n=7

Figures 3 through 5 show chromatograms obtained for 2.5 µg/L perchlorate in three different matrices. Figure 3 shows the chromatogram of 2.5 µg/L perchlorate in deionized water. Figure 4 shows 2.5 µg/L perchlorate in tap water. Note that all other anions present in tap water elute in the void volume and do not interfere with perchlorate determination. Some environmental samples may contain low levels of perchlorate in the presence of a large amount of sulfate. Figure 5 shows the determination of 2.5 µg/L perchlorate in the presence of 700 mg/L sulfate. The high concentration of sulfate does not affect perchlorate recovery or the detection limit.

### *SUMMARY*

The method outlined in this Application Note allows the determination of low-µg/L (ppb) levels of perchlorate. Linear concentration ranges have been established to accurately quantify perchlorate in drinking water and ground water samples.

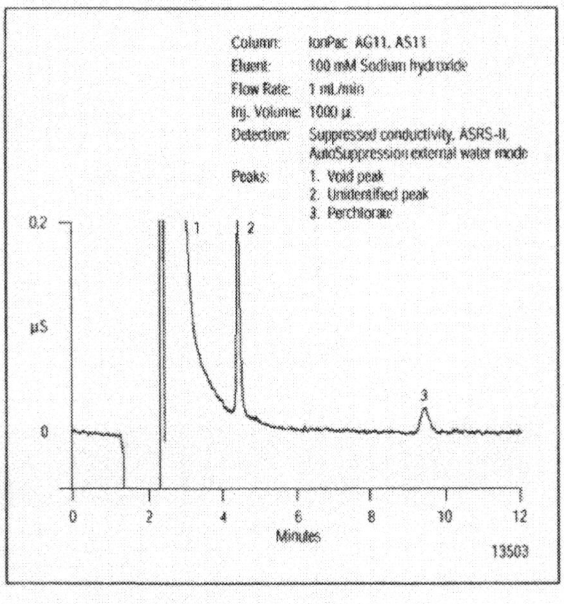


*Figure 3* 2.5 µg/L Perchlorate standard

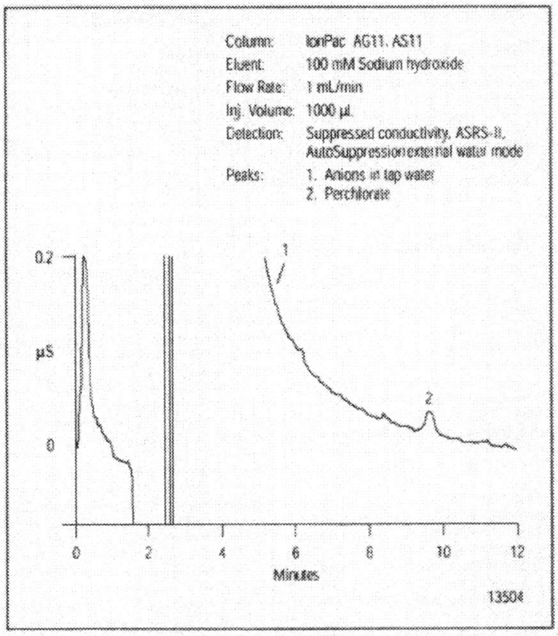


*Figure 4* 2.5 µg/L Perchlorate in tap water

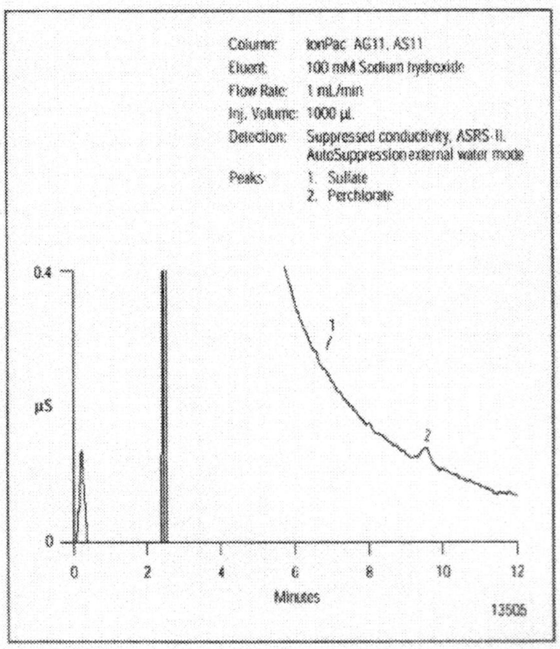


*Figure 5* 2.5 µg/L Perchlorate and 700 mg/L Sulfate

## REFERENCES

1. ìPerchlorate in California Drinking Waterî; California Department of Health Services, September 1997.
2. Correspondence from Joan S. Dollarhide, National Center for Environmental Assessment, Office of Research and Development, to Mike Girrard, Chairman, Perchlorate Study Group, U.S. EPA, 1995.

## LIST OF SUPPLIERS

Aldrich Chemical Company, Inc., 1001 West Saint Paul Avenue, P.O. Box 355, Milwaukee, Wisconsin, 53233, USA. Tel: 1-800-558-9160.

Fisher Scientific, 711 Forbes Ave., Pittsburgh, Pennsylvania, 15219-4785, USA. Tel: 1-800-766-7000.

Ultra Scientific, 250 Smith Street, North Kingstown, Rhode Island, 02852, USA. Tel: 401-294-9400.

# APPENDIX D
# RAW DATA FOR SALINITY TESTS

**Table D.1**
**The interference of different salinity levels on perchlorate biodegradation**
**(Raw data—Figure 6.53)**

| | Transmittance (%) | | | | | | | | | |
|---|---|---|---|---|---|---|---|---|---|---|
| Time (hour) | NaCl 0a | NaCl 0.5b | NaCl 1a | NaCl 1.5b | NaCl 2b | NaCl 2.5b | NaCl 3b | NaCl 3.5c | NaCl 4b | NaCl 6a |
| 0 | 89.5 | 89.5 | 89.5 | 89 | 89 | 89 | 89 | 89 | 89.5 | 89.5 |
| 3 | 89 | 89.5 | 89.5 | 89 | 88.5 | 89 | 88.5 | 89 | 89.5 | 89.5 |
| 19 | 87.5 | 89 | 89.5 | 89 | 89 | 89.5 | 89 | 89 | 90 | 90 |
| 27 | 87.5 | 88.5 | 89 | 89 | 89 | 89 | 88.5 | 89 | 88.5 | 90 |
| 46 | 78.5 | 82 | 86 | 88.5 | 89 | 89.5 | 89 | 89 | 89.5 | 90 |
| 54.5 | 76 | 79 | 80.5 | 86.5 | 89 | 89.5 | 89 | 89 | 89 | 90 |
| 69 | 76.5 | 79 | 80.5 | 81 | 86.5 | 88.5 | 89 | 89.5 | 89.5 | 90 |
| 78 | 76.5 | 79 | 80.5 | 81.5 | 86 | 88.5 | 89 | 90 | 90 | 90 |
| 93 | 76.5 | 78.5 | 80.5 | 81.5 | 83 | 85.5 | 89.5 | 90 | 89.5 | 89.5 |
| 101 | 76.5 | 79 | 80.5 | 81.5 | 83 | 84 | 87.5 | 90.5 | 89 | 90.5 |
| 117 | 76.5 | 79 | 81.5 | 82 | 83 | 80.5 | 86 | 90.5 | 91 | 91 |
| 127 | 76 | 79.5 | 81.5 | 82.5 | 84 | 80.5 | 85.5 | 90 | 90 | 91 |
| 145 | 50.5 | 78 | 82 | 82.5 | 84 | 81.5 | 84 | 89 | 91 | 90 |
| 147 | 50.5 | 78 | 82 | 82.5 | 84 | 81.5 | 84 | 89 | 91 | 90 |
| 166 | 55 | 58 | 82 | 83.5 | 85.5 | 84 | 82 | 87 | 90 | 90.5 |
| 171 | 54.5 | 53 | 82 | 84.5 | 85.5 | 84 | 82.5 | 87.5 | 91.5 | 91.5 |
| 190 | 42.5 | 57.5 | 82 | 84.5 | 86 | 84 | 83 | 86 | 90 | 91.5 |
| 199 | 40.5 | 61 | 81.5 | 84.5 | 87 | 85.5 | 84.5 | 87 | 92 | 91.5 |
| 214.5 | 28 | 70.5 | 57.5 | 84.5 | 87 | 85.5 | 84.5 | 85.5 | 91.5 | 91.5 |
| 222 | 24.5 | 71.5 | 58 | 85 | 87 | 85 | 83.5 | 85.5 | 91 | 91 |
| 235 | 28.5 | 71.5 | 62 | 83.5 | 85.5 | 85.5 | 85.5 | 83.5 | 90.5 | 91 |
| 247 | 28.5 | 71.5 | 63.5 | 76.5 | 86.5 | 86.5 | 85.5 | 85 | 91 | 91 |
| 261 | 29 | 71.5 | 73.5 | 65.5 | 87 | 86 | 85.5 | 86.5 | 92.5 | 92 |
| 268.5 | 30 | 71.5 | 75 | 68 | 87 | 86.5 | 86 | 86.5 | 92 | 92 |
| 285 | 30.5 | 72 | 74.5 | 71 | 86.5 | 86.5 | 86 | 87 | 92.5 | 91.5 |
| 295 | 30.5 | 71 | 75 | 72 | 85 | 86.5 | 85.5 | 86.5 | 91 | 92.5 |
| 310 | 31.5 | 72 | 75 | 73.5 | 86 | 85 | 85.5 | 86.5 | 91.5 | 92 |
| 334 | 32 | 72 | 75.5 | 77.5 | 86.5 | 86 | 86 | 88 | 91 | 91 |
| 359 | 32.5 | 72.5 | 75.5 | 78.5 | 86 | 86.5 | 87 | 88 | 91 | 91.5 |
| 385 | 34 | 72.5 | 76 | 80 | 86 | 87 | 87 | 88 | 91 | 91 |
| 410 | 34 | 73 | 78 | 81.5 | 86.5 | 87 | 87 | 89 | 91 | 91 |
| 435 | 35 | 73.5 | 78 | 82 | 87 | 87.5 | 87.5 | 89 | 92 | 91 |
| 454 | 36.5 | 74 | 78.5 | 83 | 87.5 | 87 | 87.5 | 89 | 92.5 | 92.5 |
| 482 | 37 | 74 | 78 | 82.5 | 87.5 | 86.5 | 87 | 89 | 91 | 91.5 |
| 512 | 38.5 | 74.5 | 80 | 82 | 88 | 87.5 | 87.5 | 88.5 | 91.5 | 92 |
| 529 | 39 | 74.5 | 78 | 81.5 | 86.5 | 87.5 | 87.5 | 89.5 | 91.5 | 91 |
| 552 | 40 | 75.5 | 79.5 | 82 | 87 | 87 | 88 | 89.5 | 91 | 91.5 |
| 573 | 41 | 75.5 | 79.5 | 82 | 87.5 | 87 | 87.5 | 89 | 91 | 91 |
| 598 | 41.5 | 75 | 79 | 81 | 87 | 87 | 88 | 87 | 91.5 | 91.5 |
| 622 | 42.5 | 76 | 80 | 81 | 87.5 | 87.5 | 88.5 | 88 | 92 | 93 |
| 650 | 42.5 | 76 | 80 | 80.5 | 88 | 87.5 | 87.5 | 88 | 92 | 93 |
| 671.5 | 43.5 | 76 | 79 | 80 | 86.5 | 87.5 | 87.5 | 88 | 91.5 | 93.5 |

**Table D.2**
**The interference of salinity on perchlorate biodegradation (0%)**
**(Raw data—Figure 6.54)**

| Time | NaCl 0a | | | NaCl 0b | | | NaCl 0c | | | NaCl 0d | NaCl 0e |
|---|---|---|---|---|---|---|---|---|---|---|---|
| (hr) | Trans. | Abs. | Ln(Abs.) | Trans. | Abs. | Ln(Abs.) | Trans. | Abs. | Ln(Abs.) | Trans. | Trans. |
| 0 | 89.5 | **0.1109** | **-2.1988** | 88 | **0.1278** | **-2.057** | 89 | 0.11653 | -2.15 | 100 | 89.5 |
| 3 | 89 | **0.1165** | **-2.1496** | 88 | **0.1278** | **-2.057** | 89 | 0.11653 | -2.15 | 100 | 89 |
| 19 | 87.5 | **0.1335** | **-2.0134** | 88 | **0.1278** | **-2.057** | 89 | 0.11653 | -2.15 | 100 | 89 |
| 27 | 87.5 | **0.1335** | **-2.0134** | 88 | **0.1278** | **-2.057** | 89 | 0.11653 | -2.15 | 100 | 89 |
| 46 | 78.5 | **0.2421** | **-1.4185** | 76.5 | **0.2679** | **-1.3172** | 79 | 0.23572 | -1.445 | 100 | 89 |
| 54.5 | 76 | **0.2744** | **-1.293** | 76 | **0.2744** | **-1.293** | 79.5 | 0.22941 | -1.472 | 100 | 89 |
| 69 | 76.5 | **0.2679** | **-1.3172** | 76 | **0.2744** | **-1.293** | 79.5 | 0.22941 | -1.472 | 100 | 89.5 |
| 78 | 76.5 | **0.2679** | **-1.3172** | 76 | **0.2744** | **-1.293** | 79 | 0.23572 | -1.445 | 100 | 89.5 |
| 93 | 76.5 | **0.2679** | **-1.3172** | 76 | **0.2744** | **-1.293** | 79 | 0.23572 | -1.445 | 100 | 89.5 |
| 101 | 76.5 | **0.2679** | **-1.3172** | 76.5 | **0.2679** | **-1.3172** | 79 | 0.23572 | -1.445 | 100 | 89.5 |
| 117 | 76.5 | **0.2679** | **-1.3172** | 75 | **0.2877** | **-1.2459** | 78.5 | 0.24207 | -1.419 | 100 | 89 |
| 127 | 76 | **0.2744** | **-1.293** | 75 | **0.2877** | **-1.2459** | 78 | 0.24846 | -1.392 | 100 | 90.5 |
| 145 | 50.5 | **0.6832** | **-0.381** | 52 | **0.6539** | **-0.4248** | 72 | 0.3285 | -1.113 | 100 | 90 |
| 147 | 50.5 | **0.6832** | **-0.381** | 52 | **0.6539** | **-0.4248** | 72 | 0.3285 | -1.113 | 100 | 90 |
| 166 | 55 | **0.5978** | **-0.5144** | 55 | **0.5978** | **-0.5144** | 55 | 0.59784 | -0.514 | 100 | 89.5 |
| 171 | 54.5 | **0.607** | **-0.4993** | 52 | **0.6539** | **-0.4248** | 55 | 0.59784 | -0.514 | 100 | 90 |
| 190 | 42.5 | **0.8557** | **-0.1559** | 41.5 | **0.8795** | **-0.1284** | 55.5 | 0.58879 | -0.53 | 100 | 89 |
| 199 | 40.5 | **0.9039** | **-0.1011** | 35 | **1.0498** | **0.0486** | 57 | 0.56212 | -0.576 | 100 | 89.5 |
| 214.5 | 28 | **1.273** | **0.2413** | 25.5 | **1.3665** | **0.3122** | 55.5 | 0.58879 | -0.53 | 100 | 89 |
| 222 | 24.5 | **1.4065** | **0.3411** | 27.5 | 1.291 | 0.2554 | 55.5 | 0.58879 | -0.53 | 100 | 89 |
| 235 | 28.5 | 1.2553 | 0.2273 | 28 | 1.273 | 0.2413 | 58 | 0.54473 | -0.607 | 100 | 89 |
| 247 | 28.5 | 1.2553 | 0.2273 | 28 | 1.273 | 0.2413 | 60 | 0.51083 | -0.672 | 100 | 89 |
| 261 | 29 | 1.2379 | 0.2134 | 29 | 1.2379 | 0.2134 | 62 | 0.47804 | -0.738 | 100 | 89 |
| 268.5 | 30 | 1.204 | 0.1856 | 29.5 | 1.2208 | 0.1995 | 63 | 0.46204 | -0.772 | 100 | 89 |
| 285 | 30.5 | 1.1874 | 0.1718 | 29.5 | 1.2208 | 0.1995 | 64 | 0.44629 | -0.807 | 100 | 89 |
| 295 | 30.5 | 1.1874 | 0.1718 | 29.5 | 1.2208 | 0.1995 | 63 | 0.46204 | -0.772 | 100 | 89 |
| 310 | 31.5 | 1.1552 | 0.1443 | 30 | 1.204 | 0.1856 | 59.5 | 0.51919 | -0.655 | 100 | 88.5 |
| 334 | 32 | 1.1394 | 0.1305 | 31 | 1.1712 | 0.158 | 59 | 0.52763 | -0.639 | 100 | 88 |
| 359 | 32.5 | 1.1239 | 0.1168 | 32 | 1.1394 | 0.1305 | 60.5 | 0.50253 | -0.688 | 100 | 88 |
| 385 | 34 | 1.0788 | 0.0759 | 33 | 1.1087 | 0.1032 | 62.5 | 0.47 | -0.755 | 100 | 88 |
| 410 | 34 | 1.0788 | 0.0759 | 33.5 | 1.0936 | 0.0895 | 62.5 | 0.47 | -0.755 | 100 | 87.5 |
| 435 | 35 | 1.0498 | 0.0486 | 35.5 | 1.0356 | 0.035 | 40.5 | 0.90387 | -0.101 | 100 | 88 |
| 454 | 36.5 | 1.0079 | 0.0078 | 35.5 | 1.0356 | 0.035 | 34 | 1.07881 | 0.0759 | 100 | 88 |
| 482 | 37 | 0.9943 | -0.0058 | 36 | 1.0217 | 0.0214 | 35 | 1.04982 | 0.0486 | 100 | 88 |
| 512 | 38.5 | 0.9545 | -0.0466 | 37.5 | 0.9808 | -0.0194 | 36 | 1.02165 | 0.0214 | 100 | 89 |
| 529 | 39 | 0.9416 | -0.0602 | 38 | 0.9676 | -0.033 | 36 | 1.02165 | 0.0214 | 100 | 88 |
| 552 | 40 | 0.9163 | -0.0874 | 38.5 | 0.9545 | -0.0466 | 37 | 0.99425 | -0.006 | 100 | 88 |
| 573 | 41 | 0.8916 | -0.1147 | 40 | 0.9163 | -0.0874 | 37.5 | 0.98083 | -0.019 | 100 | 88 |
| 598 | 41.5 | 0.8795 | -0.1284 | 40.5 | 0.9039 | -0.1011 | 37.5 | 0.98083 | -0.019 | 100 | 88 |
| 622 | 42.5 | 0.8557 | -0.1559 | 42 | 0.8675 | -0.1421 | 38.5 | 0.95451 | -0.047 | 100 | 88 |
| 650 | 42.5 | 0.8557 | -0.1559 | 42 | 0.8675 | -0.1421 | 39 | 0.94161 | -0.06 | 100 | 88 |
| 671.5 | 43.5 | 0.8324 | -0.1834 | 42.5 | 0.8557 | -0.1559 | 39.5 | 0.92887 | -0.074 | 100 | 88 |

**Table D.3**
**The interference of salinity on perchlorate biodegradation (0.5%)**
**(Raw data—Figure 6.55)**

| Time | NaCl 0.5a | | | NaCl 0.5b | | | NaCl 0.5c | | | NaCl 0.5d | NaCl 0.5e |
|---|---|---|---|---|---|---|---|---|---|---|---|
| (hr) | Trans. | Abs. | Ln(Abs.) | Trans. | Abs. | Ln(Abs.) | Trans. | Abs. | Ln(Abs.) | Trans. | Trans. |
| 0 | 90 | 0.11 | -2.199 | 90 | **0.111** | **-2.2** | 89 | **0.117** | **-2.15** | 100 | 89.5 |
| 3 | 90 | 0.11 | -2.199 | 90 | **0.111** | **-2.2** | 88 | **0.128** | **-2.057** | 100 | 89.5 |
| 19 | 89 | 0.12 | -2.15 | 89 | **0.117** | **-2.15** | 89 | **0.122** | **-2.102** | 100 | 89.5 |
| 27 | 89 | 0.12 | -2.15 | 89 | **0.122** | **-2.1** | 89 | **0.122** | **-2.102** | 100 | 89.5 |
| 46 | 81 | 0.21 | -1.557 | 82 | **0.198** | **-1.62** | 82 | **0.198** | **-1.617** | 100 | 88.5 |
| 54.5 | 80 | 0.23 | -1.472 | 79 | **0.236** | **-1.45** | 82 | **0.205** | **-1.587** | 100 | 88.5 |
| 69 | 80 | 0.23 | -1.472 | 79 | **0.236** | **-1.45** | 82 | **0.205** | **-1.587** | 100 | 88.5 |
| 78 | 80 | 0.22 | -1.5 | 79 | **0.236** | **-1.45** | 82 | **0.198** | **-1.617** | 100 | 88.5 |
| 93 | 80 | 0.23 | -1.472 | 79 | **0.242** | **-1.42** | 82 | **0.205** | **-1.587** | 100 | 88 |
| 101 | 79 | 0.24 | -1.445 | 79 | **0.236** | **-1.45** | 82 | **0.205** | **-1.587** | 100 | 88 |
| 117 | 80 | 0.23 | -1.472 | 79 | **0.236** | **-1.45** | 81 | **0.211** | **-1.557** | 100 | 88 |
| 127 | 80 | 0.23 | -1.472 | 80 | **0.229** | **-1.47** | 81 | **0.211** | **-1.557** | 100 | 88 |
| 145 | 78 | 0.25 | -1.392 | 78 | **0.248** | **-1.39** | 80 | **0.223** | **-1.5** | 100 | 87.5 |
| 147 | 78 | 0.25 | -1.392 | 78 | **0.248** | **-1.39** | 80 | **0.223** | **-1.5** | 100 | 87.5 |
| 166 | 52 | 0.66 | -0.41 | 58 | **0.545** | **-0.61** | 64 | **0.454** | **-0.789** | 100 | 87.5 |
| 171 | 52 | 0.65 | -0.425 | 53 | **0.635** | **-0.45** | 53 | **0.635** | **-0.454** | 100 | 87.5 |
| 190 | 58 | 0.54 | -0.607 | 58 | 0.553 | -0.59 | 58 | 0.545 | -0.607 | 100 | 88 |
| 199 | 61 | 0.5 | -0.688 | 61 | 0.494 | -0.7 | 62 | 0.486 | -0.721 | 100 | 88 |
| 214.5 | 71 | 0.35 | -1.051 | 71 | 0.35 | -1.05 | 72 | 0.335 | -1.092 | 100 | 88 |
| 222 | 72 | 0.34 | -1.092 | 72 | 0.335 | -1.09 | 73 | 0.315 | -1.156 | 100 | 88.5 |
| 235 | 71 | 0.34 | -1.072 | 72 | 0.335 | -1.09 | 73 | 0.315 | -1.156 | 100 | 88 |
| 247 | 72 | 0.34 | -1.092 | 72 | 0.335 | -1.09 | 73 | 0.315 | -1.156 | 100 | 88 |
| 261 | 71 | 0.34 | -1.072 | 72 | 0.335 | -1.09 | 73 | 0.315 | -1.156 | 100 | 88 |
| 268.5 | 71 | 0.35 | -1.051 | 72 | 0.335 | -1.09 | 73 | 0.315 | -1.156 | 100 | 89 |
| 285 | 71 | 0.35 | -1.051 | 72 | 0.329 | -1.11 | 74 | 0.308 | -1.178 | 100 | 88.5 |
| 295 | 71 | 0.35 | -1.051 | 71 | 0.342 | -1.07 | 72 | 0.329 | -1.113 | 100 | 87 |
| 310 | 72 | 0.33 | -1.113 | 72 | 0.329 | -1.11 | 73 | 0.322 | -1.134 | 100 | 87.5 |
| 334 | 66 | 0.42 | -0.878 | 72 | 0.329 | -1.11 | 73 | 0.322 | -1.134 | 100 | 88 |
| 359 | 64 | 0.45 | -0.807 | 73 | 0.322 | -1.13 | 74 | 0.308 | -1.178 | 100 | 89 |
| 385 | 65 | 0.44 | -0.824 | 73 | 0.322 | -1.13 | 75 | 0.294 | -1.223 | 100 | 88 |
| 410 | 67 | 0.41 | -0.897 | 73 | 0.315 | -1.16 | 75 | 0.294 | -1.223 | 100 | 88 |
| 435 | 69 | 0.38 | -0.972 | 74 | 0.308 | -1.18 | 75 | 0.294 | -1.223 | 100 | 88 |
| 454 | 70 | 0.36 | -1.011 | 74 | 0.301 | -1.2 | 76 | 0.281 | -1.269 | 100 | 88 |
| 482 | 70 | 0.36 | -1.011 | 74 | 0.301 | -1.2 | 75 | 0.294 | -1.223 | 100 | 88 |
| 512 | 73 | 0.31 | -1.156 | 75 | 0.294 | -1.22 | 76 | 0.281 | -1.269 | 100 | 88.5 |
| 529 | 72 | 0.34 | -1.092 | 75 | 0.294 | -1.22 | 75 | 0.288 | -1.246 | 100 | 88 |
| 552 | 66 | 0.42 | -0.878 | 76 | 0.281 | -1.27 | 76 | 0.281 | -1.269 | 100 | 88 |
| 573 | 33 | 1.11 | 0.103 | 76 | 0.281 | -1.27 | 76 | 0.274 | -1.293 | 100 | 88 |
| 598 | 33 | 1.11 | 0.103 | 75 | 0.288 | -1.25 | 76 | 0.281 | -1.269 | 100 | 88 |
| 622 | 35 | 1.05 | 0.049 | 76 | 0.274 | -1.29 | 76 | 0.281 | -1.269 | 100 | 88 |
| 650 | 35 | 1.05 | 0.049 | 76 | 0.274 | -1.29 | 76 | 0.274 | -1.293 | 100 | 87 |
| 671.5 | 35 | 1.05 | 0.049 | 76 | 0.274 | -1.29 | 77 | 0.268 | -1.317 | 100 | 87 |

**Table D.4**
**The interference of salinity on perchlorate biodegradation (1%)**
**(Raw data—Figure 6.56)**

| Time | NaCl 1a | | | NaCl 1b | | | NaCl 1c | | | NaCl 1d | NaCl 1e |
|---|---|---|---|---|---|---|---|---|---|---|---|
| (hr) | Trans. | Abs. | Ln(Abs.) | Trans. | Abs. | Ln(Abs.) | Trans. | Abs. | Ln(Abs.) | Trans. | Trans. |
| 0 | 90 | **0.111** | **-2.199** | 89 | **0.117** | **-2.15** | 89 | **0.117** | **-2.15** | 100 | 89 |
| 3 | 90 | **0.111** | **-2.199** | 89 | **0.117** | **-2.15** | 89 | **0.117** | **-2.15** | 100 | 89 |
| 19 | 90 | **0.111** | **-2.199** | 89 | **0.117** | **-2.15** | 89 | **0.117** | **-2.15** | 100 | 89.5 |
| 27 | 89 | **0.117** | **-2.15** | 89 | **0.117** | **-2.15** | 88.5 | **0.122** | **-2.1** | 100 | 89 |
| 46 | 86 | **0.151** | **-1.892** | 85 | **0.163** | **-1.817** | 86 | **0.151** | **-1.89** | 100 | 89 |
| 54.5 | 81 | **0.217** | **-1.528** | 80.5 | **0.217** | **-1.528** | 81.5 | **0.205** | **-1.59** | 100 | 89 |
| 69 | 81 | **0.217** | **-1.528** | 80.5 | **0.217** | **-1.528** | 81.5 | **0.205** | **-1.59** | 100 | 90 |
| 78 | 81 | **0.217** | **-1.528** | 80.5 | **0.217** | **-1.528** | 81 | **0.211** | **-1.56** | 100 | 89.5 |
| 93 | 81 | **0.217** | **-1.528** | 80.5 | **0.217** | **-1.528** | 81 | **0.211** | **-1.56** | 100 | 88.5 |
| 101 | 81 | **0.217** | **-1.528** | 80 | **0.223** | **-1.5** | 80.5 | **0.217** | **-1.53** | 100 | 88 |
| 117 | 82 | **0.205** | **-1.587** | 81 | **0.211** | **-1.557** | 81.5 | **0.205** | **-1.59** | 100 | 89 |
| 127 | 82 | **0.205** | **-1.587** | 81.5 | **0.205** | **-1.587** | 81.5 | **0.205** | **-1.59** | 100 | 88 |
| 145 | 82 | **0.198** | **-1.617** | 81.5 | **0.205** | **-1.587** | 82 | **0.198** | **-1.62** | 100 | 88.5 |
| 147 | 82 | **0.198** | **-1.617** | 81.5 | **0.205** | **-1.587** | 82 | **0.198** | **-1.62** | 100 | 88.5 |
| 166 | 82 | **0.198** | **-1.617** | 82.5 | **0.192** | **-1.648** | 82.5 | **0.192** | **-1.65** | 100 | 88 |
| 171 | 82 | **0.198** | **-1.617** | 82.5 | **0.192** | **-1.648** | 82.5 | **0.192** | **-1.65** | 100 | 88 |
| 190 | 82 | **0.198** | **-1.617** | 81.5 | **0.205** | **-1.587** | 82.5 | **0.192** | **-1.65** | 100 | 88 |
| 199 | 82 | **0.205** | **-1.587** | 79 | **0.236** | **-1.445** | 82 | **0.198** | **-1.62** | 100 | 88 |
| 214.5 | 58 | **0.553** | **-0.592** | 55.5 | **0.589** | **-0.53** | 79.5 | **0.229** | **-1.47** | 100 | 88.5 |
| 222 | 58 | 0.545 | -0.607 | 59.5 | 0.519 | -0.655 | 73 | **0.315** | **-1.16** | 100 | 89 |
| 235 | 62 | 0.478 | -0.738 | 62 | 0.478 | -0.738 | 57.5 | **0.553** | **-0.59** | 100 | 88.5 |
| 247 | 64 | 0.454 | -0.789 | 65.5 | 0.423 | -0.86 | 60.5 | 0.503 | -0.69 | 100 | 88.5 |
| 261 | 74 | 0.308 | -1.178 | 75.5 | 0.281 | -1.269 | 63.5 | 0.454 | -0.79 | 100 | 88.5 |
| 268.5 | 75 | 0.288 | -1.246 | 75.5 | 0.281 | -1.269 | 68 | 0.386 | -0.95 | 100 | 89 |
| 285 | 75 | 0.294 | -1.223 | 75 | 0.288 | -1.246 | 75 | 0.288 | -1.25 | 100 | 87.5 |
| 295 | 75 | 0.288 | -1.246 | 75 | 0.288 | -1.246 | 75 | 0.288 | -1.25 | 100 | 86.5 |
| 310 | 75 | 0.288 | -1.246 | 75.5 | 0.281 | -1.269 | 75.5 | 0.281 | -1.27 | 100 | 86.5 |
| 334 | 76 | 0.281 | -1.269 | 76 | 0.274 | -1.293 | 75 | 0.288 | -1.25 | 100 | 87.5 |
| 359 | 76 | 0.281 | -1.269 | 75.5 | 0.281 | -1.269 | 75.5 | 0.281 | -1.27 | 100 | 86.5 |
| 385 | 76 | 0.274 | -1.293 | 76.5 | 0.268 | -1.317 | 75.5 | 0.281 | -1.27 | 100 | 86.5 |
| 410 | 78 | 0.248 | -1.392 | 78 | 0.248 | -1.392 | 76 | 0.274 | -1.29 | 100 | 88 |
| 435 | 78 | 0.248 | -1.392 | 78.5 | 0.242 | -1.419 | 77 | 0.261 | -1.34 | 100 | 87.5 |
| 454 | 79 | 0.242 | -1.419 | 78.5 | 0.242 | -1.419 | 77.5 | 0.255 | -1.37 | 100 | 88 |
| 482 | 78 | 0.248 | -1.392 | 78.5 | 0.242 | -1.419 | 76 | 0.274 | -1.29 | 100 | 88.5 |
| 512 | 80 | 0.223 | -1.5 | 80 | 0.223 | -1.5 | 78 | 0.248 | -1.39 | 100 | 88.5 |
| 529 | 78 | 0.248 | -1.392 | 78.5 | 0.242 | -1.419 | 77 | 0.261 | -1.34 | 100 | 87.5 |
| 552 | 80 | 0.229 | -1.472 | 80 | 0.223 | -1.5 | 79 | 0.236 | -1.45 | 100 | 88 |
| 573 | 80 | 0.229 | -1.472 | 80 | 0.223 | -1.5 | 78.5 | 0.242 | -1.42 | 100 | 88 |
| 598 | 79 | 0.236 | -1.445 | 80 | 0.223 | -1.5 | 78 | 0.248 | -1.39 | 100 | 87.5 |
| 622 | 80 | 0.223 | -1.5 | 80.5 | 0.217 | -1.528 | 78.5 | 0.242 | -1.42 | 100 | 88 |
| 650 | 80 | 0.223 | -1.5 | 80 | 0.223 | -1.5 | 78 | 0.248 | -1.39 | 100 | 88 |
| 671.5 | 79 | 0.236 | -1.445 | 80.5 | 0.217 | -1.528 | 79 | 0.236 | -1.45 | 100 | 87.5 |

**Table D.5**
**The interference of salinity on perchlorate biodegradation (1.5%)**
**(Raw data—Figure 6.57)**

| Time | NaCl 1.5a | | | NaCl 1.5b | | | NaCl 1.5c | | | NaCl 1.5d | NaCl 1.5e |
|---|---|---|---|---|---|---|---|---|---|---|---|
| (hr) | Trans. | Abs. | Ln(Abs.) | Trans. | Abs. | Ln(Abs.) | Trans. | Abs. | Ln(Abs.) | Trans. | Trans. |
| 0 | 89 | 0.12 | -2.15 | 89 | 0.1165 | -2.15 | 89 | 0.12 | -2.14957 | 100 | 89 |
| 3 | 89 | 0.12 | -2.15 | 89 | 0.1165 | -2.15 | 89 | 0.12 | -2.14957 | 100 | 89 |
| 19 | 89 | 0.12 | -2.15 | 89 | 0.1165 | -2.15 | 89 | 0.12 | -2.14957 | 100 | 89.5 |
| 27 | 89 | 0.12 | -2.15 | 89 | 0.1165 | -2.15 | 89 | 0.12 | -2.14957 | 100 | 89 |
| 46 | 88.5 | 0.12 | -2.102 | 88.5 | 0.1222 | -2.102 | 88.5 | 0.12 | -2.10236 | 100 | 89 |
| 54.5 | 86.5 | 0.15 | -1.931 | 86.5 | 0.145 | -1.931 | 88 | 0.13 | -2.05703 | 100 | 89.5 |
| 69 | 82 | 0.2 | -1.617 | 81 | 0.2107 | -1.557 | 81 | 0.21 | -1.55722 | 100 | 89.5 |
| 78 | 82 | 0.2 | -1.617 | 81.5 | 0.2046 | -1.587 | 81.5 | 0.2 | -1.58686 | 100 | 89.5 |
| 93 | 82 | 0.2 | -1.617 | 81.5 | 0.2046 | -1.587 | 81.5 | 0.2 | -1.58686 | 100 | 89 |
| 101 | 82 | 0.2 | -1.617 | 81.5 | 0.2046 | -1.587 | 81.5 | 0.2 | -1.58686 | 100 | 88 |
| 117 | 82 | 0.2 | -1.617 | 82 | 0.1985 | -1.617 | 82 | 0.2 | -1.61721 | 100 | 88.5 |
| 127 | 82 | 0.2 | -1.617 | 82.5 | 0.1924 | -1.648 | 83 | 0.19 | -1.68024 | 100 | 88 |
| 145 | 81.5 | 0.2 | -1.587 | 82.5 | 0.1924 | -1.648 | 83 | 0.19 | -1.68024 | 100 | 88 |
| 147 | 81.5 | 0.2 | -1.587 | 82.5 | 0.1924 | -1.648 | 83 | 0.19 | -1.68024 | 100 | 88 |
| 166 | 83 | 0.19 | -1.68 | 83.5 | 0.1803 | -1.713 | 83.5 | 0.18 | -1.713 | 100 | 88 |
| 171 | 83.5 | 0.18 | -1.713 | 84.5 | 0.1684 | -1.781 | 84 | 0.17 | -1.74667 | 100 | 88 |
| 190 | 83.5 | 0.18 | -1.713 | 84.5 | 0.1684 | -1.781 | 84 | 0.17 | -1.74667 | 100 | 88 |
| 199 | 84 | 0.17 | -1.747 | 84.5 | 0.1684 | -1.781 | 84.5 | 0.17 | -1.7813 | 100 | 88.5 |
| 214.5 | 84.5 | 0.17 | -1.781 | 84.5 | 0.1684 | -1.781 | 84 | 0.17 | -1.74667 | 100 | 88 |
| 222 | 85 | 0.16 | -1.817 | 85 | 0.1625 | -1.817 | 84.5 | 0.17 | -1.7813 | 100 | 88.5 |
| 235 | 84.5 | 0.17 | -1.781 | 83.5 | 0.1803 | -1.713 | 81 | 0.21 | -1.55722 | 100 | 88 |
| 247 | 84 | 0.17 | -1.747 | 76.5 | 0.2679 | -1.317 | 72 | 0.33 | -1.11321 | 100 | 88 |
| 261 | 84.5 | 0.17 | -1.781 | 65.5 | 0.4231 | -0.86 | 65 | 0.43 | -0.84215 | 100 | 88 |
| 268.5 | 86 | 0.15 | -1.892 | 68 | 0.3857 | -0.953 | 68 | 0.39 | -0.95279 | 100 | 88.5 |
| 285 | 85.5 | 0.16 | -1.854 | 71 | 0.3425 | -1.072 | 72 | 0.33 | -1.11321 | 100 | 88.5 |
| 295 | 84.5 | 0.17 | -1.781 | 72 | 0.3285 | -1.113 | 72 | 0.33 | -1.11321 | 100 | 88 |
| 310 | 85 | 0.16 | -1.817 | 73.5 | 0.3079 | -1.178 | 73.5 | 0.31 | -1.17803 | 100 | 88 |
| 334 | 84.5 | 0.17 | -1.781 | 77.5 | 0.2549 | -1.367 | 77 | 0.26 | -1.34184 | 100 | 89 |
| 359 | 84.5 | 0.17 | -1.781 | 78.5 | 0.2421 | -1.419 | 78 | 0.25 | -1.39247 | 100 | 88.5 |
| 385 | 85 | 0.16 | -1.817 | 80 | 0.2231 | -1.5 | 79 | 0.24 | -1.4451 | 100 | 88 |
| 410 | 85 | 0.16 | -1.817 | 81.5 | 0.2046 | -1.587 | 80 | 0.22 | -1.49994 | 100 | 88 |
| 435 | 85.5 | 0.16 | -1.854 | 82 | 0.1985 | -1.617 | 81.5 | 0.2 | -1.58686 | 100 | 89 |
| 454 | 86 | 0.15 | -1.892 | 83 | 0.1863 | -1.68 | 81.5 | 0.2 | -1.58686 | 100 | 89 |
| 482 | 83.5 | 0.18 | -1.713 | 82.5 | 0.1924 | -1.648 | 81 | 0.21 | -1.55722 | 100 | 88 |
| 512 | 84 | 0.17 | -1.747 | 82 | 0.1985 | -1.617 | 81 | 0.21 | -1.55722 | 100 | 88.5 |
| 529 | 82 | 0.2 | -1.617 | 81.5 | 0.2046 | -1.587 | 81 | 0.21 | -1.55722 | 100 | 88 |
| 552 | 82 | 0.2 | -1.617 | 82 | 0.1985 | -1.617 | 80.5 | 0.22 | -1.52826 | 100 | 88 |
| 573 | 80 | 0.22 | -1.5 | 82 | 0.1985 | -1.617 | 81 | 0.21 | -1.55722 | 100 | 88 |
| 598 | 76 | 0.27 | -1.293 | 81 | 0.2107 | -1.557 | 80.5 | 0.22 | -1.52826 | 100 | 88 |
| 622 | 71 | 0.34 | -1.072 | 81 | 0.2107 | -1.557 | 80.5 | 0.22 | -1.52826 | 100 | 88 |
| 650 | 66.5 | 0.41 | -0.897 | 80.5 | 0.2169 | -1.528 | 79.5 | 0.23 | -1.47223 | 100 | 88 |
| 671.5 | 64.5 | 0.44 | -0.824 | 80 | 0.2231 | -1.5 | 79.5 | 0.23 | -1.47223 | 100 | 88.5 |

## Table D.6
## The interference of salinity on perchlorate biodegradation (2%)
## (Raw data—Figure 6.58)

| Time | NaCl 2a | | | NaCl 2b | | | NaCl 2c | | | NaCl 2d | NaCl 2e |
|---|---|---|---|---|---|---|---|---|---|---|---|
| (hr) | Trans. | Abs. | Ln(Abs.) | Trans. | Abs. | Ln(Abs.) | Trans. | Abs. | Ln(Abs.) | Trans. | Trans. |
| 0 | 88 | 0.1278 | -2.05703 | 89 | 0.1165 | -2.14957 | 89 | 0.117 | -2.15 | 100 | 89 |
| 3 | 88.5 | **0.1222** | **-2.10236** | 89 | **0.1222** | **-2.10236** | 89 | **0.117** | **-2.15** | 100 | 89 |
| 19 | 88.5 | **0.1222** | **-2.10236** | 89 | **0.1165** | **-2.14957** | 89 | **0.117** | **-2.15** | 100 | 89 |
| 27 | 88.5 | **0.1222** | **-2.10236** | 89 | **0.1165** | **-2.14957** | 89 | **0.117** | **-2.15** | 100 | 89 |
| 46 | 88 | **0.1278** | **-2.05703** | 89 | **0.1165** | **-2.14957** | 89 | **0.117** | **-2.15** | 100 | 89.5 |
| 54.5 | 88 | **0.1278** | **-2.05703** | 89 | **0.1165** | **-2.14957** | 89 | **0.117** | **-2.15** | 100 | 89.5 |
| 69 | 85 | **0.1625** | **-1.81696** | 87 | **0.145** | **-1.93084** | 86.5 | **0.145** | **-1.931** | 100 | 89 |
| 78 | 84 | **0.1744** | **-1.74667** | 86 | **0.1508** | **-1.89165** | 86 | **0.151** | **-1.892** | 100 | 89.5 |
| 93 | 81.5 | **0.2046** | **-1.58686** | 83 | **0.1863** | **-1.68024** | 83 | **0.186** | **-1.68** | 100 | 89 |
| 101 | 79 | **0.2357** | **-1.4451** | 83 | 0.1863 | -1.68024 | 83 | 0.186 | -1.68 | 100 | 89 |
| 117 | 81.5 | 0.2046 | -1.58686 | 83 | 0.1863 | -1.68024 | 84 | 0.174 | -1.747 | 100 | 88 |
| 127 | 81.5 | 0.2046 | -1.58686 | 84 | 0.1744 | -1.74667 | 84 | 0.174 | -1.747 | 100 | 89 |
| 145 | 81.5 | 0.2046 | -1.58686 | 84 | 0.1744 | -1.74667 | 82 | 0.198 | -1.617 | 100 | 88 |
| 147 | 81.5 | 0.2046 | -1.58686 | 84 | 0.1744 | -1.74667 | 82 | 0.198 | -1.617 | 100 | 88 |
| 166 | 83.5 | 0.1803 | -1.713 | 86 | 0.1567 | -1.85372 | 83.5 | 0.18 | -1.713 | 100 | 88 |
| 171 | 83.5 | 0.1803 | -1.713 | 86 | 0.1567 | -1.85372 | 84.5 | 0.168 | -1.781 | 100 | 88.5 |
| 190 | 84 | 0.1744 | -1.74667 | 86 | 0.1508 | -1.89165 | 84.5 | 0.168 | -1.781 | 100 | 88 |
| 199 | 84.5 | 0.1684 | -1.7813 | 87 | 0.1393 | -1.9714 | 86 | 0.151 | -1.892 | 100 | 88 |
| 214.5 | 85 | 0.1625 | -1.81696 | 87 | 0.1393 | -1.9714 | 86 | 0.151 | -1.892 | 100 | 88 |
| 222 | 84.5 | 0.1684 | -1.7813 | 87 | 0.1393 | -1.9714 | 85 | 0.163 | -1.817 | 100 | 88 |
| 235 | 84.5 | 0.1684 | -1.7813 | 86 | 0.1567 | -1.85372 | 85.5 | 0.157 | -1.854 | 100 | 88 |
| 247 | 86.5 | 0.145 | -1.93084 | 87 | 0.145 | -1.93084 | 86.5 | 0.145 | -1.931 | 100 | 88 |
| 261 | 85.5 | 0.1567 | -1.85372 | 87 | 0.1393 | -1.9714 | 87 | 0.139 | -1.971 | 100 | 88 |
| 268.5 | 87 | 0.1393 | -1.9714 | 87 | 0.1393 | -1.9714 | 87 | 0.139 | -1.971 | 100 | 88 |
| 285 | 86.5 | 0.145 | -1.93084 | 87 | 0.145 | -1.93084 | 87 | 0.139 | -1.971 | 100 | 88 |
| 295 | 85.5 | 0.1567 | -1.85372 | 85 | 0.1625 | -1.81696 | 86.5 | 0.145 | -1.931 | 100 | 87.5 |
| 310 | 85.5 | 0.1567 | -1.85372 | 86 | 0.1508 | -1.89165 | 85.5 | 0.157 | -1.854 | 100 | 88 |
| 334 | 86.5 | 0.145 | -1.93084 | 87 | 0.145 | -1.93084 | 86.5 | 0.145 | -1.931 | 100 | 89 |
| 359 | 86.5 | 0.145 | -1.93084 | 86 | 0.1508 | -1.89165 | 87 | 0.139 | -1.971 | 100 | 88 |
| 385 | 86 | 0.1508 | -1.89165 | 86 | 0.1508 | -1.89165 | 87 | 0.139 | -1.971 | 100 | 88 |
| 410 | 87 | 0.1393 | -1.9714 | 87 | 0.145 | -1.93084 | 88 | 0.128 | -2.057 | 100 | 88.5 |
| 435 | 88 | 0.1278 | -2.05703 | 87 | 0.1393 | -1.9714 | 88 | 0.128 | -2.057 | 100 | 88.5 |
| 454 | 88.5 | 0.1222 | -2.10236 | 88 | 0.1335 | -2.01342 | 87.5 | 0.134 | -2.013 | 100 | 89 |
| 482 | 87 | 0.1393 | -1.9714 | 88 | 0.1335 | -2.01342 | 87.5 | 0.134 | -2.013 | 100 | 88.5 |
| 512 | 87.5 | 0.1335 | -2.01342 | 88 | 0.1278 | -2.05703 | 88 | 0.128 | -2.057 | 100 | 88.5 |
| 529 | 86.5 | 0.145 | -1.93084 | 87 | 0.145 | -1.93084 | 87 | 0.139 | -1.971 | 100 | 88 |
| 552 | 86.5 | 0.145 | -1.93084 | 87 | 0.1393 | -1.9714 | 87.5 | 0.134 | -2.013 | 100 | 88 |
| 573 | 86 | 0.1508 | -1.89165 | 88 | 0.1335 | -2.01342 | 86.5 | 0.145 | -1.931 | 100 | 89 |
| 598 | 85.5 | 0.1567 | -1.85372 | 87 | 0.1393 | -1.9714 | 86.5 | 0.145 | -1.931 | 100 | 88 |
| 622 | 85.5 | 0.1567 | -1.85372 | 88 | 0.1335 | -2.01342 | 86.5 | 0.145 | -1.931 | 100 | 88.5 |
| 650 | 84.5 | 0.1684 | -1.7813 | 88 | 0.1278 | -2.05703 | 87 | 0.139 | -1.971 | 100 | 88.5 |
| 671.5 | 84.5 | 0.1684 | -1.7813 | 87 | 0.145 | -1.93084 | 86.5 | 0.145 | -1.931 | 100 | 88 |

**Table D.7**
**The interference of salinity on perchlorate biodegradation (2.5%)**
**(Raw data—Figure 6.59)**

| Time | NaCl 2.5a | | | NaCl 2.5b | | | NaCl 2.5c | | | NaCl 2.5d | NaCl 2.5e |
|---|---|---|---|---|---|---|---|---|---|---|---|
| (hr) | Trans. | Abs. | Ln(Abs.) | Trans. | Abs. | Ln(Abs.) | Trans. | Abs. | Ln(Abs.) | Trans. | Trans. |
| 0 | 89 | **0.1165** | **-2.1496** | 89 | **0.11653** | **-2.14957** | 89.5 | **0.111** | **-2.1988** | 100 | 89.5 |
| 3 | 89 | **0.1165** | **-2.1496** | 89 | **0.11653** | **-2.14957** | 89 | **0.117** | **-2.1496** | 100 | 89 |
| 19 | 89 | **0.1165** | **-2.1496** | 89.5 | **0.11093** | **-2.19884** | 89 | **0.117** | **-2.1496** | 100 | 89.5 |
| 27 | 88.5 | **0.1222** | **-2.1024** | 89 | **0.11653** | **-2.14957** | 89 | **0.117** | **-2.1496** | 100 | 89 |
| 46 | 88.5 | **0.1222** | **-2.1024** | 89.5 | **0.11093** | **-2.19884** | 89 | **0.117** | **-2.1496** | 100 | 89 |
| 54.5 | 88.5 | **0.1222** | **-2.1024** | 89.5 | **0.11093** | **-2.19884** | 89.5 | **0.111** | **-2.1988** | 100 | 89.5 |
| 69 | 88.5 | **0.1222** | **-2.1024** | 88.5 | **0.12217** | **-2.10236** | 89 | **0.117** | **-2.1496** | 100 | 89.5 |
| 78 | 88.5 | **0.1222** | **-2.1024** | 88.5 | **0.12217** | **-2.10236** | 89 | **0.117** | **-2.1496** | 100 | 90.5 |
| 93 | 85.5 | **0.1567** | **-1.8537** | 85.5 | **0.15665** | **-1.85372** | 86 | **0.151** | **-1.8916** | 100 | 89.5 |
| 101 | 84.5 | **0.1684** | **-1.7813** | 84 | **0.17435** | **-1.74667** | 86 | **0.151** | **-1.8916** | 100 | 88.5 |
| 117 | 82 | **0.1985** | **-1.6172** | 80.5 | **0.21691** | **-1.52826** | 83.5 | **0.18** | **-1.713** | 100 | 88 |
| 127 | 82 | 0.1985 | -1.6172 | 80.5 | **0.21691** | **-1.52826** | 82 | **0.198** | **-1.6172** | 100 | 89 |
| 145 | 83.5 | 0.1803 | -1.713 | 81.5 | 0.20457 | -1.58686 | 82 | 0.198 | -1.6172 | 100 | 89 |
| 147 | 83.5 | 0.1803 | -1.713 | 81.5 | 0.20457 | -1.58686 | 82 | 0.198 | -1.6172 | 100 | 89 |
| 166 | 85 | 0.1625 | -1.817 | 84 | 0.17435 | -1.74667 | 84 | 0.174 | -1.7467 | 100 | 89 |
| 171 | 85 | 0.1625 | -1.817 | 84 | 0.17435 | -1.74667 | 84.5 | 0.168 | -1.7813 | 100 | 88.5 |
| 190 | 85.5 | 0.1567 | -1.8537 | 84 | 0.17435 | -1.74667 | 84.5 | 0.168 | -1.7813 | 100 | 89 |
| 199 | 85.5 | 0.1567 | -1.8537 | 85.5 | 0.15665 | -1.85372 | 85 | 0.163 | -1.817 | 100 | 88.5 |
| 214.5 | 85.5 | 0.1567 | -1.8537 | 85.5 | 0.15665 | -1.85372 | 85 | 0.163 | -1.817 | 100 | 87.5 |
| 222 | 85.5 | 0.1567 | -1.8537 | 85 | 0.16252 | -1.81696 | 85 | 0.163 | -1.817 | 100 | 87 |
| 235 | 86 | 0.1508 | -1.8916 | 85.5 | 0.15665 | -1.85372 | 85.5 | 0.157 | -1.8537 | 100 | 86 |
| 247 | 86.5 | 0.145 | -1.9308 | 86.5 | 0.14503 | -1.93084 | 86 | 0.151 | -1.8916 | 100 | 88 |
| 261 | 86.5 | 0.145 | -1.9308 | 86 | 0.15082 | -1.89165 | 85 | 0.163 | -1.817 | 100 | 88 |
| 268.5 | 87 | 0.1393 | -1.9714 | 86.5 | 0.14503 | -1.93084 | 85 | 0.163 | -1.817 | 100 | 87 |
| 285 | 87 | 0.1393 | -1.9714 | 86.5 | 0.14503 | -1.93084 | 84.5 | 0.168 | -1.7813 | 100 | 87.5 |
| 295 | 86.5 | 0.145 | -1.9308 | 86.5 | 0.14503 | -1.93084 | 84.5 | 0.168 | -1.7813 | 100 | 87 |
| 310 | 86.5 | 0.145 | -1.9308 | 85 | 0.16252 | -1.81696 | 84.5 | 0.168 | -1.7813 | 100 | 87.5 |
| 334 | 86.5 | 0.145 | -1.9308 | 86 | 0.15082 | -1.89165 | 84 | 0.174 | -1.7467 | 100 | 88 |
| 359 | 88 | 0.1278 | -2.057 | 86.5 | 0.14503 | -1.93084 | 85.5 | 0.157 | -1.8537 | 100 | 88 |
| 385 | 87.5 | 0.1335 | -2.0134 | 87 | 0.13926 | -1.9714 | 84.5 | 0.168 | -1.7813 | 100 | 87.5 |
| 410 | 87 | 0.1393 | -1.9714 | 87 | 0.13926 | -1.9714 | 85.5 | 0.157 | -1.8537 | 100 | 89.5 |
| 435 | 87 | 0.1393 | -1.9714 | 87.5 | 0.13353 | -2.01342 | 86 | 0.151 | -1.8916 | 100 | 89 |
| 454 | 87 | 0.1393 | -1.9714 | 87 | 0.13926 | -1.9714 | 85 | 0.163 | -1.817 | 100 | 88.5 |
| 482 | 86.5 | 0.145 | -1.9308 | 86.5 | 0.14503 | -1.93084 | 85 | 0.163 | -1.817 | 100 | 89 |
| 512 | 87.5 | 0.1335 | -2.0134 | 87.5 | 0.13353 | -2.01342 | 86.5 | 0.145 | -1.9308 | 100 | 89.5 |
| 529 | 87.5 | 0.1335 | -2.0134 | 87.5 | 0.13353 | -2.01342 | 86.5 | 0.145 | -1.9308 | 100 | 89.5 |
| 552 | 87 | 0.1393 | -1.9714 | 87 | 0.13926 | -1.9714 | 86.5 | 0.145 | -1.9308 | 100 | 88.5 |
| 573 | 87 | 0.1393 | -1.9714 | 87 | 0.13926 | -1.9714 | 87 | 0.139 | -1.9714 | 100 | 88.5 |
| 598 | 86.5 | 0.145 | -1.9308 | 87 | 0.13926 | -1.9714 | 86.5 | 0.145 | -1.9308 | 100 | 88.5 |
| 622 | 87.5 | 0.1335 | -2.0134 | 87.5 | 0.13353 | -2.01342 | 86 | 0.151 | -1.8916 | 100 | 89.5 |
| 650 | 87.5 | 0.1335 | -2.0134 | 87.5 | 0.13353 | -2.01342 | 86 | 0.151 | -1.8916 | 100 | 88.5 |
| 671.5 | 87 | 0.1393 | -1.9714 | 87.5 | 0.13353 | -2.01342 | 86 | 0.151 | -1.8916 | 100 | 89 |

**Table D.8**
**The interference of salinity on perchlorate biodegradation (3%)**
**(Raw data—Figure 6.60)**

| Time (hr) | NaCl 3a Trans. | NaCl 3a Abs. | NaCl 3a Ln(Abs) | NaCl 3b Trans. | NaCl 3b Abs. | NaCl 3b Ln(Abs.) | NaCl 3c Trans. | NaCl 3c Abs. | NaCl 3c Ln(Abs.) | NaCl 3d Trans. | NaCl 3e Trans. |
|---|---|---|---|---|---|---|---|---|---|---|---|
| 0 | 89 | **0.1165** | **-2.14957** | 89 | **0.1165** | **-2.14957** | 89 | **0.1165** | **-2.1496** | 100 | 89 |
| 3 | 88.5 | **0.1221** | **-2.10236** | 88.5 | **0.1221** | **-2.10236** | 88.5 | **0.1222** | **-2.1024** | 100 | 89 |
| 19 | 89 | **0.1165** | **-2.14957** | 89 | **0.1165** | **-2.14957** | 89 | **0.1165** | **-2.1496** | 100 | 89 |
| 27 | 89 | **0.1165** | **-2.14957** | 88.5 | **0.1221** | **-2.10236** | 89 | **0.1165** | **-2.1496** | 100 | 89 |
| 46 | 89 | **0.1165** | **-2.14957** | 89 | **0.1165** | **-2.14957** | 89 | **0.1165** | **-2.1496** | 100 | 89.5 |
| 54.5 | 89 | **0.1165** | **-2.14957** | 89 | **0.1165** | **-2.14957** | 89 | **0.1165** | **-2.1496** | 100 | 89.5 |
| 69 | 89 | **0.1165** | **-2.14957** | 89 | **0.1165** | **-2.14957** | 89 | **0.1165** | **-2.1496** | 100 | 90 |
| 78 | 89 | **0.1165** | **-2.14957** | 89 | **0.1165** | **-2.14957** | 89 | **0.1165** | **-2.1496** | 100 | 90 |
| 93 | 89 | **0.1165** | **-2.14957** | 89.5 | **0.1109** | **-2.19884** | 89 | **0.1165** | **-2.1496** | 100 | 90 |
| 101 | 85.5 | **0.1566** | **-1.85372** | 87.5 | **0.1335** | **-2.01342** | 88 | **0.1278** | **-2.057** | 100 | 90.5 |
| 117 | 84.5 | **0.1684** | **-1.7813** | 86 | **0.1508** | **-1.89165** | 86.5 | **0.145** | **-1.9308** | 100 | 90 |
| 127 | 83.5 | **0.1803** | **-1.713** | 85.5 | **0.1566** | **-1.85372** | 86 | **0.1508** | **-1.8916** | 100 | 90 |
| 145 | 82 | **0.1984** | **-1.61721** | 84 | **0.1743** | **-1.74667** | 84.5 | **0.1684** | **-1.7813** | 100 | 90 |
| 147 | 82 | **0.1984** | **-1.61721** | 84 | **0.1743** | **-1.74667** | 84.5 | **0.1684** | **-1.7813** | 100 | 90 |
| 166 | 78.5 | **0.2420** | **-1.41852** | 82 | **0.1984** | **-1.61721** | 83 | **0.1863** | **-1.6802** | 100 | 89 |
| 171 | 80.5 | 0.2169 | -1.52826 | 82.5 | 0.1923 | -1.64832 | 84 | 0.1744 | -1.7467 | 100 | 90.5 |
| 190 | 80 | 0.2231 | -1.49994 | 83 | 0.1863 | -1.68024 | 84.5 | 0.1684 | -1.7813 | 100 | 89 |
| 199 | 80.5 | 0.2169 | -1.52826 | 84.5 | 0.1684 | -1.7813 | 85.5 | 0.1567 | -1.8537 | 100 | 90.5 |
| 214.5 | 81 | 0.2107 | -1.55722 | 84.5 | 0.1684 | -1.7813 | 86 | 0.1508 | -1.8916 | 100 | 90.5 |
| 222 | 82.5 | 0.1923 | -1.64832 | 83.5 | 0.1803 | -1.713 | 84.5 | 0.1684 | -1.7813 | 100 | 89 |
| 235 | 82.5 | 0.1923 | -1.64832 | 85.5 | 0.1566 | -1.85372 | 85.5 | 0.1567 | -1.8537 | 100 | 89 |
| 247 | 82.5 | 0.1923 | -1.64832 | 85.5 | 0.1566 | -1.85372 | 85.5 | 0.1567 | -1.8537 | 100 | 89 |
| 261 | 82.5 | 0.1923 | -1.64832 | 85.5 | 0.1566 | -1.85372 | 87 | 0.1393 | -1.9714 | 100 | 89 |
| 268.5 | 84 | 0.1743 | -1.74667 | 86 | 0.1508 | -1.89165 | 86.5 | 0.145 | -1.9308 | 100 | 89.5 |
| 285 | 84.5 | 0.1684 | -1.7813 | 86 | 0.1508 | -1.89165 | 86.5 | 0.145 | -1.9308 | 100 | 89 |
| 295 | 84.5 | 0.1684 | -1.7813 | 85.5 | 0.1566 | -1.85372 | 85.5 | 0.1567 | -1.8537 | 100 | 88.5 |
| 310 | 84.5 | 0.1684 | -1.7813 | 85.5 | 0.1566 | -1.85372 | 86 | 0.1508 | -1.8916 | 100 | 88 |
| 334 | 84.5 | 0.1684 | -1.7813 | 86 | 0.1508 | -1.89165 | 86.5 | 0.145 | -1.9308 | 100 | 89 |
| 359 | 84.5 | 0.1684 | -1.7813 | 87 | 0.1392 | -1.9714 | 86.5 | 0.145 | -1.9308 | 100 | 89 |
| 385 | 85.5 | 0.1566 | -1.85372 | 87 | 0.1392 | -1.9714 | 87 | 0.1393 | -1.9714 | 100 | 88.8 |
| 410 | 85 | 0.1625 | -1.81696 | 87 | 0.1392 | -1.9714 | 87 | 0.1393 | -1.9714 | 100 | 89.5 |
| 435 | 85 | 0.1625 | -1.81696 | 87.5 | 0.1335 | -2.01342 | 87 | 0.1393 | -1.9714 | 100 | 89.5 |
| 454 | 86.5 | 0.1450 | -1.93084 | 87.5 | 0.1335 | -2.01342 | 87 | 0.1393 | -1.9714 | 100 | 89.5 |
| 482 | 85 | 0.1625 | -1.81696 | 87 | 0.1392 | -1.9714 | 88 | 0.1278 | -2.057 | 100 | 90 |
| 512 | 85.5 | 0.1566 | -1.85372 | 87.5 | 0.1335 | -2.01342 | 87.5 | 0.1335 | -2.0134 | 100 | 89 |
| 529 | 86.5 | 0.1450 | -1.93084 | 87.5 | 0.1335 | -2.01342 | 87.5 | 0.1335 | -2.0134 | 100 | 90.5 |
| 552 | 86.5 | 0.1450 | -1.93084 | 88 | 0.1278 | -2.05703 | 88 | 0.1278 | -2.057 | 100 | 90.5 |
| 573 | 87.5 | 0.1335 | -2.01342 | 87.5 | 0.1335 | -2.01342 | 88 | 0.1278 | -2.057 | 100 | 90.5 |
| 598 | 86 | 0.1508 | -1.89165 | 88 | 0.1278 | -2.05703 | 87 | 0.1393 | -1.9714 | 100 | 90.5 |
| 622 | 86.5 | 0.1450 | -1.93084 | 88.5 | 0.1221 | -2.10236 | 86.5 | 0.145 | -1.9308 | 100 | 91 |
| 650 | 86.5 | 0.1450 | -1.93084 | 87.5 | 0.1335 | -2.01342 | 86.5 | 0.145 | -1.9308 | 100 | 90 |
| 671.5 | 86.5 | 0.1450 | -1.93084 | 87.5 | 0.1335 | -2.01342 | 86.5 | 0.145 | -1.9308 | 100 | 90.5 |

## Table D.9
## The interference of salinity on perchlorate biodegradation (3.5%)
## (Raw data—Figure 6.61)

| Time | NaCl 3.5a | | | NaCl 3.5b | | | NaCl 3.5c | | | NaCl 3.5d | NaCl 3.5e |
|---|---|---|---|---|---|---|---|---|---|---|---|
| (hr) | Trans. | Abs. | Ln(Abs.) | Trans. | Abs. | Ln(Abs.) | Trans. | Abs. | Ln(Abs) | Trans. | Trans. |
| 0 | 89 | 0.12 | -2.15 | 89 | **0.12** | **-2.1496** | 89 | **0.117** | **-2.15** | 100 | 89 |
| 3 | 89 | 0.12 | -2.15 | 89 | **0.12** | **-2.1496** | 89 | **0.117** | **-2.15** | 100 | 89 |
| 19 | 89 | 0.12 | -2.15 | 89.5 | **0.11** | **-2.1988** | 89 | **0.117** | **-2.15** | 100 | 89.5 |
| 27 | 89 | 0.12 | -2.15 | 89.5 | **0.11** | **-2.1988** | 89 | **0.117** | **-2.15** | 100 | 89 |
| 46 | 89 | 0.12 | -2.15 | 89.5 | **0.11** | **-2.1988** | 89 | **0.117** | **-2.15** | 100 | 89 |
| 54.5 | 89.5 | 0.11 | -2.199 | 89.5 | **0.11** | **-2.1988** | 89 | **0.117** | **-2.15** | 100 | 89.5 |
| 69 | 90 | 0.11 | -2.25 | 89.5 | **0.11** | **-2.1988** | 89.5 | **0.111** | **-2.199** | 100 | 89.5 |
| 78 | 90 | 0.11 | -2.25 | 89.5 | **0.11** | **-2.1988** | 90 | **0.105** | **-2.25** | 100 | 89.5 |
| 93 | 90.5 | 0.1 | -2.304 | 89.5 | **0.11** | **-2.1988** | 90 | **0.105** | **-2.25** | 100 | 90 |
| 101 | 90.5 | 0.1 | -2.304 | 90.5 | **0.1** | **-2.3044** | 90.5 | **0.1** | **-2.304** | 100 | 90 |
| 117 | 91 | 0.09 | -2.361 | 90.5 | **0.1** | **-2.3044** | 90.5 | **0.1** | **-2.304** | 100 | 91 |
| 127 | 90 | 0.11 | -2.25 | 89.5 | **0.11** | **-2.1988** | 90 | **0.105** | **-2.25** | 100 | 91.5 |
| 145 | 90 | 0.11 | -2.25 | 87.5 | **0.13** | **-2.0134** | 89 | **0.117** | **-2.15** | 100 | 91.5 |
| 147 | 90 | 0.11 | -2.25 | 87.5 | **0.13** | **-2.0134** | 89 | **0.117** | **-2.15** | 100 | 91.5 |
| 166 | 88.5 | 0.12 | -2.102 | 86 | **0.15** | **-1.8916** | 87 | **0.139** | **-1.971** | 100 | 91.5 |
| 171 | 88 | 0.13 | -2.057 | 85.5 | **0.16** | **-1.8537** | 87.5 | **0.134** | **-2.013** | 100 | 92 |
| 190 | 86.5 | 0.15 | -1.931 | 85.5 | **0.16** | **-1.8537** | 86 | **0.151** | **-1.892** | 100 | 91.5 |
| 199 | 87 | 0.14 | -1.971 | 86 | **0.15** | **-1.8916** | 87 | **0.139** | **-1.971** | 100 | 92 |
| 214.5 | 86 | 0.15 | -1.892 | 85 | **0.16** | **-1.817** | 85.5 | **0.157** | **-1.854** | 100 | 92 |
| 222 | 86.5 | 0.15 | -1.931 | 84 | **0.17** | **-1.7467** | 85.5 | **0.157** | **-1.854** | 100 | 92.5 |
| 235 | 85 | 0.16 | -1.817 | 82.5 | **0.19** | **-1.6483** | 83.5 | **0.18** | **-1.713** | 100 | 92.5 |
| 247 | 85.5 | 0.16 | -1.854 | 83 | 0.19 | -1.6802 | 85 | 0.163 | -1.817 | 100 | 93 |
| 261 | 85.5 | 0.16 | -1.854 | 83 | 0.19 | -1.6802 | 86.5 | 0.145 | -1.931 | 100 | 93.5 |
| 268.5 | 85.5 | 0.16 | -1.854 | 84 | 0.17 | -1.7467 | 86.5 | 0.145 | -1.931 | 100 | 93 |
| 285 | 86 | 0.15 | -1.892 | 85 | 0.16 | -1.817 | 87 | 0.139 | -1.971 | 100 | 93 |
| 295 | 85.5 | 0.16 | -1.854 | 84.5 | 0.17 | -1.7813 | 86.5 | 0.145 | -1.931 | 100 | 92 |
| 310 | 85.5 | 0.16 | -1.854 | 85.5 | 0.16 | -1.8537 | 86.5 | 0.145 | -1.931 | 100 | 92 |
| 334 | 86.5 | 0.15 | -1.931 | 86.5 | 0.15 | -1.9308 | 88 | 0.128 | -2.057 | 100 | 92 |
| 359 | 88.5 | 0.12 | -2.102 | 87.5 | 0.13 | -2.0134 | 88 | 0.128 | -2.057 | 100 | 92.5 |
| 385 | 89.5 | 0.11 | -2.199 | 87.5 | 0.13 | -2.0134 | 88 | 0.128 | -2.057 | 100 | 92.5 |
| 410 | 89 | 0.12 | -2.15 | 88 | 0.13 | -2.057 | 89 | 0.117 | -2.15 | 100 | 93.5 |
| 435 | 89 | 0.12 | -2.15 | 89 | 0.12 | -2.1496 | 89 | 0.117 | -2.15 | 100 | 93.5 |
| 454 | 90 | 0.11 | -2.25 | 89 | 0.12 | -2.1496 | 89 | 0.117 | -2.15 | 100 | 93 |
| 482 | 88.5 | 0.12 | -2.102 | 87.5 | 0.13 | -2.0134 | 89 | 0.117 | -2.15 | 100 | 92 |
| 512 | 89 | 0.12 | -2.15 | 88 | 0.13 | -2.057 | 88.5 | 0.122 | -2.102 | 100 | 93 |
| 529 | 89 | 0.12 | -2.15 | 89 | 0.12 | -2.1496 | 89.5 | 0.111 | -2.199 | 100 | 92 |
| 552 | 90.5 | 0.1 | -2.304 | 89.5 | 0.11 | -2.1988 | 89.5 | 0.111 | -2.199 | 100 | 93 |
| 573 | 89.5 | 0.11 | -2.199 | 89.5 | 0.11 | -2.1988 | 89 | 0.117 | -2.15 | 100 | 92.5 |
| 598 | 89 | 0.12 | -2.15 | 89 | 0.12 | -2.1496 | 87 | 0.139 | -1.971 | 100 | 92 |
| 622 | 90 | 0.11 | -2.25 | 89 | 0.12 | -2.1496 | 88 | 0.128 | -2.057 | 100 | 92.5 |
| 650 | 90 | 0.11 | -2.25 | 89 | 0.12 | -2.1496 | 88 | 0.128 | -2.057 | 100 | 93 |
| 671.5 | 90 | 0.11 | -2.25 | 89.5 | 0.11 | -2.1988 | 88 | 0.128 | -2.057 | 100 | 92.5 |

**Table D.10**
**The interference of salinity on perchlorate biodegradation (4%)**
**(Raw data—Figure 6.62)**

| Time | NaCl 4a | | | NaCl 4b | | | NaCl 4c | | | NaCl 4d | NaCl 4e |
|---|---|---|---|---|---|---|---|---|---|---|---|
| (hr) | Trans. | Abs. | Ln(Abs.) | Trans. | Abs. | Ln(Abs.) | Trans. | Abs. | Ln(Abs.) | Trans. | Trans. |
| 0 | 89 | 0.1165 | -2.1495 | 89.5 | 0.1109 | -2.19884 | 89 | 0.11653 | -2.14957 | 100 | 88.5 |
| 3 | 89 | 0.1165 | -2.1495 | 89.5 | 0.1109 | -2.19884 | 89 | 0.11653 | -2.14957 | 100 | 88 |
| 19 | 90 | 0.1053 | -2.2503 | 90 | 0.1053 | -2.25037 | 90 | 0.10536 | -2.25037 | 100 | 88.5 |
| 27 | 89.5 | 0.1109 | -2.1988 | 88.5 | 0.1221 | -2.10236 | 89.5 | 0.11093 | -2.19884 | 100 | 88.5 |
| 46 | 89.5 | 0.1109 | -2.1988 | 89.5 | 0.1109 | -2.19884 | 89 | 0.11653 | -2.14957 | 100 | 88.5 |
| 54.5 | 89 | 0.1165 | -2.1495 | 89 | 0.1165 | -2.14957 | 89 | 0.11653 | -2.14957 | 100 | 89 |
| 69 | 89 | 0.1165 | -2.1495 | 89.5 | 0.1109 | -2.19884 | 89.5 | 0.11093 | -2.19884 | 100 | 89 |
| 78 | 89.5 | 0.1109 | -2.198 | 90 | 0.1053 | -2.25037 | 89.5 | 0.11093 | -2.19884 | 100 | 89.5 |
| 93 | 89.5 | 0.1109 | -2.1988 | 89.5 | 0.1109 | -2.19884 | 89.5 | 0.11093 | -2.19884 | 100 | 89 |
| 101 | 89.5 | 0.1109 | -2.1988 | 89 | 0.1165 | -2.14957 | 89.5 | 0.11093 | -2.19884 | 100 | 89.5 |
| 117 | 90.5 | 0.0998 | -2.3043 | 91 | 0.0943 | -2.36116 | 90.5 | 0.09982 | -2.30438 | 100 | 90 |
| 127 | 90.5 | 0.0998 | -2.3043 | 90 | 0.1053 | -2.25037 | 90.5 | 0.09982 | -2.30438 | 100 | 90 |
| 145 | 90.5 | 0.0998 | -2.3043 | 91 | 0.0943 | -2.36116 | 91 | 0.09431 | -2.36116 | 100 | 90 |
| 147 | 90.5 | 0.0998 | -2.3043 | 91 | 0.0943 | -2.36116 | 91 | 0.09431 | -2.36116 | 100 | 90 |
| 166 | 90 | 0.1053 | -2.2503 | 90 | 0.1053 | -2.25037 | 90 | 0.10536 | -2.25037 | 100 | 90 |
| 171 | 91 | 0.0943 | -2.3611 | 91.5 | 0.0888 | -2.42102 | 90 | 0.10536 | -2.25037 | 100 | 90 |
| 190 | 90 | 0.1053 | -2.2503 | 90 | 0.1053 | -2.25037 | 90 | 0.10536 | -2.25037 | 100 | 90 |
| 199 | 91.5 | 0.0888 | -2.4210 | 92 | 0.0833 | -2.48433 | 91 | 0.09431 | -2.36116 | 100 | 92 |
| 214.5 | 91.5 | 0.0888 | -2.4210 | 91.5 | 0.0888 | -2.42102 | 90.5 | 0.09982 | -2.30438 | 100 | 92 |
| 222 | 91 | 0.0943 | -2.3611 | 91 | 0.0943 | -2.36116 | 91.5 | 0.08883 | -2.42102 | 100 | 92 |
| 235 | 91 | 0.0943 | -2.3611 | 90.5 | 0.0998 | -2.30438 | 91.5 | 0.08883 | -2.42102 | 100 | 91.5 |
| 247 | 91.5 | 0.0888 | -2.4210 | 91 | 0.0943 | -2.36116 | 92 | 0.08338 | -2.48433 | 100 | 92 |
| 261 | 91 | 0.0943 | -2.3611 | 92.5 | 0.0779 | -2.55154 | 92 | 0.08338 | -2.48433 | 100 | 92.5 |
| 268.5 | 92.5 | 0.0779 | -2.5515 | 92 | 0.0833 | -2.48433 | 92.5 | 0.07796 | -2.55154 | 100 | 92.5 |
| 285 | 93 | 0.0725 | -2.6231 | 92.5 | 0.0779 | -2.55154 | 92.5 | 0.07796 | -2.55154 | 100 | 92.5 |
| 295 | 91 | 0.0943 | -2.3611 | 91 | 0.0943 | -2.36116 | 90.5 | 0.09982 | -2.30438 | 100 | 91.5 |
| 310 | 91 | 0.0943 | -2.3611 | 91.5 | 0.0888 | -2.42102 | 90 | 0.10536 | -2.25037 | 100 | 92 |
| 334 | 91 | 0.0943 | -2.3611 | 91 | 0.0943 | -2.36116 | 92 | 0.08338 | -2.48433 | 100 | 92 |
| 359 | 90.5 | 0.0998 | -2.3043 | 91 | 0.0943 | -2.36116 | 91 | 0.09431 | -2.36116 | 100 | 91 |
| 385 | 90.5 | 0.0998 | -2.3043 | 91 | 0.0943 | -2.36116 | 92 | 0.08338 | -2.48433 | 100 | 91 |
| 410 | 91 | 0.0943 | -2.3611 | 91 | 0.0943 | -2.36116 | 91.5 | 0.08883 | -2.42102 | 100 | 91.5 |
| 435 | 91.5 | 0.0888 | -2.4210 | 92 | 0.0833 | -2.48433 | 91 | 0.09431 | -2.36116 | 100 | 92 |
| 454 | 92 | 0.0833 | -2.4843 | 92.5 | 0.0779 | -2.55154 | 92 | 0.08338 | -2.48433 | 100 | 93 |
| 482 | 92 | 0.0833 | -2.4843 | 91 | 0.0943 | -2.36116 | 91 | 0.09431 | -2.36116 | 100 | 91 |
| 512 | 92 | 0.0833 | -2.4843 | 91.5 | 0.0888 | -2.42102 | 91.5 | 0.08883 | -2.42102 | 100 | 92.5 |
| 529 | 92 | 0.0833 | -2.4843 | 91.5 | 0.0888 | -2.42102 | 91.5 | 0.08883 | -2.42102 | 100 | 92.5 |
| 552 | 91 | 0.0943 | -2.3611 | 91 | 0.0943 | -2.36116 | 90.5 | 0.09982 | -2.30438 | 100 | 92 |
| 573 | 91 | 0.0943 | -2.3611 | 91 | 0.0943 | -2.36116 | 90.5 | 0.09982 | -2.30438 | 100 | 91.5 |
| 598 | 91 | 0.094 | -2.3611 | 91.5 | 0.0888 | -2.42102 | 91 | 0.09431 | -2.36116 | 100 | 91.5 |
| 622 | 92 | 0.0833 | -2.4843 | 92 | 0.0833 | -2.48433 | 91 | 0.09431 | -2.36116 | 100 | 91 |
| 650 | 92 | 0.0833 | -2.4843 | 92 | 0.0833 | -2.48433 | 90.5 | 0.09982 | -2.30438 | 100 | 92 |
| 671.5 | 92 | 0.0833 | -2.4843 | 91.5 | 0.0888 | -2.42102 | 91.5 | 0.08883 | -2.42102 | 100 | 91 |

**Table D.11**
**The interference of salinity on perchlorate biodegradation (6%)**
**(Raw data—Figure 6.63)**

| Time (hour) | Transmittance (%) NaCl 6a | NaCl 6b | NaCl 6c | NaCl 6d | NaCl 6e |
|---|---|---|---|---|---|
| 0 | 89.5 | 89 | 88 | 100 | 90 |
| 3 | 89.5 | 89 | 88 | 100 | 89.5 |
| 19 | 90 | 90 | 89.5 | 100 | 91 |
| 27 | 90 | 90 | 88.5 | 100 | 90 |
| 46 | 90 | 90 | 88.5 | 100 | 90 |
| 54.5 | 90 | 90 | 88.5 | 100 | 90.5 |
| 69 | 90 | 90 | 89.5 | 100 | 90.5 |
| 78 | 90 | 90 | 89.5 | 100 | 90 |
| 93 | 89.5 | 89.5 | 88.5 | 100 | 89.5 |
| 101 | 90.5 | 90.5 | 89 | 100 | 90 |
| 117 | 91 | 91 | 89.5 | 100 | 90 |
| 127 | 91 | 91 | 91 | 100 | 91 |
| 145 | 90 | 91 | 88.5 | 100 | 92 |
| 147 | 90 | 91 | 88.5 | 100 | 92 |
| 166 | 90.5 | 91 | 89 | 100 | 92 |
| 171 | 91.5 | 92 | 90.5 | 100 | 92.5 |
| 190 | 91.5 | 92 | 90 | 100 | 92 |
| 199 | 91.5 | 92 | 90.5 | 100 | 92.5 |
| 214.5 | 91.5 | 92 | 90 | 100 | 92 |
| 222 | 91 | 92 | 90.5 | 100 | 92 |
| 235 | 91 | 92 | 90 | 100 | 92 |
| 247 | 91 | 92 | 91 | 100 | 92.5 |
| 261 | 92 | 93 | 92.5 | 100 | 92.5 |
| 268.5 | 92 | 92.5 | 92.5 | 100 | 93 |
| 285 | 91.5 | 92.5 | 91 | 100 | 92.5 |
| 295 | 92.5 | 92 | 92 | 100 | 92.5 |
| 310 | 92 | 92 | 91.5 | 100 | 92 |
| 334 | 91 | 91.5 | 90 | 100 | 92 |
| 359 | 91.5 | 91 | 91 | 100 | 92.5 |
| 385 | 91 | 92.5 | 92 | 100 | 92.5 |
| 410 | 91 | 92 | 92 | 100 | 92.5 |
| 435 | 91 | 91.5 | 90 | 100 | 92.5 |
| 454 | 92.5 | 92.5 | 91 | 100 | 92.5 |
| 482 | 91.5 | 93 | 90.5 | 100 | 92.5 |
| 512 | 92 | 93 | 90.5 | 100 | 92.5 |
| 529 | 91 | 92 | 91 | 100 | 91 |
| 552 | 91.5 | 93 | 91 | 100 | 91.5 |
| 573 | 91 | 91.5 | 90.5 | 100 | 91 |
| 598 | 91.5 | 92.5 | 90 | 100 | 92 |
| 622 | 93 | 93 | 90.5 | 100 | 92.5 |
| 650 | 93 | 93.5 | 91.5 | 100 | 92.5 |
| 671.5 | 93.5 | 93 | 91.5 | 100 | 92.5 |

# APPENDIX E
# EXAMPLE CALCULATIONS FOR MEMBRANE-IMMOBILIZED BIOFILM REACTOR

**EXAMPLE 1: Moles of perchlorate biodegraded per day per $m^2$ of a membrane through a BTS-55 membrane-immobilized biofilm (data from Tables 6.18 and 6.19)**

At day 1, the initial amount of perchlorate in the DR and BR reactors was:
$(231.12 \pm 0.194)$ mg/L * 5 L =1156.57 mg

At day 7, the perchlorate concentration in both reactors was zero. Therefore, the perchlorate biodegraded per day was:
1156.57 / 6 = 192. 76 mg/day = 0.19276 g/day
since the molecular weight of perchlorate is: 99.5 g/mole, so
$0.19276 / 99.5 = 1.94 \times 10^{-3}$ mole/day

The diameter of the membrane = 10 cm = 0.1 m
Thus, the area of the membrane = $3.14 \times 0.1^2 /4 = 7.85 \times 10^{-3}$ $m^2$

Then, the perchlorate biodegradation rate was:
$1.94 \times 10^{-3} / (7.85 \times 10^{-3}) = 0.25$ mole/$m^2$-day

**EXAMPLE 2: Calculation for perchlorate removal after acclimation in KC3 to KC5 (data from Table 6.38)**

Perchlorate removal in KC3
At day 0, the initial concentration of perchlorate in KC3 = 63.65 mg/L
After acclimation, at day 3, the concentration of perchlorate = 17.54 mg/L
Thus, perchlorate removed = (63.65-17.54)/63.65 = 72.4%

Perchlorate removal in KC4
At day 0, the initial concentration of perchlorate = 108.72 mg/L
After acclimation, at day 5, the concentration of perchlorate = 52.17 mg/L
Thus, perchlorate removed = (108.72-52.17)/108.72 = 52%

Perchlorate removal in KC5
At day 0, the initial concentration of perchlorate = 232.2 mg/L
After acclimation, at day 5, the concentration of perchlorate = 166.68 mg/L
Thus, perchlorate removed = (232.2-166.68)/232.2 = 28.2%

**EXAMPLE 3: Calculation for perchlorate removal rate (data from Tables 6.18 and 6.19)**

Perchlorate biodegradation rate experiments were performed in triplicate with the BTS-55 membrane. The perchlorate removal rate for the 1st test, after 4 days is calculated as:

$(231.12 \pm 0.194 - 3.21)$ mg $ClO_4^-$ / (3 days $\times$ 24 hours/day)

= 3.168 mg $ClO_4^-$ /hr/L

The area of membrane covered by the biofilm is $7.85 \times 10^{-3}$ m$^2$
Assuming the density of microbes in the biofilm as 1.07 g/mL. The measured thickness of the biofilm was approximately 25μm, thus the mass of microbes attached to the membrane is:

Mass of microbes = area * thickness * density
= $7.85 \times 10^{-3}$ m$^2$ $\times$ $25 \times 10^{-6}$ m x 1.07 g/mL $\times$ $10^6$ mL/m$^3$
= 210 mg microbes

Thus, the perchlorate removal rate is:

3.168 mg $ClO_4^-$ /hr/L / 210 mg microbes
= 0.251 μg $ClO_4^-$ /min/L/mg microbes
**= 251 μg $ClO_4^-$ /min/L/μg microbes**

For the replicated tests, the perchlorate removal rates were calculated using the data from Tables 6.26-6.29 and were found to be 110 and 219 μg $ClO_4^-$/min/L/μg microbes.

# APPENDIX F
# 16S rRNA ANALYSIS OF ISOLATE OF THE "BALI" MIXED CULTURE

## Alignment Report

Customer: Kesterson

11/9/2000

MIDI LABS

### Partial 16S rRNA Gene Alignment with GenBank

| sample # | closest GenBank match | % ID |
|---|---|---|
| C4299 KLI-A con | Dechlorisoma suilla | 98% |

**C#** - Customer number assigned by MIDI Labs.

**sample #** - Sample number assigned by MIDI Labs, followed by name assigned by customer.

**closest GenBank match** - Results from blast search of GenBank database.
% ID - percent identity; this is essentially the percent similarity.

See GenBank web page (www.ncbi.nlm.nih.gov/BLAST/blast_help.html)
for more detailed information.

MIDI Labs claims no responsibility for the validity of sequences found in GenBank or RDP.

**For research use only**

# REFERENCES


APHA, AWWA, and WEF (American Public Health Association, American Water Works Association, and Water Environment Federation). 1995. *Standard Methods for the Examination of Water and Wastewater.* 19th ed. Washington, D.C.: APHA.

APHA, AWWA, and WEF (American Public Health Association, American Water Works Association, and Water Environment Federation). 1995. *Standard Methods for the Examination of Water and Wastewater.* 19th ed. Washington, D.C.: APHA.

Attaway, H., and M. Smith. 1993. Reduction of Perchlorate by an Anaerobic Enrichment Culture. *Journal of Industrial Microbiology,* 12:408-412.

AwwaRF (American Water Works Association Research Foundation). 1997. Final report of the Perchlorate Research Issue Group Workshop, September 30 to October 2. Denver, CO: AwwaRF.

Ballapragada, B.S., H.D. Stensel, J.A. Puhakka, and J.F. Ferguson. 1997. Effect of Hydrogen on Reductive Dechlorination of Chlorinated Ethenes. *Environmental Science and Technology,* 31(6):1728-1734.

Basye, L., and S. Swaminathan. 1997. *Hydrogen Production Costs-A Survey.* Produced for the U.S. Department of Energy, Washington D.C.

Batista, J. 1998. Verbal communication.

*Batista, J.A., and J. Liu. 2001. Biological Perchlorate Removal from Drinking Waters Incorporating Microporous Membranes. Abstract submitted to the In Situ and On-site Bioremediation: Sixth Batelle International Symposium, San Diego, Calif., June 4-7, 2001. Accepted for presentation on November 21, 2000.

*Batista, J.A., J. Liu, and K. Kersterson. 2001. Application of Membrane-Immobilized Biofilm System to Perchlorate Contaminated Drinking Water. AWWA Conference, Washington, DC, June 17-21. Abstract approved for presentation. On December 22, 2000.

*Batista, J.R., F.X. McGarvey, and A.R. Vieira. 2000. The Removal of Perchlorate From Waters Using Ion-Exchange Resins. In *Perchlorate in the Environment.* Chap. 13, pp. 135-145. New York: Plenum Publisher.

*Batista, J., A. Vieira, and F. McGarvey. 2000. The Influence of Humic Substances and Sulfate on the Removal of Perchlorate from a Groundwater by Ion-Exchange Resins. Proceedings of the 5th Annual Joint Services Pollution Prevention and Hazardous Waste Management. August 21-24, 2000, San Antonio, Texas.

Bliven, A.R. 1996. Chlorate Respiring Bacteria: Isolation, Identification, and Effects on Environmentally Significant Substrates. Master's thesis. Department of Chemical and Environmental Engineering, University of Arizona, Tucson, AZ.

*Boralessa, R., and J. Batista. 2000. Historical Perchlorate levels in the Las Vegas Wash and Lake Mead. Proceedings of the 2000 AWWA Inorganic Contaminants Conference, February 27-29, 2000, Albuquerque, NM.

Bowien, B., and H. Schlegel. 1981. Physiology and Biochemistry of Aerobic Hydrogen-Oxidizing Bacteria. *Annual Reviews in Microbiology,* 35:405-452.

Bryan, E.H. 1966. Application of the Chlorate BOD Procedure to Routine Measurement of Wastewater Strength. *Journal of the Water Pollution Control Federation,* 35:1350-1362.

Bryan, E.H., and G.A. Rohlich. 1954. Biological Reduction of Sodium Chlorate as Applied to Measurement of Sewage B.O.D. *Sewage and Industrial Wastes,* 26:1315-1324.

Catts, J.G. 1999. The Biochemical Removal of Perchlorate from San Gabriel Basin Groundwater and Potable Uses of the Treated Water. In *Extended Abstracts*, 218th ACS Meeting, New Orleans, August 22-24, 39:107-109.

Catts, J.G., and M.L. McCullough. 1998. Draft Final Phase 1 Treatability Study Report Perchlorate in Groundwater, Baldwin Park Operable Unit, San Gabriel Basin. Harding Lawson Associates, Irvine, CA. HLA Project No. 39860355.

Chiovato, L., F. Santini, and A. Pinchera. 1997. *Treatment of Hyperthyroidism.* Pisa, Italy. http://www.thyrolink.com/. October 2000.

Coates, J.R., U. Michaelidou, R.A. Bruce, S.M. O'Connor, J.N. Crespi, and L.A. Achenbach. 1999. Ubiquity and Diversity of Dissimilatory (Per)chlorate Reducing Bacteria. *Applied Environmental Microbiology,* 65(12):5234-5241.

Coppola, Edward N. 1998. Perchlorate Biodegradation Technology: Multiple Applications, NGWA Southwest Focused Ground Water Conference, Anaheim, CA.

Cox, K.E., and K.D. Williamson, Jr., (eds.). 1997. *Hydrogen: Its Technology and Implications.* Vol. 1. Cleveland, OH: CRC Press. pp. 3–41.

DeGroot, G.N., and A.H. Stouthamer. 1969. Regulation of Reductase Formation in Proteus Mirabilis I, Formation of Rreductases and Enzymes of the Formic Hydrogenase Complex in the Wild Type and in Chlorate-Resistant Mutants. *Archives in Microbiology,* 66:220-233.

Dionex Corporation. 1998. Verbal communication.

Dollarhide, J.S. 1992. Provisional Non-Cancer and Cancer Toxicity Values for Potassium Perchlorate. Memorandum to Dan Stralka, U.S. Environmental Protection Agency.

Espenson, J. 1997. The Problem and Perversity of Perchlorate. In *Perchlorate in the Environment.* Edited by E.T. Urbansky. Plenum, 1-8.

Foye, W.O., ed. 1989. Principles of Medicinal Chemistry, 3rd ed. , Philadelphia, Pa.: Lea & Febiger. pp. 612-613.

Friedrich, C., and B. Friedrich. 1983. Regulation of Hydrogenase Formation is Temperature-Sensitive and Plasmid-Encoded in *Alcaligenes eutrophus. Journal of Bacteriology,* 153:176-181.

Friedrich, B., and E. Schwartz. 1993. Molecular Biology of Hydrogen Utilization in Aerobic Chemolithotrophs. *Annual Reviews of Microbiology,* 43:351-383.

Giblin T., D. Herman, M.A. Deshusses, and W.T. Frankenberger Jr. 2000. Removal of Perchlorate in Ground Water with a Flow-Through Bioreactor. *Journal of Environmental Quality,* 29:578-583.

Giblin, Tara, David Herman, and William T. Frankenberger Jr. 2000. An Autotrophic System For the Removal of Perchlorate From Groundwater. In *Perchlorate in the Environment.* Edited by E.T. Urbansky. New York: Kluwer Academic/Plenum Publishers. pp 199-213.

Grady, C.P., and H.C. Lim. 1980. Biological Wastewater Treatment, Theory and Applications. New York: Dekker. p. 290.

Greene, M., and M.P. Pitre. 1999. Treatment of Groundwater Containing Perchlorate Using Biological Fluidized Bed Reactors with GAC or Sand Media. In *Exteneded Abstract.* 218th ACS Meeting, New Orleans, August 22-24. 39:105-107.

Greene, M., and M.P. Pitre. 2000. Treatment of Groundwater Containing Perchlorate Using Biological Fluidized Bed Reactors with GAC or Sand Media. In *Perchlorate in the Environment.* Edited by E.T. Urbansky. Plenum. 241-256.

Gros, H., G. Schnoor, and P. Rutten. 1986. Nitrate Removal from Groundwater by Autotrohpic Microorganisms. *Water Supply,* 4:11-21.

Harris, D.C. 1991. Quantitative Chemical Analysis, 3rd ed. New York: Freedman. pp.146, 722-723.

Herman, D.C., and W.T. Frankenberger Jr. 1998. Microbial-Mediated Reduction of Perchlorate in Groundwater. *Journal of Environmental Quality,* 27:750-754.

Herman, D.C., and W.T. Frankenberger Jr. 1999. Bacterial Reduction of Perchlorate and Nitrate in Water. *Journal of Environmental Quality,* 28:1018-1024.

Hijnen, W.A.M, R. Voogt, H.R. Veenendaal, H. van der Jagt, and D. van der Kooij. 1995. Bromate Reduction by Denitrifying Bacteria. *Applied and Environmental Microbiology,* 61(1):239-244.

Hogrefe, C., D. Romermann, and B. Friedrich. 1984. *Alcaligenes eutrophus* Hydrogenase genes (hox). *Journal of Bacteriology,* 158:43-48.

Kim, Kijung. 1999. Microbial Treatment of Perchlorate-Contaminated Water. Master's thesis. Department of Civil and Environmental Engineering, The Pennsylvania State University.

*Kim, K., and B.E. Logan. 2000. Fixed-Bed Bioreactor Treating Perchlorate-Contaminated Waters. *Environmental Engineering and Science,* 17(5):257-265.

*Kim, K., and B.E. Logan. 2001. Microbial Reduction of Perchlorate in Pure and Mixed Culture Packed-Bed Bioreactors. *Water Research,* In press.

Komen, R., and K. Schmidt. 1992. Hydrogenase Mutants of *Alcaligenes Eutrophus* H16 Show Alterations in the Electron Transport System. *FEMS Microbiology Letters,* 96:173-178.

Korenkov, V.N., V. Ivanovich, S.I. Kuznetsov, and J.V. Vorenov. 1976. Process for Purification of Industrial Waste Waters from Perchlorates and Chlorates. U.S. Patent No. 3,943,055, March 9.

Kurt, M., I.J. Dunn, and J.R. Bourne. 1987. Biological Denitrification of Drinking Water Using Autotrophic Organisms with $H_2$ in a Fluidized-Bed Biofilm Reactor. *Biotechnology and Bioengineering,* 29:493-501.

LaGrega, M.D., P.L. Buckinham, and J.C. Evans. 1994. *Hazardous Waste Management.* New York: McGraw Hill.

LaPoint, D.M. 2000. Factors Affecting Perchlorate Removal Rates in Autotropic Packed Bed Reactors. Master's thesis. Department of Civil and Environmental Engineering. The Pennsylvania State University.

Lemoine, D., T. Jouenne, and G.A. Junter. 1988. Reduction of Nitrate by Pseudomonas putrefaciens Entrapped in Composite Agar Layer/Micro-porous Membrane Structures. *Biotechnology Letters,* 10(6):399-402.

Lemoine, D., T. Jouenne, and G.A. Junter. 1991a. Biological Denitrification of Water in a Two-Chambered Immobilized-Cell Bioreactor. *Applied Microbiology and Biotechnology,* 36(2):257-264.

Lemoine, D., G.A. Junter, and T. Jouenne. 1991b. Denitrification of Water by Pseudomonas Denitrificans Entrapped in Agar Layer/Membrane Filter Structures II: Double Flow Reactor. In *Biochemical Engineering-Stuttgart.* Edited by M. Reuss, H. Chmiel, E-D Gilles, and H-J Knackmuss. Stuttart, Germany: Gustaf Fisher.

*Liu, J., and J. Batista. 2000a. Removal of Perchlorate From Water Using a Membrane-Immobilized Biofilm. Proceedings of the 5th Annual Joint Services Pollution Prevention and Hazardous Waste Management. August 21-24, San Antonio, Texas.

*Liu, J., and J. Batista. 2000b. Removal of Perchlorate from Waters by a Membrane- Immobilized Biofilm. Proceedings of the 2000 AWWA Inorganic Contaminants Conference, February 27-29, 2000, Albuquerque, NM.

*Liu, J., and J. Batista. 2000c. Perchlorate Removal from Waters by a Membrane-immobilized Biofilm. Proc. AWWA Inorganics Conference, Albuquerque, New Mexico, Feb 28.

Logan, B.E. 1993. Oxygen Transfer in Trickling Filters. *Journal of Environmental Engineering*, 119(6):1059-1076.

Logan, B.E. 1998a. A Review of Chlorate and Perchlorate Respiring Microorganisms. *Bioremediation Journal,* 2(2):69-79.

Logan, B.E. 1998b. Process for Treating Perchlorate Contaminated Drinking Water. Patent pending.

Logan, B.E. 1999. *Environmental Transport Processes*. New York: John Wiley & Sons.

*Logan, B.E. 2000. Evaluation of Biological Reactors to Degrade Perchlorate to Levels Suitable for Drinking Water. In *Perchlorate in the Environment.* Edited by E.T. Urbansky. Plenum. 189-197.

*Logan, B.E. 2001. Analysis of Overall Perchlorate Removal Rates in Packed-Bed Reactors. *Journal of Environmental Engineering.* In press.

*Logan, B.E., A.R. Bliven, S.R. Olsen, and R. Patnaik. 1998. Growth Kinetics of Mixed Cultures under Chlorate-Reducing Conditions. *Journal of Environmental Engineering*, 124(10):1008-1011.

Logan, B.E., and K. Kim. 1998. Microbiological Treatment of Perchlorate Contaminated Ground Water. Proceedings of The Southwest Focused Ground Water Conference: Discussing the Issue of MTBE and Perchlorate in the Ground Water. National Ground Water Association, Anaheim, CA, June 3-4.

*Logan, B.E., K. Kim, P. Mulvaney, J. Miller, and R. Unz. 1999a. Biological Treatment of Perchlorate Contaminated Waters. In: Bioremediation of Metals and Inorganic Compounds. In *Proc. Fifth International Symposium of In-situ and On-site Bioremediation*, Edited by A. Leeson and B.C. Alleman. Columbus, OH: Battelle. 5(4):147–151.

*Logan, B.E., K. Kim, J. Miller, P. Mulvaney, H. Zhang, R. Unz. 1999b. Factors Affecting Biodegradation of Perchlorate Contaminated Waters. In *Extended Abstracts*, 218th ACS Meeting, New Orleans, August 22-24, 39:112-114.

*Logan, B.E., J. Wu, and R.F. Unz. 2001. Biological Perchlorate Reduction in High Salinity Solutions. *Water Research.* In press.

*Logan, B.E., H. Zhang, P. Mulvaney, M.G. Milner, I. Head, and R.F. Unz. 2000. Kinetics of Chlorate- and Perchlorate-Respiring Bacteria. Submitted.

Madigan, M.T., J.M. Martinko, and J. Parker. 1997. Brock Biology of Microorganisms, 8th ed. New Jersey: Prentice Hall.

Malmqvist, A., and T. Welander. 1992. Anaerobic Removal of Chlorate from Bleach Effluents. *Water Science and Technology,* 25(7):237-242.

Malmqvist, A., T. Welander, and L. Gunnarsson. 1991. Anaerobic Growth of Microorganisms with Chlorate as an Election Acceptor. *Applied and Environmental Microbiology,* 57:2229-2232.

Malmqvist, A., T. Welander, E. Moore, A. Ternstrom, G. Molin, and I.-M. Stenstrom. 1994. Ideonella Dechloratans Gen. Nov., sp. Nov., a New Bacterium Capable of Growing Anaerobically with Chlorate as an Electron Acceptor. *Systems and Applied Microbiology* 17:58-64.

Mateju, V., S. Cizinska, J. Krejci, and T. Janocj. 1992 . Biological Water Denitrification: A Review. *Enzyme Microbiology and Technology,* 14:170-182.

Maymó-Gatell, X., Y-t. Chien, J.M. Gossett, and S.H. Zinder. 1997. Isolation of a Bacterium that Reductively Dechlorinates Tetrachloroethene to Ethene. *Science,* 276:1568-1571.

McCleaf, Philip R., and Edward D. Schroeder. 1995. Denitrification Using a Membrane-Immobilized Biofilm, *Jour. AWWA*, 89:77-86.

Milazzo, G., and S. Caroli. 1978. *Tables of Standard Electrode Potentials.* New York: Wiley.

Miller, Joel. 1999. Bioreduction of Perchlorate Under Hydrogen-Oxidizing Conditions. Master's thesis. Department of Civil and Environmental Engineering, The Pennsylvania State University.

Miller, J.P., and B.E. Logan. 2000. Sustained Perchlorate Degradation in an Autotrophic, Gas Phase, Packed Bed Bioreactor. *Environmental Science and Technology,* 34(14):3018-3022.

Mulvaney, Peter. 1999. Perchlorate and Chlorate Reduction by Axenic Cultures. Master's thesis. Environmental Pollution Control, The Pennsylvania State University.

NDEP (Nevada Department of Environmental Protection). 2000. NPDES Discharge Permit Application. No. NV0023060.

Nilsson, I., and S. Ohlson. 1982. Immobilized Cells in Microbial Nitrate Reduction. *Applied Biochemistry and Biotechnology,* 7(1):39-41.

Nilsson, I., S. Ohlson, L. Haggstrom, L.N. Molin, and K. Mosbach. 1980. Denitrification of Water Using Immobilized Pseudomonas Denitrificans Cells. *European Journal of Applied Microbiology and Biotechnology,* 10(4):261-274.

NREL (National Renewable Energy Laboratory). 1995. *Advanced Hydrogen Production Technologies.* Produced for the U.S. Department of Energy, Washington, D.C.

Nzengung, V.A., C. Wang, and G. Harvey. 1999. Plant-Mediated Transformation of Perchlorate into Chloride. *Environmental Science and Technology,* 33:1470-1478.

Olsen, S. 1997. Consortium and Pure Culture Kinetics of Chlorate-Reducing Microorganisms. Master's thesis. Department of Chemical and Environmental Engineering, University of Arizona, Tucson, AZ.

Oltmann, L.P, W.N.M. Reijnders, and A.H. Stouthamer. 1976. The Correlation Between the Protein Composition of Cytoplasmic Membranes and the Formation of Nitrate Reductase A, Chlorate Reductase C and Tetrathionate Reductase in Proteus Mirabilis Wild Type and Some Chlorate Resistant Mutants. *Archives in Microbiology,* 111:37-43.

Orgiassi, J., and R. Mornex. 1990. Hyperthyroidism. In *The Thyroid Gland.* Edited by M.A. Greer. New York: Raver Press. pp. 405-495.

Patnaik, R. 1996. Oxygen Demand Tests and Mixed-Culture Bioreactor Studies Using Chlorate (ClO3-) as an Electron Acceptor. Master's thesis. Department of Chemical and Environmental Engineering, University of Arizona, Tucson, AZ.

Perry, J., and J. Staley. 1997. *Microbiology Dynamics & Diversity.* Fort Worth, TX: Saunders College Publishing.

Ray, B.T. 1995. *Environmental Engineering.* New York: PWS Publishing Company.

Reid, R.C., J.M. Prausnitz, and B.E. Poling. 1987. *The Properties of Gases and Liquids,* 4th ed. McGraw-Hill, Inc.

Rikken, G.B., A.G. Kroon, and C.G. van Ginkel. 1996. Transformation of (Per)chlorate into Chloride by a Newly Isolated Bacterium: Reduction Dismutation. *Applied Microbiology and Biotechnology,* 45:420426.

Rittmann, B.E. 1995. Fundamentals and Application of Biofilm Processes in Drinking-Water Treatment. In *Handbook of Environmental Chemistry, Quality and Treatment of Drinking Water.* Editor in chief, O. Hutzinger. 5B:61-87.

Rittmann, Bruce E., and Issam Najm. 2000. Autohydrogenotrophic Perchlorate Reduction. Proceedings at AWWA 2000 Inorganic Contaminants Workshop. Albuquerque, New Mexico. February 26-27.

Rittmann, Bruce E., and P.L. McCarty. 2001. *Environmental Biotechnology: Principles and Applications.* New York McGraw-Hill. p. 609.

Sawyer, C.N., P.L. McCarty, and G.F. Parkin. 1994. *Chemistry for Environmental Engineers*, 4th ed. New York: McGraw-Hill.

Schilt, A.A. 1979. *Perchloric Acid and Perchlorates.* Columbus, OH: The G. Frederick Smith Chemical Company.

Siantar, D.P., C.G. Schreier, C.-S. Chou, and M. Reinhard. 1997. Treatment of 1,2-Dibromo-3-Chloropropane and Nitrate-Contaminated Water with Zero-Valent Iron or Hydrogen/palladium Catalysts. *Water Research,* 30(10):2315-2322.

Simes, G.F. 1991. *Preparation Aids for the Development of Category IV Quality Assurance Project Plants.* Risk Reduction Engineering Laboratory, Office of Research and Development. Cincinnati, OH: U.S. Environmental Protection Agency.

Stanbury, J.B., and J.B. Wyngaarden. 1952. Effect of Perchlorate on the Human Thyroid Gland. *Metabolism*, 1:533-539.

Tramposch, W.G. 2001. ISEP+™ Technology for Nitrate and Perchlorate Treatment. Presentation dated December 6, 2001 provided by Charles Drewry at Calgon.

Urbansky, E.T. 1998. Review and Discussion of Perchlorate Chemistry as Related to Analysis and Remediation. *Bioremediation Journal,* 2:81–95.

Urbansky, E.T. 2000. *Perchlorate in the Environment.* New York: Plenum, Kluwer Academic/Plenum Publishers.

Urbansky, E.T., and M.R. Schock. 1999. Issues in Managing the Risks Associated with Perchlorate in Drinking Water. *Journal of Environmental Management,* 56:79-95.

US EPA (US Environmental Protection Agency). 1979. *Methods for Chemical Analysis of Water and Wastes.* Washington, D.C.: US EPA.

US EPA. EPA TNNweb 2000. Unified Air Toxics Website. Methanol. http://www.epa.gov/ttn/uatw/hlthef/methanol.html.

US EPA. 2002. Perchlorate Environmental Contamination: Toxicological Review and Risk Characterization, External Review Draft, NCEA-1-0503, January 16, 2002.

van Ginkel, G.C., A.G.M. Kroon, G.B. Rikken, and S.W.M. Kengen. 1998. Microbial Conversion of Perchlorate, Chlorate, and Chlorite. In *Proceedings of the Southwest Focused Groundwater Conference: Discussing the Issue of MTBE and Perchlorate in the Ground Water*, pp. 92-95. National Ground Water Association.

van Ginkel, G.C., C.M. Plugge, and C.A. Stroo. 1995. Reduction of Chlorate with Various Energy Substrates and Innocula Under Anaerobic Conditions. *Chemosphere,* 31:40574066.

van Ginkel, G.C., G.B. Rikken, A.G.M. Kroon, and S.W.M. Kengen. 1996. Purification and Characterization of Chlorite Dismutase: A Novel Oxygen Generating Enzyme. *Archives in Microbiology,* 166:321-326.

*Vieira, A., and J. Batista. 2000a. The Influence of Humic Substances on the Removal of Perchlorate from Waters using Ion-Exchange Resins. In *Invited Speakers*, Nevada Chapter of the Air and Waste Management meeting, June 7, 2000.

*Vieira, A., and J. Batista. 2000b. Removal of Perchlorate from Contaminated Groundwater in the Las Vegas Valley Using Ion-Exchange Resins. In *Proceedings AWWA Inorganics Conference*, Albuquerque, New Mexico, Feb. 28, 2000 (9 pages).

Wallace, W., T. Ward, A. Breen, and H. Attaway. 1996. Identification of an Anaerobic Bacterium which Reduces Perchlorate and Chlorate as Wolinella Succinogenes. *Journal of Industrial Microbiology,* 16:68-72.

Wallace, W., S. Beshear, D. Williams, S. Hospadar, and M. Owens. 1998. Perchlorate Reduction by a Mixed Culture in an Up-Flow Anaerobic Fixed Bed Reactor. *Journal of Industrial Microbiology and Biotechnology,* 20:26-131.

Wirt, K., M. Laikhtman, J. Rohrer, and P.E. Jackson. 1998. Low-level Perchlorate Analysis in Drinking Water and Groundwater by Ion Chromatography. *Amer. Environ. Lab.*, 10(3):1-5.

*Wu, Jun. 2000. The Abundance and Growth of (Per)chlorate Reducing Bacteria Using Different Substrates. Master's thesis. Department of Civil and Environmental Engineering, The Pennsylvania State University.

---

* Publications by the PI's resulting totally or partially from AWWA support.

# ABBREVIATIONS

| | |
|---|---|
| AFBF | anaerobic fixed bed filler |
| ANOVA | analysis of variance |
| APHA | American Public Health Association |
| atm | atmospheres |
| atm•mL/mmol | atmospheres times milliliters per millimole |
| ATP | adenosine triphosphate |
| AWWA | American Water Works Association |
| AwwaRF | Awwa Research Foundation |
| | |
| B-1168 | bacterial strain |
| BALI | culture named after Batista (BA) and Liu (LI) |
| Biodeg. | mass of lactate biodegraded after 1 day |
| BOD | biochemical oxygen demand |
| BPOU | Baldwin Park Operable Unit |
| BR | bioreactor side of membrane bioreactor |
| | |
| °C | degrees Celsius |
| °C/min | change in degrees Celsius per minute |
| CA | California |
| cc | cubic centimeter |
| CCSD | Clark County Sanitation District |
| CDHS | California Department of Health Services |
| CDM | Camp, Dresser and McKee, Inc. |
| CI | confidence interval |
| cm | centimeters |
| cm/min | centimeters per minute |
| $cm^2$ | centimeters squared |
| $cm^2/sec$ | centimeters squared per second |
| $cm^3/g.mole$ | centimeters cubed per gram mole |
| $cm^3/min$ | centimeters cubed per minute |
| $cm^3/mol$ | centimeters cubed per mole |
| cmmol/mL | centimeters times moles per milliliter |
| CO | carbon monoxide |
| CRMs | chlorate-respiring microorganisms |
| | |
| $d^{-1}$ | per day |
| DELCD | dry electrolytic conductivity detector |
| DHS | Department of Health Services |
| DI | deionized |
| DNA | deoxyribonucleic acid |
| DO | dissolved oxygen |
| DOC | dissolved organic carbon |
| DR | diffusion reactor side of membrane bioreactor |

| | |
|---|---|
| E° | electron volts, standard conditions |
| EA | electron acceptor |
| EBCT | empty bed contact time |
| EP | electrophoresis |
| EPA | Environmental Protection Agency |
| ETC | electron transport chain |
| | |
| FBR | fluidized bed reactor |
| FGLP | type of membrane used in membrane bioreactor |
| ft | feet |
| | |
| g | gram |
| g/cm$^3$ | gram per centimeter cubed |
| g/L | gram per liter |
| g/L-min | gram per liter per minute |
| g/m$^3$ | gram per meter cubed |
| g/ml | gram per milliliter |
| g/mole | gram per mole |
| GAC | granular activated carbon |
| gal/day | gallons per day |
| gal/min | gallons per minute |
| g-cell/g-acetate | grams of cells per gram of acetate |
| g-cell/g-substrate | grams of cells per gram of substrate |
| GF/C | glass fiber, type C |
| g-perchlorate/m$^3$reactor/h | grams of perchlorate per meter cubed of reactor volume per hour |
| gpm/ft$^2$ | gallons per minute per square foot |
| GR-1 | strain of bacteria |
| g-VSS | grams of volatile suspended solids |
| GWR | Groundwater Rule |
| | |
| H$^{-1}$ | per hour |
| HACH DR/2000 | model of instrument used |
| HAP-1 | strain of bacteria |
| HDPE | high density polyethylene membrane |
| HDT | hydraulic detention time |
| HLR | hydraulic loading rate |
| hr | hour |
| | |
| IC | ion chromatography |
| Inc. | incorporated |
| Isol. | isolate |
| | |
| KC1-5 | abbreviation used for different perchlorate concentrations used in batch studies |
| kg/m$^3$ | kilogram per meter cubed |
| KJ | strain of bacteria |

| | |
|---|---|
| kJ/mol-acetate | kilojoules per mole of acetate |
| KL1-5 | abbreviation used for different lactate concentrations used in batch studies |
| $K_s$ | half saturation's constant |
| L | liters |
| L/hr | liters per hour |
| $L^{-1}$ | per liter |
| LCR | Lead and Copper Rule |
| M | moles |
| $m^2/m^3$ | meters squared per cubic meter |
| $m^3$ | cubic meters |
| $m^3$/hr | cubic meters per hour |
| MA | Massachusetts |
| MCL | maximum contaminant level |
| MCLG | maximum contaminent level goal |
| MCR | Master Culture Reactor |
| MDL | minimum detection limit |
| mg | milligrams |
| $mg^{-1}$ | per milligram |
| mg/day | milligrams per day |
| mg/g | milligrams per gram |
| mg/hr | milligrams per hour |
| mg/kg/day | milligrams per kilogram per day |
| mg/L | milligrams per liter |
| mg/L-min | milligrams per liter per minute |
| MGD | million gallons per day |
| mg-DW/g-sand | milligrams dryweight per gram of sand |
| mg-$NO^{3-}$/min | milligrams of nitrate per minute |
| MI | Michigan |
| MilliQ | Milli-Q water |
| min | minute |
| $min^{-1}$ | per minute |
| mL | milliliters |
| mL/L | milliliters per liter |
| mL/min | milliliters per minute |
| mM | millimolar |
| mm | millimeter |
| mmol | millimoles |
| MN | Minnesota |
| moles/day-$m^2$ | moles per day per square meter |
| mS | milli-Siemens |
| MS | measured perchlorate mass in the reactor |
| mS/cm | milli-Siemens per centimeter |
| mv | millivolts |

| | |
|---|---|
| NAD± | nicotinamide-adenine dinucleotide |
| NADH | hydrogyen carrier of NAD± |
| NADP± | NAD phosphate |
| NADPH | hydrogen carrier of NADP± |
| NCPO | noncatalytic partial oxidation |
| NDEP | Nevada Department of Environmental Protection |
| NJ | New Jersey |
| nm | nanometers |
| NREL | Nation Renewable Energy Lab |
| NY | New York |
| | |
| O&M | operating and maintenance |
| OK | Oklahoma |
| ORP | oxidation reduction potentional |
| | |
| PA | Pennsylvania |
| PCE | perchloroethylene |
| PCR | polymerase chain reaction |
| PDA | strain of bacteria |
| PDB | strain of bacteria |
| PDX | strain of bacteria |
| PEC | photoelectrochemical |
| PEM | proton exchange membrane |
| Penn-4, Penn-5 | sample identification numbers |
| pH | units of hydrogen ion concentration |
| PHB | polyhydroxybutryrate |
| POTW | publicly owned treatment works |
| ppb | parts per billion |
| ppm | parts per million |
| PRMs | perchlorate respiring microorganisms |
| PSU | Penn State University |
| p-values | statistical test, p value |
| PVDF | membrane material used in membrane bioreactor |
| | |
| QA | quality assurance |
| QA/QC | quality assurance and quality control |
| QAPP | Quality Assurance Project Plan |
| QC | quality control |
| | |
| RAS-CCSD | bacterial isolate identification letters |
| RPD | relative percent deviation |
| rRNA | ribosomal ribonucleic acid |
| | |
| s | seconds |
| SDWA | Safe Drinking Water Act |
| sec | seconds |

| | |
|---|---|
| SLR | surface loading rate |
| spp | species |
| SRI | instrument manufacturer: SRI, Inc. |
| SRT | solids retention time |
| SS | suspended solids |
| SWTR | Surface Water Treatment Rule |
| | |
| TBD | to be determined |
| TCE | trichloroethylene |
| TCR | Total Coliform Rule |
| TDS | total dissolved solids |
| TE | terminal electron |
| TEA | terminal electron acceptor |
| TEMP | temperature |
| TOC | total organic carbon |
| | |
| UNLV | University of Nevada, Las Vegas |
| U.S. | United States |
| U.S.A. | United States of America |
| US DOE | United States Department of Energy |
| USEPA | United States Environmental Protection Agency |
| | |
| VG | van Ginkel |
| VOC | volatile organic carbon |
| vol/vol | volume per volume |
| VSS | volatile suspended solids |
| v/v | volume per volume |
| | |
| w/w | weight per weight |
| | |
| YSI | Yellow Springs Instruments, Inc. |
| | |
| µg | microgram |
| µg perchlorate/L-min | micrograms of perchlorate per liter per minute |
| µg/L | micrograms per liter |
| µg/L-min | micrograms per liter per minute |
| µg-DW/mL | micrograms dry weight per milliliter |
| µL | microliters |
| µm | micrometers |
| µmhos/cm | per centimeter per micro-ohm |

IWA Publishing's authorised EU representative for General Product Safety Regulations is Diane D'Arras, 15 rue Duret, 75116 Paris, France, e-mail: safety@iwap.co.uk.

Printed and bound by CPI Group (UK) Ltd, Croydon, CR0 4YY
17/09/2026
02218808-0007